Butterflies of Australia

I.F.B. Common and D.F. Waterhouse

BUTTERFLIES OF AUSTRALIA

Field Edition

Angus & Robertson Publishers

ANGUS & ROBERTSON PUBLISHERS
London · Sydney · Melbourne

First published by Angus & Robertson Publishers, Australia, 1972
Revised edition 1981
This abridged field guide edition 1982

National Library of Australia
Cataloguing-in-publication data.

Common, I. F. B. (Ian Francis Bell).
   Butterflies of Australia.

   Abridged ed.
   Abridged version of rev. ed.: Sydney: Angus & Robertson, 1981.
   Bibliography.
   Includes index.
   ISBN 0 207 14498 2.

   1. Butterflies—Australia.   I. Waterhouse, D. F. (Douglas Frew).   II. Title.

595.78′9′0994

Typeset in 10/11pt Baskerville by Asco Trade Typesetting Ltd, Hong Kong
Printed in Hong Kong

Contents

CONTENTS

Illustrations

(All plates appear after p. xiv)

PLATES

LINE DRAWINGS

MAPS

Foreword

THIS abridged version of the second (1981) edition of *Butterflies of Australia* is intended to provide butterfly collectors and other naturalists with a field guide to the Australian butterflies. Ninety-eight per cent (373) of the 382 species recorded from Australia are figured; those not figured are either indistinguishable from those already represented, except on the basis of genitalia, or are known from only very few specimens from remote northern areas. Although it should be possible to identify most species from the colour figures, readers should confirm their identifications by checking the distribution maps and the relevant text. Although we have included the common names, these are not featured at the expense of the scientific names. Many of the coined common names are in fact more cumbersome and difficult to remember than the scientific names and, of course, seldom provide any clue to the relationships of the species or groups which the generic name often does.

The terminology is covered in the introductory sections and technical terms have usually been explained on the first occasion they are used in the text, or their meanings are given in the glossary. In the distribution maps the black spots represent the positions of Cooktown, Cairns, Townsville, Rockhampton, Brisbane and Mitchell in Queensland, Port Macquarie, Sydney and Batemans Bay in New South Wales, Adelaide in South Australia, Albany, Perth, Geraldton,

Roebourne, Derby and Wyndham in Western Australia, and Darwin and Alice Springs in the Northern Territory.

Probably more than ninety per cent of Australian butterflies are already known, and few collectors will have an opportunity of discovering new species. Nevertheless, the assembling of a collection of adult butterflies is undeniably an absorbing, instructive and rewarding activity. Much still remains to be learnt and recorded about the precise distribution of our species, their life histories and other aspects of their biology and behaviour. This field guide provides an up-to-date summary of the available information on the eggs, larvae and pupae, and the recorded food plants; life history stages of representative species are also figured. Careful observations on most species will greatly extend our knowledge on these subjects, and the discovery of new facts will bring additional pleasure and satisfaction to the study of our butterflies.

Acknowledgements

IT is difficult in a book of this kind to give adequate credit to all those who have provided technical information or who have helped in many other ways. We have drawn heavily on G. A. Waterhouse's *What Butterfly is That?* (1932) and its predecessor, *The Butterflies of Australia* by G. A. Waterhouse and G. Lyell (1914), and the many papers by Waterhouse and others. Special thanks are due to Mr A. F. Atkins, Mr D. Bell, Mr D. Binns, Mr A. D. Bishop, Mrs Kay Breeden, Mr Stanley Breeden, Mr A. N. Burns, Dr B. A. Conroy, Mr G. Daniels, Mr J. W. C. d'Apice, Mr M. De Baar, Mr E. D. Edwards, the late Mr E. O. Edwards, Lt.-Col. J. N. Eliot, Mr D. Ferguson, Dr R. P. Field, Mr R. H. Fisher, the late Mr C. W. Frazier, Mr W. Graham, Mr G. Griffin, Mr R. Grund, Dr R. W. Guard, the late Mr T. H. Guthrie, Mrs Jean Harslett, Mr W. R. Hindson, Mr T. G. Howarth, Mr M. Hutchison, Mr S. J. Johnson, Professor J. F. R. Kerr, Dr R. L. Kitching, Mr R. Lachlan, Mr J. C. Le Souef, Mr C. McCubbin, Mr S. McEvey, Mr J. Macqueen, Mr R. C. Manskie, Mr G. Marriott, Dr C. G. Miller, Dr G. B. Monteith, Mr D. E. A. Morton, Mr M. S. Moulds, Mr J. Olive, Mr J. V. Peters, Mr W. N. B. Quick, Dr D. P. A. Sands, Mr G. Sankowsky, Dr A. Sibatani, Mr E. Slater, Dr C. N. Smithers, Mr A. Sundholm, Mr J. R. Turner, Mr M. S. Upton, Mr R. I. Vane-Wright, Ms J. Waddy, Mr S. Wallace, and Mr J. Williams.

All but three of the colour plates were by the late Neville W. Cayley; some colour deterioration and inaccuracies in these plates have been rectified by I. F. B. Common. Plates 13 and 23 are by Mrs Ninon Geier, Plate 14 by Mrs Sybil Monteith, and Plates I, IV, VII, VIII, XI, XIII, XIV, XVII and XX by Mrs D. North. Photographs of immature stages were kindly provided by Mr E. D. Edwards (Plate II, figs 1–5, 7, 8; Plate V, figs 1–4; Plate X, figs 2, 4; Plate XII, figs 5–9; Plate XV, figs 6, 7; Plate XVI, figs 3, 4, 7; Plate XIX, fig 7), Mr A. F. Atkins (Plate III, figs 1, 4, 6–10; Plate VI, figs 1, 2; Plate XIX, figs 1, 5, 6), the late Mr T. H. Guthrie (Plate X, figs 1, 5), Mr M. S. Upton (Plate VI, figs 3, 4, 8; Plate XII, figs 3, 10; Plate XVIII, fig 5), Dr G. B. Monteith (Plate XV, figs 1, 2), Mr S. and Mrs K. Breeden (Plate VI, fig 6), and Mr B. Jessop (Plate XV, fig 4). The other photographs of immature stages are by I. F. B. Common. The photographs in Plates IX and 29 are by Mr J. P. Green. The maps and Figs 20–25 were drawn by Mr L. A. Marshall and Mrs G. Palmer, and Figs 1–19 by I. F. B. Common.

Food plant specimens were kindly identified by the late Dr N. T. Burbidge, Mr M. Gray and Mr R. Pullen of the Herbarium Australiense, CSIRO, Canberra. Mr Gray also checked the botanical names and Dr R. W. Taylor identified many ant specimens.

The Plates

Plate 1

HESPERIIDAE

(About 0.9 times natural size)

Plate 2

HESPERIIDAE
Skippers

(About 1.2 times natural size)

1
2
3
4
4A
5
7
6
8
9
10
11
12
19
20
13
14
15
16
17
17A
18
20A

Plate 3

HESPERIIDAE
Skippers

1, *Mesodina halyzia halyzia* (p. 78)
2, *M. aeluropis* (p. 79)
3, *Croitana croites* (p. 80)
4, 4A, *Neohesperilla xanthomera* (p. 52)
5, *N. crocea* (p. 51)
6, *N. xiphiphora* (p. 51)
7, *N. senta* (p. 52)
8, *Toxidia parvula* (p. 48)
9, *T. thyrrhus* (p. 48)
10, *T. doubledayi* (p. 49)
11, *T. rietmanni rietmanni* (p. 49)
12, *T. rietmanni parasema* (p. 50)
13, *T. peron* (p. 47)
14, *T. melania* (p. 50)
15, *Hesperilla crypsigramma* (p. 55)
16, *H. sexguttata* (p. 56)
17, *Motasingha atralba atralba* (p. 76)
18, *M. atralba anaces* (p. 76)
19, 19A, *M. dirphia dilata* (p. 74)
20, *Anisynta dominula drachmophora* (p. 43)
21, *A. monticolae* (p. 42)
22, 22A, *Dispar compacta* (p. 45)
(About 1.3 times natural size)

Plate 4

HESPERIIDAE
Skippers

(About 1.2 times natural size)

Plate 5

HESPERIIDAE
Skippers

(From 1.1 to 1.4 times natural size)

1
2
2A
3
4
4A
14
15
12
13
13A
5
6
7
8
8A
9
9A
10
11
11A
16
17
17A
17B
18
18A

Plate 6

HESPERIIDAE
Skippers

(From 1.1 to 1.4 times natural size)

Plate 7

PAPILIONIDAE
Swallowtails

1, *Papilio demoleus sthenelus* (p. 123)
2, *P. anactus* (p. 118)
3, *P. ulysses joesa* (p. 124)
4, *Protographium leosthenes leosthenes* (p. 112)
5, *Graphium aristeus parmatum* (p. 118)
6, *G. macleayanum macleayanum* (p. 113)
7, *G. sarpedon choredon* (p. 115)
8, *G. eurypylus lycaon* (p. 116)
9, *G. macfarlanei macfarlanei* (p. 117)
10, *G. agamemnon ligatum* (p. 117)
11, *Pachliopta polydorus queenslandicus* (p. 126)
12, *Cressida cressida cressida* (p. 125)
(About 0.6 times natural size)

Plate 8

PAPILIONIDAE
Swallowtails

1–1C, *Papilio aegeus aegeus* (p. 120)
2, 2A, *P. ambrax egipius* (p. 123)
3, *P. fuscus capaneus* (p. 121)
4, *P. canopus canopus* (p. 122)
(About 0.6 times natural size)

1
1B
1A
1C
2
2A
4
3

Plate 9

PAPILIONIDAE
Cape York Birdwing

1, 1A, *Ornithoptera priamus pronomus* (p. 127)
(About 0.8 times natural size)

1

1A

Plate 10

PIERIDAE
Whites and Yellows

1, 1A, *Catopsilia pyranthe crokera* (p. 131)
2–2B, *C. pomona pomona* (p. 132)
3–3B, *C. gorgophone gorgophone* (p. 133)
4, *C. scylla etesia* (p. 133)
5, 5A, *Appias paulina ega* (p. 149)
6, 6A, *A. albina albina* (p. 149)
7, 7A, *A. melania* (p. 150)
8, 8A, *A. ada caria* (p. 151)
9–9B, *Anaphaeis java teutonia* (p. 147)
(About 0.6 times natural size)

1
1A
2
2A
2B
3
3A
3B
4
5
5A
6
6A
7
7A
8
8A
9
9A
9B

Plate 11

PIERIDAE
Whites and Yellows

1, *Eurema candida virgo* (p. 135)
2, 2A, *E. hecabe phoebus* (p. 135)
3, 3A, *E. smilax* (p. 137)
4, *E. sana* (p. 138)
5, *E. brigitta australis* (p. 134)
6, *E. herla* (p. 138)
7, *E. laeta lineata* (p. 138)
8, *Elodina parthia* (p. 139)
9, *E. perdita perdita* (p. 140)
10, *E. angulipennis* (p. 139)
11, 11A, *E. padusa* (p. 140)
12, 12A, *Cepora perimale scyllara* (p. 148)
13, *C. perimale latilimbata* (p. 149)
(From 1.0 to 1.2 times natural size)

2
2A
1
8
9
10
12
12A
13
11
11A
3
3A
4
5
6
7

Plate 12

PIERIDAE
Whites

1, 1A, *Delias mysis mysis* (p. 143)
2, 2A, *D. mysis waterhousei* (p. 143)
3, 3A, *D. aganippe* (p. 144)
4, *D. nysa nivira* (p. 147)
5, *D. nysa nysa* (p. 146)
6, 6A, *D. aruna inferna* (p. 145)
7, *D. ennia tindalii* (p. 144)
8, *D. ennia nigidius* (p. 144)
9, 9A, *D. argenthona argenthona* (p. 142)
10, 10A, *D. nigrina* (p. 146)
11, 11A, *D. harpalyce* (p. 145)
(About 0.6 times natural size)

1
4
9
1A
5
9A
2
6
10
2A
6A
10A
3
7
11
3A
8
11A

Plate 13

HESPERIIDAE, PIERIDAE, NYMPHALIDAE, LYCAENIDAE

1, 1A, *Trapezites macqueeni* (p. 31)
2, *Euploea algea amycus* (p. 159)
3, *E. alcathoe monilifera* (p. 158)
4, 4A, *Appias celestina* (p. 150)
5, *Taenaris artemis jamesi* (p. 201)
6, *Toxidia inornata inornata* (p. 50)
7, *Anisynta albovenata albovenata* (p. 39)
8, 8A, *Oreixenica ptunarra roonina* (p. 189)
9, *Virachola smilis dalyensis* (p. 298)
10, *Euploea usipetes usipetes* (p. 161)
11, *E. batesii belia* (p. 158)
12, 12A, *Elymnias agondas australiana* (p. 165)
(Figs 1, 6, 7, 8, 9 about 1.1 times natural size, remainder about 0.6 times natural size)

Plate 14

HESPERIIDAE, PAPILIONIDAE, PIERIDAE, NYMPHALIDAE, LYCAENIDAE

1, *Junonia erigone walkeri* (p. 218)
2, *Theclinesthes albocincta* (p. 328)
3, *Jalmenus clementi* (p. 288)
4, *Candalides geminus* (p. 307)
5, *Acrodipsas arcana* (p. 236)
6, *Hesperilla furva* (p. 54)
7, *Mimene atropatene* (p. 106)
8, *Hypolimnas anomala albula* (p. 214)
9, *Ornithoptera richmondia* (p. 129)
10, *Apaturina erminea* (New Guinea) (p. 205)
11, *Lexias aeropa* (p. 206)
12, *Rachelia extrusa* (p. 28)
13, *Vanessa cardui* (p. 216)
14, *Paralucia spinifera* (p. 239)
15, *Petrelaea dana* (p. 314)
16, *Pithecops dionisius dionisius* (p. 342)
17, *Pieris rapae rapae* (p. 151)
18, *Hesperilla sarnia* (p. 54)
19, *Allora major major* (p. 18)
(Figs 1, 8–11, 13, 17 about two-thirds natural size, remainder about natural size)

Plate 15

NYMPHALIDAE
Danaids and Glasswing

1, *Danaus plexippus plexippus* (p. 154)
2, *D. genutia alexis* (p. 155)
3, *D. philene gelanor* (p. 156)
4, *D. affinis affinis* (p. 156)
5, *D. chrysippus petilia* (p. 155)
6, *D. hamatus hamatus* (p. 156)
7, *Tellervo zoilus gelo* (p. 162)
7A, *T. zoilus zoilus* (p. 162)
8, *Euploea eichhorni* (p. 158)
9, *E. tulliolus tulliolus* (p. 160)
10, 10A, *E. darchia darchia* (p. 161)
11, *E. darchia niveata* (p. 161)
12, *E. core corinna* (p. 158)
13–13B, *E. sylvester sylvester* (p. 159)
13C, *E. sylvester pelor* (p. 160)
14, *Acraea andromacha andromacha* (p. 224)
(About 0.6 times natural size)

Plate 16

NYMPHALIDAE
Browns

1, *Tisiphone abeona rawnsleyi* (p. 199)
2, *T. abeona morrisi* (p. 198)
3, 3A, *T. abeona "joanna"* (p. 196)
4, *T. abeona abeona* (p. 196)
5, *T. abeona albifascia* (p. 197)
6, *T. helena* (p. 196)
7, 7A, *Melanitis leda bankia* (p. 164)
8, 8A, *Mycalesis terminus terminus* (p. 166)
9, 9A, *M. perseus perseus* (p. 166)
10, *M. sirius sirius* (p. 166)
11, *Orsotriaena medus moira* (p. 167)
12, 12A, *Hypocysta euphemia* (p. 167)
13, *Xois arctoa arctoa* (p. 199)
(About three-quarters natural size)

Plate 17

NYMPHALIDAE

Browns

1, *Heteronympha banksii banksii* (p. 181)
2, 2A, *H. solandri solandri* (p. 182)
3, 3A, *H. paradelpha paradelpha* (p. 177)
4, 4A, *H. penelope penelope* (p. 178)
5, 5A, *H. merope merope* (p. 176)
6, *H. merope duboulayi* (p. 177)
7, 7A, *H. mirifica* (p. 177)
8, *H. cordace cordace* (p. 183)
9, *Geitoneura acantha acantha* (p. 172)
10, *G. acantha ocrea* (p. 173)
11, *G. klugii klugii* (p. 174)
12, *G. minyas minyas* (p. 173)
13, *G. minyas mjobergi* (p. 174)
(About three-quarters natural size)

Plate 18

NYMPHALIDAE
Browns

1, 1A, *Oreixenica latialis latialis* (p. 190)
2, *O. lathoniella lathoniella* (p. 191)
3, *O. lathoniella laranda* (p. 192)
4, *O. lathoniella herceus* (p. 192)
5, 5A, *O. correae* (p. 193)
6, 6A, *O. orichora orichora* (p. 187)
7, *O. kershawi kershawi* (p. 194)
8, *O. kershawi ella* (p. 194)
9, *Argynnina hobartia hobartia* (p. 171)
10, *A. cyrila* (p. 170)
11, *A. hobartia tasmanica* (p. 171)
12, *Nesoxenica leprea leprea* (p. 185)
13, *N. leprea elia* (p. 186)
14, *Hypocysta adiante adiante* (p. 169)
15, *H. adiante antirius* (p. 170)
16, *H. irius* (p. 168)
17, *H. metirius* (p. 168)
18, *H. pseudirius* (p. 169)
19, *H. angustata angustata* (p. 170)
(About natural size)

1
1A
2
3
4
5
6
6A
9
10
11
12
13
5A
7
14
15
16
17
18
19
8

Plate 19

NYMPHALIDAE
Nymphs

1, *Pantoporia consimilis consimilis* (p. 209)
2, *P. venilia moorei* (p. 209)
3, *Neptis praslini staudingereana* (p. 207)
4, *Phaedyma shepherdi shepherdi* (p. 206)
5, 5A, *Mynes geoffroyi guerini* (p. 210)
6, *Yoma sabina parva* (p. 215)
7, 7A, *Hypolimnas bolina nerina* (p. 212)
8, 8A, *H. alimena lamina* (p. 213)
9, 9A, *H. misippus* (p. 213)
10, 10A, *Doleschallia bisaltide australis* (p. 211)
11, 11A, *Cethosia penthesilea paksha* (p. 220)
12, *Phalanta phalantha araca* (p. 222)
(About 0.6 times natural size)

1
2
3
4
5
5A
7
7A
8
8A
9
9A
10
10A
11
11A
12
6

Plate 20

NYMPHALIDAE, LIBYTHEIDAE, LYCAENIDAE
Nymphs, Beaks and Harlequin

1, *Polyura pyrrhus sempronius* (p. 202)
2, *Junonia villida calybe* (p. 219)
3, 3A, *J. orithya albicincta* (p. 219)
4, *J. hedonia zelima* (p. 218)
5, *Vanessa kershawi* (p. 216)
6, *V. itea* (p. 217)
7, 7A, *Vindula arsinoe ada* (p. 221)
8, *Cethosia cydippe chrysippe* (p. 220)
9, *Vagrans egista propinqua* (p. 221)
10, *Argyreus hyperbius inconstans* (p. 210)
11, *Cupha prosope prosope* (p. 223)
11A, *C. prosope turneri* (p. 223)
12, 12A, *Libythea geoffroy genia* (p. 226)
13, 13A, *Praetaxila segecia punctaria* (p. 345)
(About 0.6 times natural size)

Plate 21

LYCAENIDAE
Blues

1, 1A, *Hypochrysops ignitus ignitus* (p. 246)
2, *H. ignitus olliffi* (p. 248)
3, 3A, *H. cyane* (p. 249)
4, *H. epicurus* (p. 248)
5, 5A, *H. pythias euclides* (p. 252)
6, 6A, *H. byzos byzos* (p. 252)
7, 7A, *H. miskini* (p. 250)
8, 8A, *H. delicia delicia* (p. 243)
9, *H. delicia delos* (p. 244)
10, 10A, *H. narcissus narcissus* (p. 257)
11, 11A, *H. polycletus rovena* (p. 257)
12, 12A, *H. halyaetus* (p. 243)
(About 1.1 times natural size)

Plate 22

LYCAENIDAE
Blues

1, 1A, *Philiris innotata innotata* (p. 260)
2, 2A, *P. nitens nitens* (p. 259)
3, 3A, *P. fulgens kurandae* (p. 259)
4, 4A, *P. diana diana* (p. 258)
5, 5A, *Acrodipsas cuprea* (p. 233)
6, 6A, *A. myrmecophila* (p. 234)
7, 7A, *Pseudodipsas eone iole* (p. 240)
8, 8A, *P. cephenes* (p. 240)
9, 9A, *Hypochrysops digglesii* (p. 258)
10, 10A, *Celastrina tenella tenella* (p. 342)
11, *Jamides cytus claudia* (p. 335)
12, 12A, *J. aleuas coelestis* (p. 335)
13, *Megisba strongyle nigra* (p. 343)
14, 14A, *Hypochrysops apelles apelles* (p. 254)
(About natural size)

1
1A
2
2A
3
3A
4
4A
10
6
6A
5
5A
7
7A
12
12A
10A
9
9A
8
8A
11
14
14A
13

Plate 23

LYCAENIDAE
Blues

1, *Deudorix epirus agimar* (p. 297)
2, *Virachola democles* (p. 298)
3, 3A, *Hypochrysops theon medocus* (p. 242)
4, *H. hippuris* (p. 242)
5, *H. piceatus* (p. 250)
6, *Acrodipsas brisbanensis brisbanensis* (p. 232)
7, *A. illidgei* (p. 235)
8, *Catopyrops ancyra mysia* (p. 319)
9, *Candalides margarita margarita* (p. 302)
10, *C. consimilis consimilis* (p. 304)
11, *Philiris ziska titeus* (p. 260)
12, *Adaluma urumelia* (p. 312)
13, 13A, *Hypochrysops apollo apollo* (p. 255)
14, *H. elgneri barnardi* (p. 256)
14A, *H. elgneri elgneri* (p. 256)
15, *Zetona delospila* (p. 313)
16, *Ionolyce helicon hyllus* (p. 318)
17, *Catochrysops amasea amasea* (p. 337)
18, *Prosotas nora auletes* (p. 318)
(About 1.1 times natural size)

1
2
3
3A
4
5
6
7
8
9
10
11
12
13
13A
14
14A
15
16
17
18

Plate 24

LYCAENIDAE
Blues

1, 1A, *Ogyris genoveva genoveva* (p. 264)
1B, *O. zosine zosine* (p. 268)
2, 2A, *O. aenone* (p. 276)
3, 3A, *O. idmo idmo* (p. 270)
4, 4A, *O. oroetes oroetes* (p. 278)
5, 5A, *O. amaryllis amaryllis* (p. 279)
6, 6A, *O. amaryllis hewitsoni* (p. 281)
7, 7A, *O. ianthis* (p. 275)
8, 8A, *O. iphis iphis* (p. 276)
9, 9A, *O. abrota* (p. 272)
10, 10A, *O. otanes* (p. 271)
11, 11A, *O. olane ocela* (p. 273)
12, 12A, *O. barnardi barnardi* (p. 274)
(About 0.7 times natural size)

1
1A
1B
2
2A
3
3A
4
4A
5
5A
6
6A
7
7A
8A
9
9A
10
10A
11
11A
12
12A
8

Plate 25

LYCAENIDAE
Blues

1, 1A, *Arhopala centaurus centaurus* (p. 261)
2, 2A, *A. madytus* (p. 262)
3, 3A, *A. micale amphis* (p. 263)
4, 4A, *A. wildei wildei* (p. 261)
5, *Jalmenus evagoras evagoras* (p. 282)
6, *J. evagoras eubulus* (p. 283)
7, *J. eichhorni* (p. 284)
8, *J. daemeli* (p. 286)
9, *J. inous* (p. 287)
10, *J. icilius* (p. 288)
11, *J. ictinus* (p. 284)
12, *J. lithochroa* (p. 286)
13, *Pseudalmenus chlorinda zephyrus* (p. 290)
14, *P. chlorinda chloris* (p. 293)
15, 15A, *Hypolycaena phorbas phorbas* (p. 295)
16, *H. danis turneri* (p. 295)
17, 17A, *Deudorix epijarbas diovis* (p. 296)
18, *Rapala varuna simsoni* (p. 299)
19, 19A, *Bindahara phocides yurgama* (p. 300)
20, *Liphyra brassolis major* (p. 230)
(About 0.7 times natural size)

Plate 26

LYCAENIDAE
Blues

1, 1A, *Candalides absimilis* (p. 303)
2, 2A, *C. helenita helenita* (p. 303)
3, 3A, *C. gilberti* (p. 301)
4, *C. heathi heathi* (p. 310)
5, *C. heathi alpinus* (p. 311)
6, 6A, *C. hyacinthinus hyacinthinus* (p. 308)
7, *C. hyacinthinus simplex* (p. 309)
8, 8A, *C. cyprotus cyprotus* (p. 305)
9, 9A, *C. xanthospilos* (p. 309)
10, 10A, *C. erinus erinus* (p. 306)
11, 11A, *C. acastus* (p. 307)
12, 12A, *Danis danis serapis* (p. 333)
13, 13A, *D. cyanea arinia* (p. 335)
14–14B, *D. hymetus taygetus* (p. 334)
15, *Neopithecops lucifer heria* (p. 343)
(About 1.1 times natural size)

1
1A
2
2A
3
3A
4
5
6
6A
7
8
8A
9
9A
10
10A
11
11A
12
12A
13
13A
14
14B
14A
15

Plate 27

LYCAENIDAE
Blues and Coppers

1, 1A, *Jamides phaseli* (p. 336)
2, 2A, *Syntarucus plinius pseudocassius* (p. 338)
3, 3A, *Everes lacturnus australis* (p. 341)
4, *Nesolycaena albosericea* (p. 312)
5, 5A, *Paralucia aurifera* (p. 239)
6, *P. pyrodiscus pyrodiscus* (p. 238)
7, 7A, *Lucia limbaria* (p. 231)
8, 8A, *Neolucia agricola agricola* (p. 321)
9, 9A, *N. hobartensis hobartensis* (p. 322)
10, *N. hobartensis monticola* (p. 323)
11, 11A, *N. mathewi* (p. 323)
12, 12A, *Theclinesthes serpentata serpentata* (p. 331)
13, 13A, *T. sulpitius sulpitius* (p. 332)
14, *Freyeria trochylus putli* (p. 344)
15, 15A, *Lampides boeticus* (p. 337)
16, 16A, *Theclinesthes onycha onycha* (p. 325)
17, 17A, *T. scintillata* (p. 332)
(About 1.1 times natural size)

Plate 28

LYCAENIDAE
Blues

1, 1A, *Nacaduba kurava parma* (p. 315)
2, 2A, *N. berenice berenice* (p. 314)
3, 3A, *Catopyrops florinda halys* (p. 319)
4, *Theclinesthes miskini miskini* (p. 326)
5, 5A, *Erysichton lineata lineata* (p. 320)
6, 6A, *E. palmyra tasmanicus* (p. 320)
7, 7A, *Nacaduba biocellata biocellata* (p. 316)
8, 8A, *Prosotas dubiosa dubiosa* (p. 317)
9, 9A, *P. felderi* (p. 317)
10, *Euchrysops cnejus cnidus* (p. 343)
11, 11A, *Anthene seltuttus affinis* (p. 300)
12, 12A, *A. lycaenoides godeffroyi* (p. 301)
13, 13A, *Catochrysops panormus platissa* (p. 337)
14, 14A, *Zizina labradus labradus* (p. 340)
15, 15A, *Zizeeria karsandra* (p. 339)
16, *Zizula hylax attenuata* (p. 341)
17, *Famegana alsulus alsulus* (p. 340)
(About 1.1 times natural size)

Plate 29

HESPERIIDAE, NYMPHALIDAE, LYCAENIDAE

1, *Croitana arenaria* (p. 81)
2, *Theclinesthes hesperia* (p. 329)
3, 4, *Acrodipsas hirtipes* (p. 236)
5, *A. illidgei* (p. 235)
6, *Croitana arenaria* (p. 81)
7, *Paralucia spinifera* (p. 239)
8, *Charaxes latona* (p. 203)
9, *Hypochrysops hippuris* (p. 242)
10, *Croitana aestiva* (p. 81)
11, *Hypochrysops cleon* (p. 251)
12, *Acrodipsas arcana* (p. 236)
13, 14, *A. melania* (p. 237)
(Fig 8 about 0.9, Fig 9 about 1.7 and remainder about 1.9 times natural size)

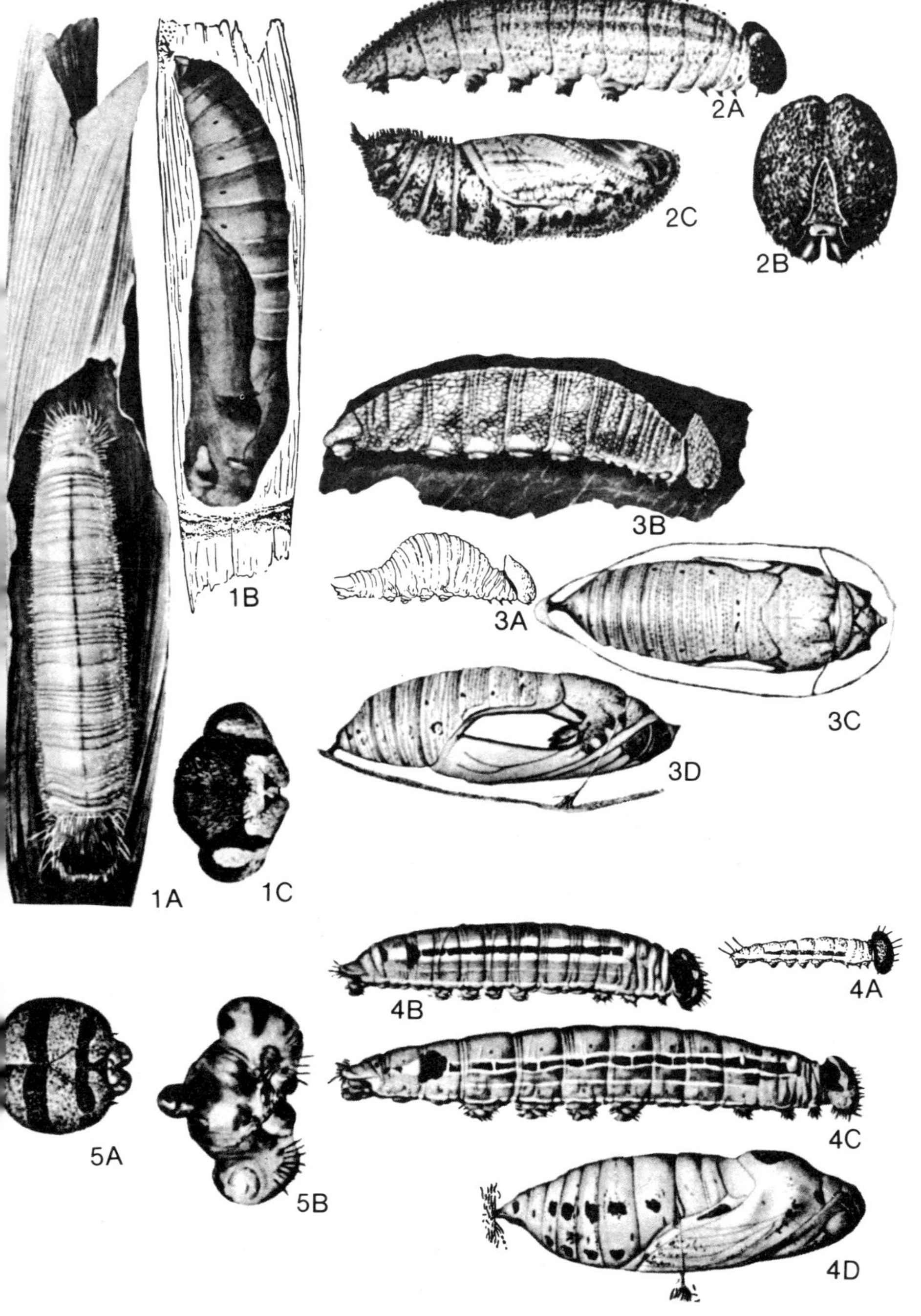

1A
1B
1C
2A
2B
2C
3A
3B
3C
3D
4A
4B
4C
4D
5A
5B

Plate II

HESPERIIDAE

1, 2, Mature larva and pupa of *Hasora discolor mastusia* (p. 18)
3, Mature larva of *Netrocoryne repanda repanda* (p. 24)
4, Late instar larva of *Trapezites luteus leucus* (p. 37)
5–8, Mature larvae of *Anisynta dominula draco* (p. 43), *Hesperilla chaostola chares* (p. 68), *Hesperilla mastersi mastersi* (p. 69), and *Motasingha dirphia dea* (p. 75)

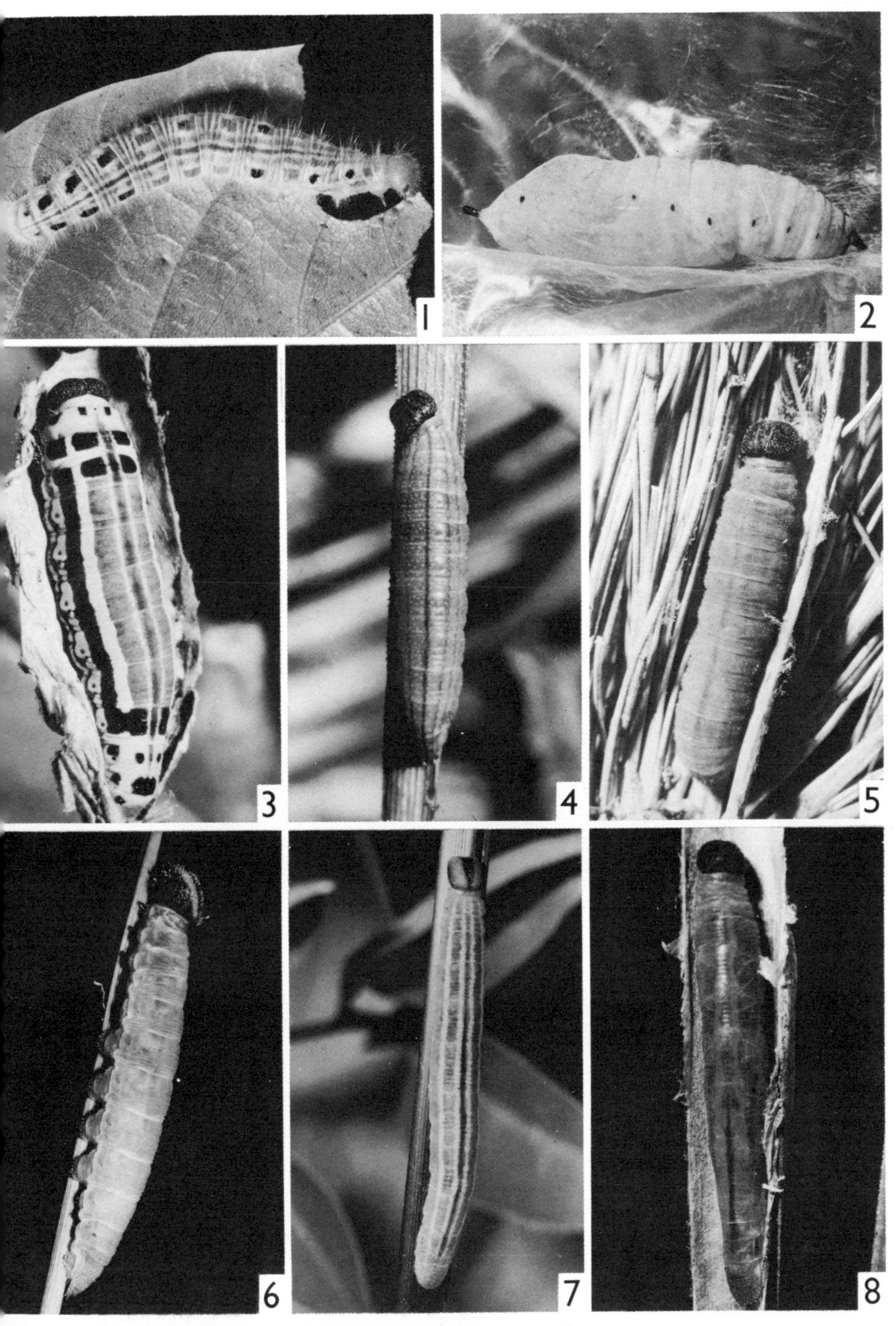

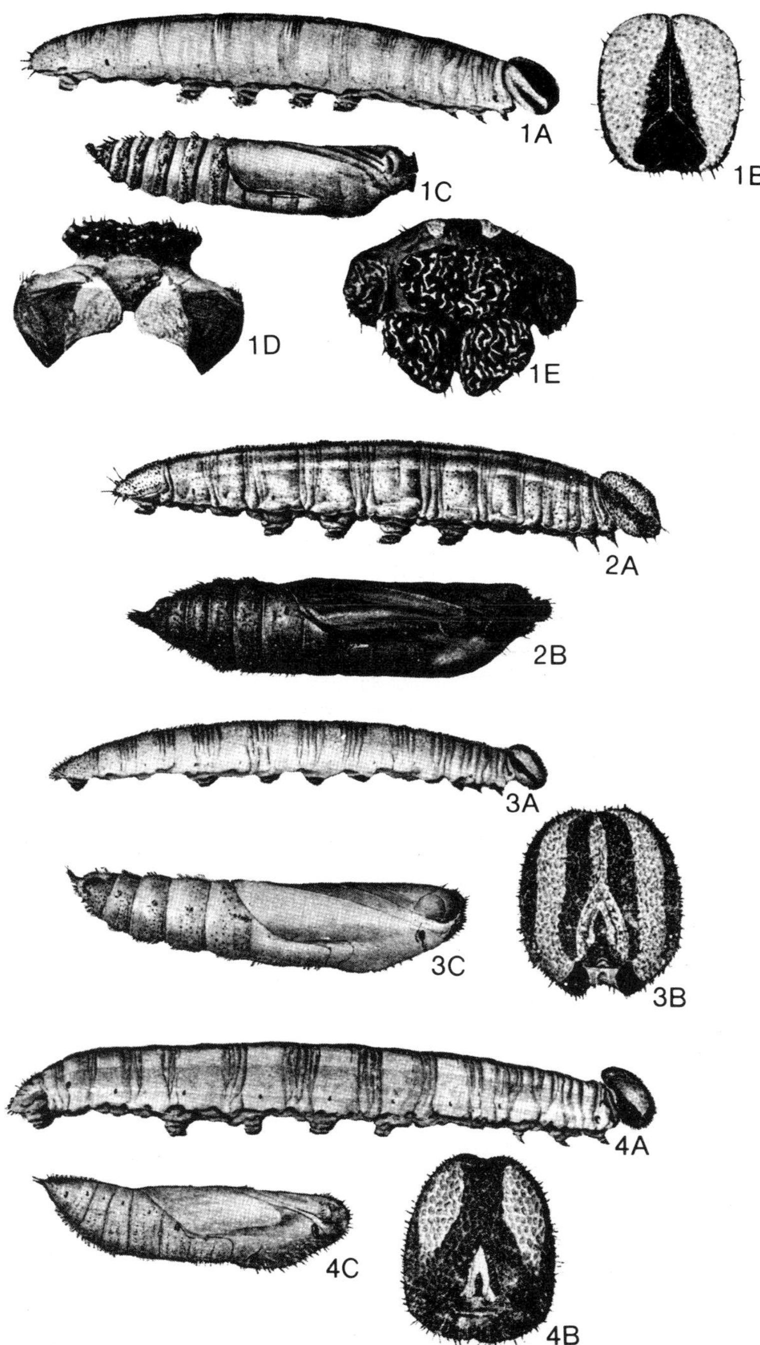

Plate V

HESPERIIDAE AND PAPILIONIDAE

1–3, Eggs of *Euschemon rafflesia rafflesia* (p. 21), *Mesodina aeluropis* (p. 79) and
Suniana lascivia lascivia (p. 93)
4, 5, Mature larvae of *Telicota anisodesma* (p. 101) and *Cephrenes trichopepla*
(p. 104)
6, 7, Mature larva and pupa of *Pelopidas lyelli lyelli* (p. 109)
8, 9, Mature larvae (mottled and green forms) of *Papilio canopus canopus*
(p. 122)
10, 11, Mature larvae of *Papilio anactus* (p. 118) and *Ornithoptera priamus
euphorion* (p. 128)

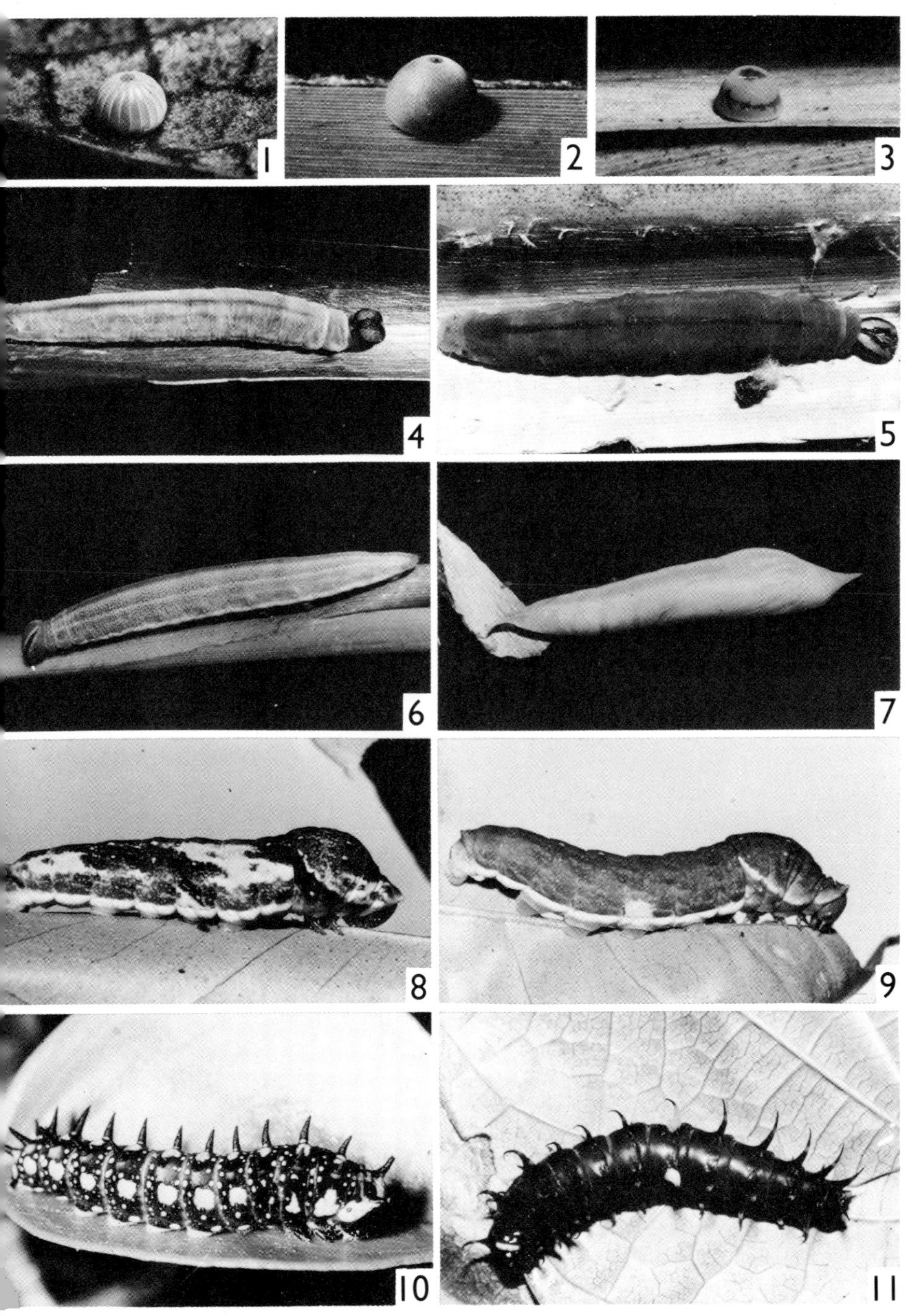

Plate VII

PAPILIONIDAE

1A–1D, Larvae and pupa of *Papilio aegeus aegeus* (p. 120)
2A–2E, Egg, larvae and pupa of *Cressida cressida cressida* (p. 125)
3A–3C, Larvae and pupa of *Graphium sarpedon choredon* (p. 115)
4, Pupa of *Protographium leosthenes leosthenes* (p. 112)
(Enlarged)

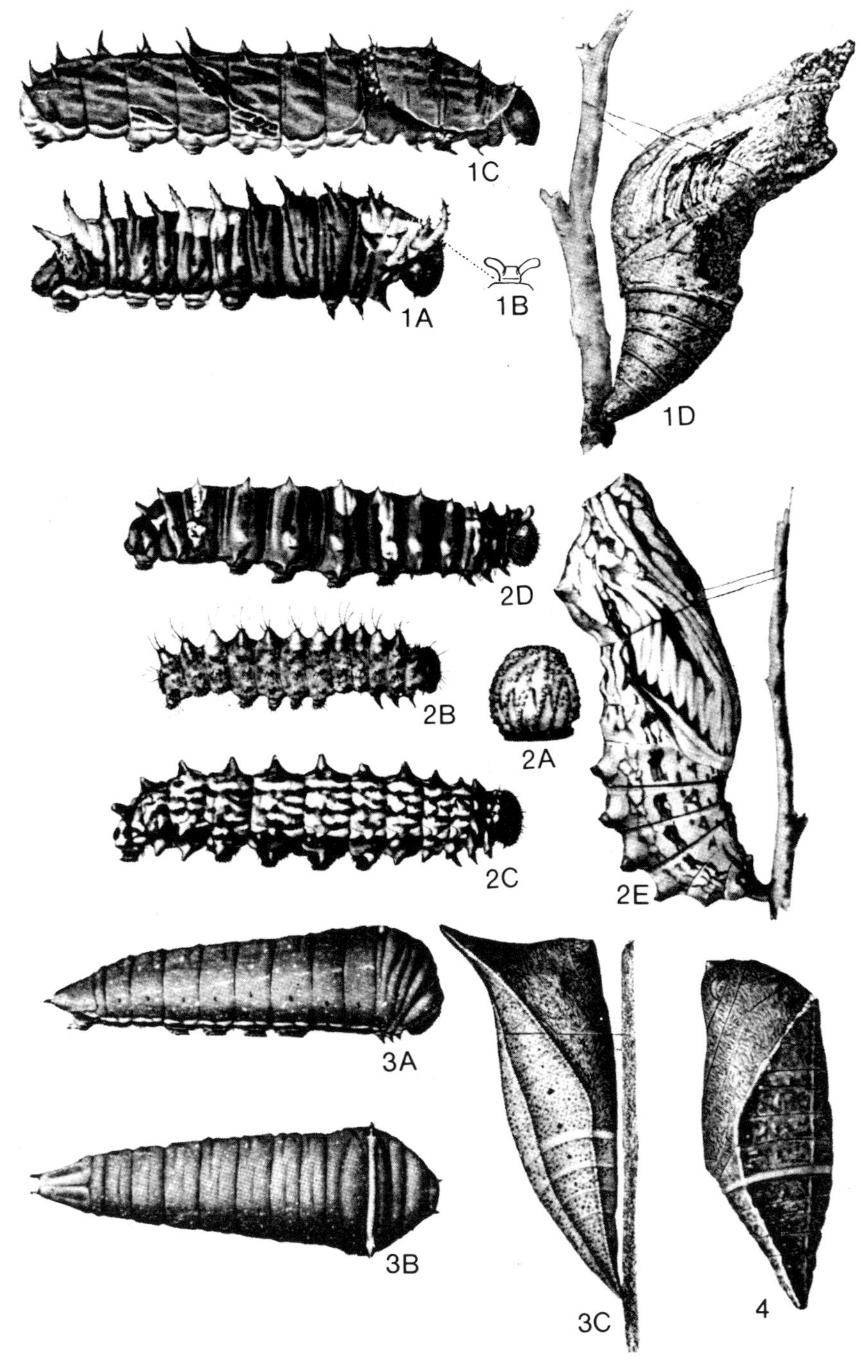

1C
1A
1B
1D
2D
2B
2A
2C
2E
3A
3B
3C
4

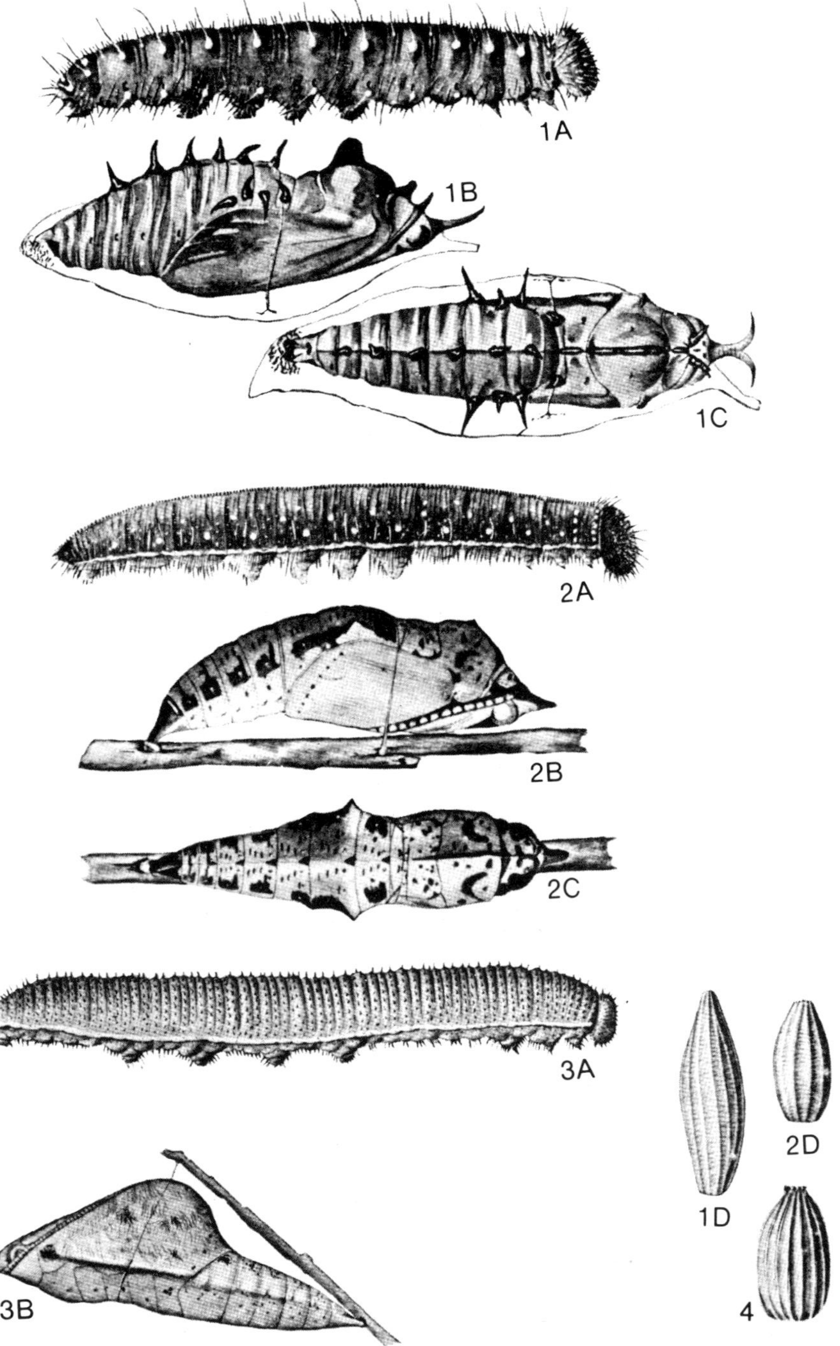

1A
1B
1C
2A
2B
2C
3A
3B
1D
2D
4

Plate X

PIERIDAE AND NYMPHALIDAE

1, 2, Mature larvae of *Delias mysis waterhousei* (p. 143) and *D. aganippe* (p. 144)
3, 4, Pupae of *D. mysis waterhousei* (p. 143) and *D. aganippe* (p. 144)
5, 6, Mature larvae of *Tellervo zoilus gelo* (p. 162) and *Polyura pyrrhus sempronius* (p. 202)
7–9, Pupa, lateral and dorsal, and mature larva of *Geitoneura klugii klugii* (p. 174)
10, Mature larva of *Vanessa kershawi* (p. 216)

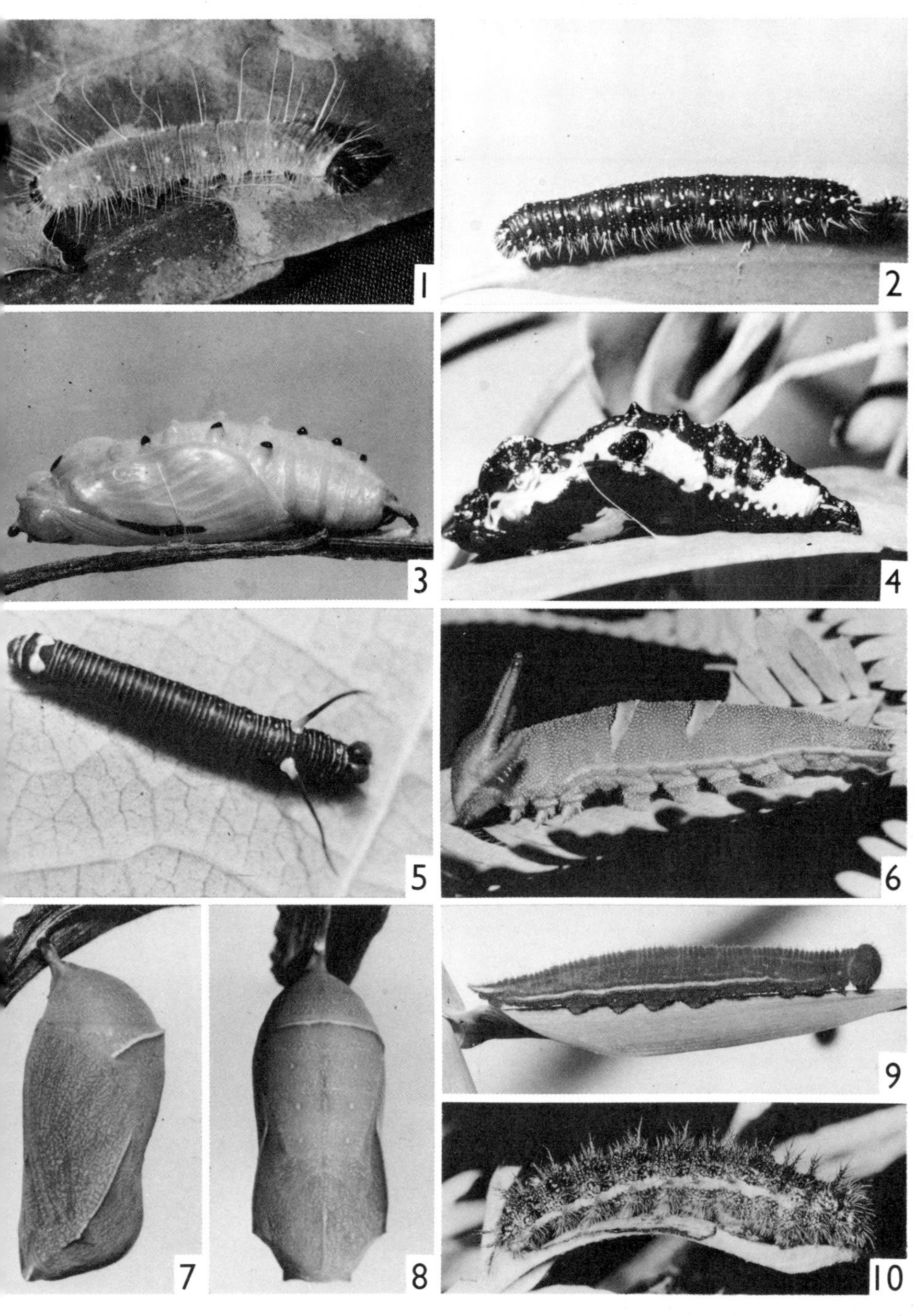

Plate XI

NYMPHALIDAE

1A–1C, Larvae and pupa of *Euploea core corinna* (p. 158)
2A, 2B, Larva and pupa of *Junonia villida calybe* (p. 219)
3A, 3B, Larva and pupa of *Hypolimnas bolina nerina* (p. 212)
(Enlarged)

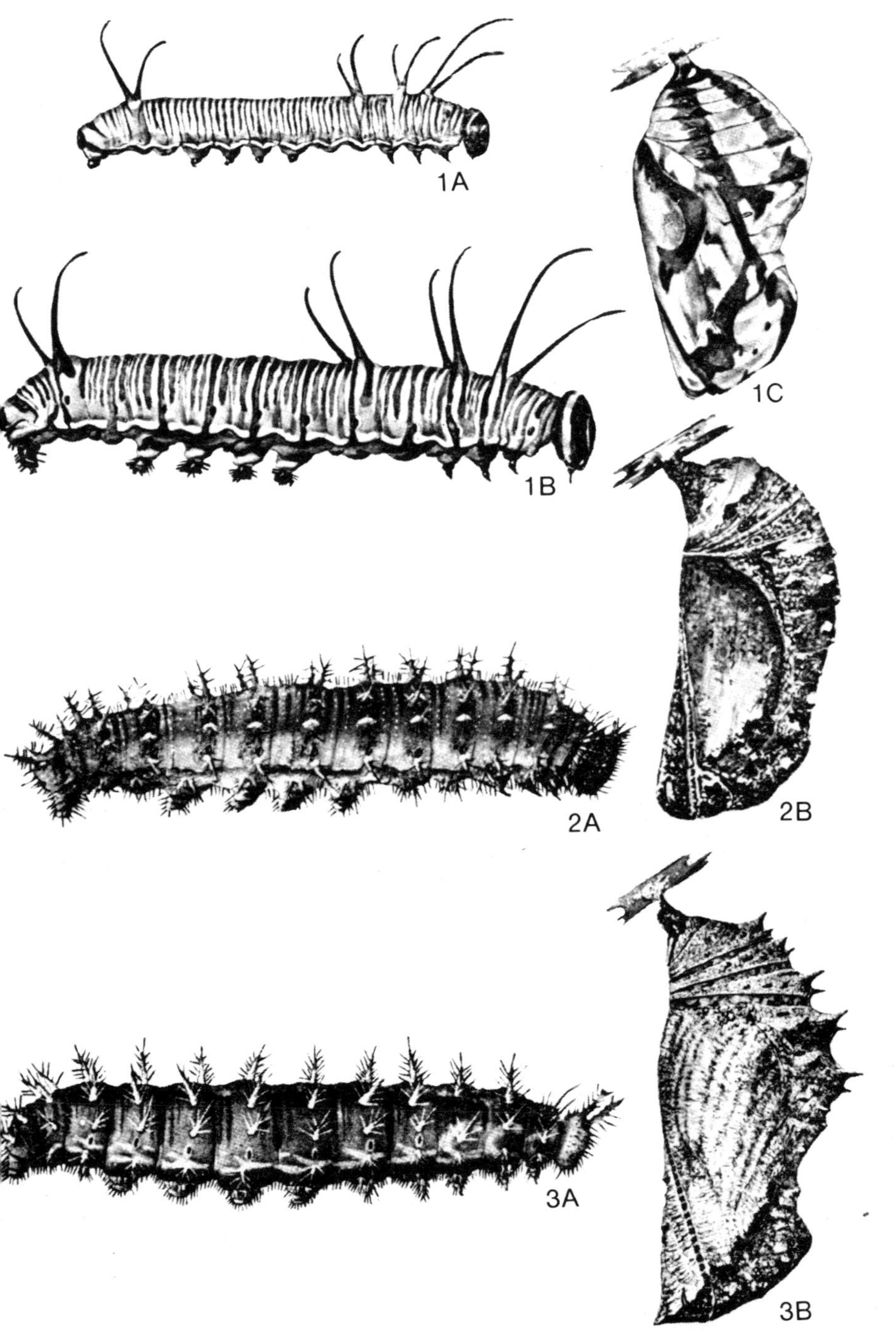
1A
1B
1C
2A
2B
3A
3B

Plate XII

NYMPHALIDAE

1, 2, Mature larva and pupa of *Danaus hamatus hamatus* (p. 156)
3, Pupa of *Polyura pyrrhus sempronius* (p. 202)
4, Mature larva of *Melanitis leda bankia* (p. 164)
5, 6, Mature larva and pupa of *Heteronympha penelope penelope* (p. 178)
7, Pupa of *Oreixenica latialis latialis* (p. 190)
8, 9, Mature larva and pupa of *Oreixenica lathoniella herceus* (p. 192)
10, Mature larva of *Tisiphone abeona regalis* (p. 199)

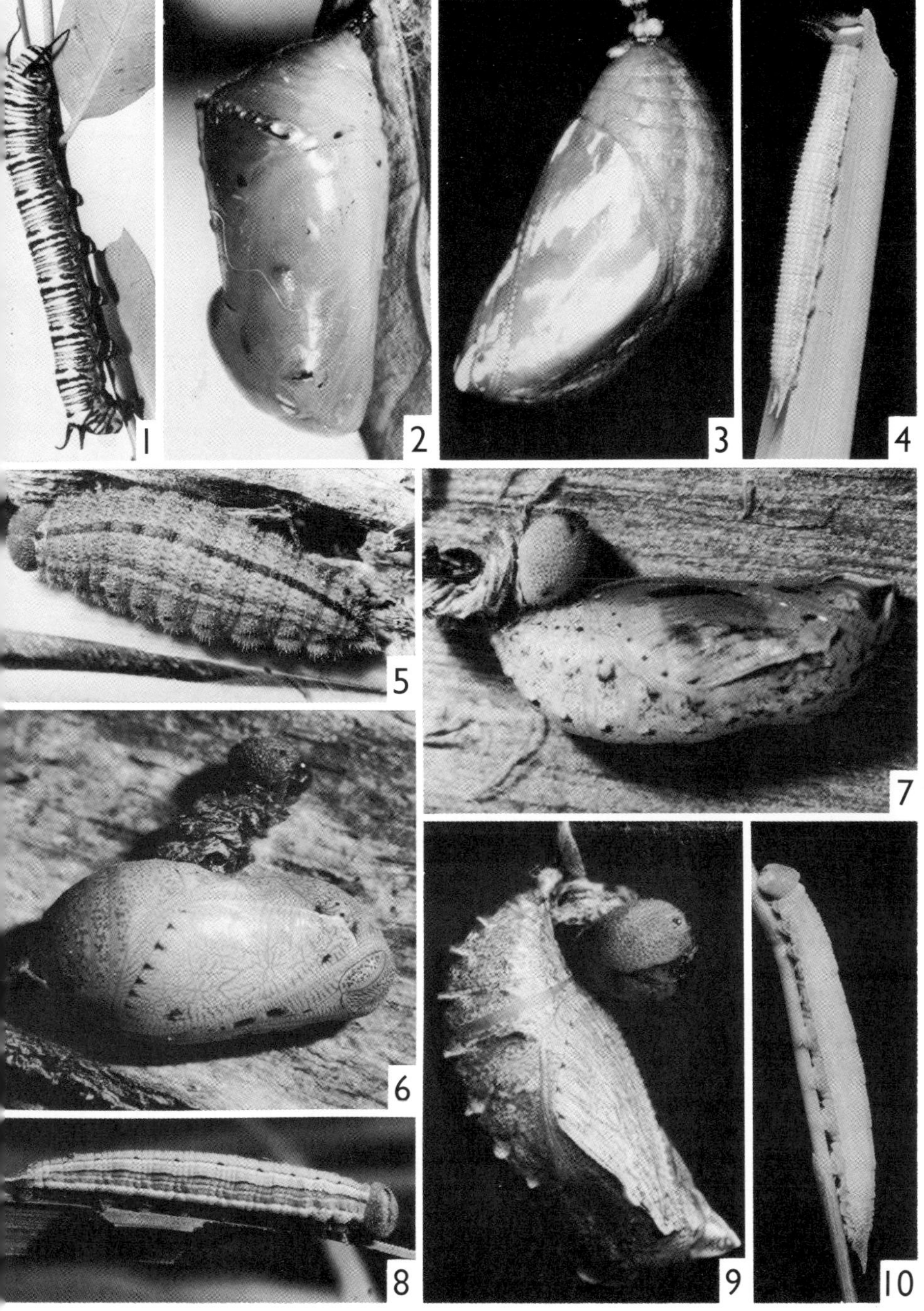

Plate XIII

NYMPHALIDAE

1A–1D, Larvae, larval head and pupa of *Heteronympha paradelpha paradelpha* (p. 177)
2A–2E, Larvae, larval heads and pupa of *H. mirifica* (p. 177)
3A–3F, Larva, larval heads and pupae of *Geitoneura acantha acantha* (p. 172)
4A–4E, Larva, larval head and pupae of *Hypocysta pseudirius* (p. 169)
(Enlarged)

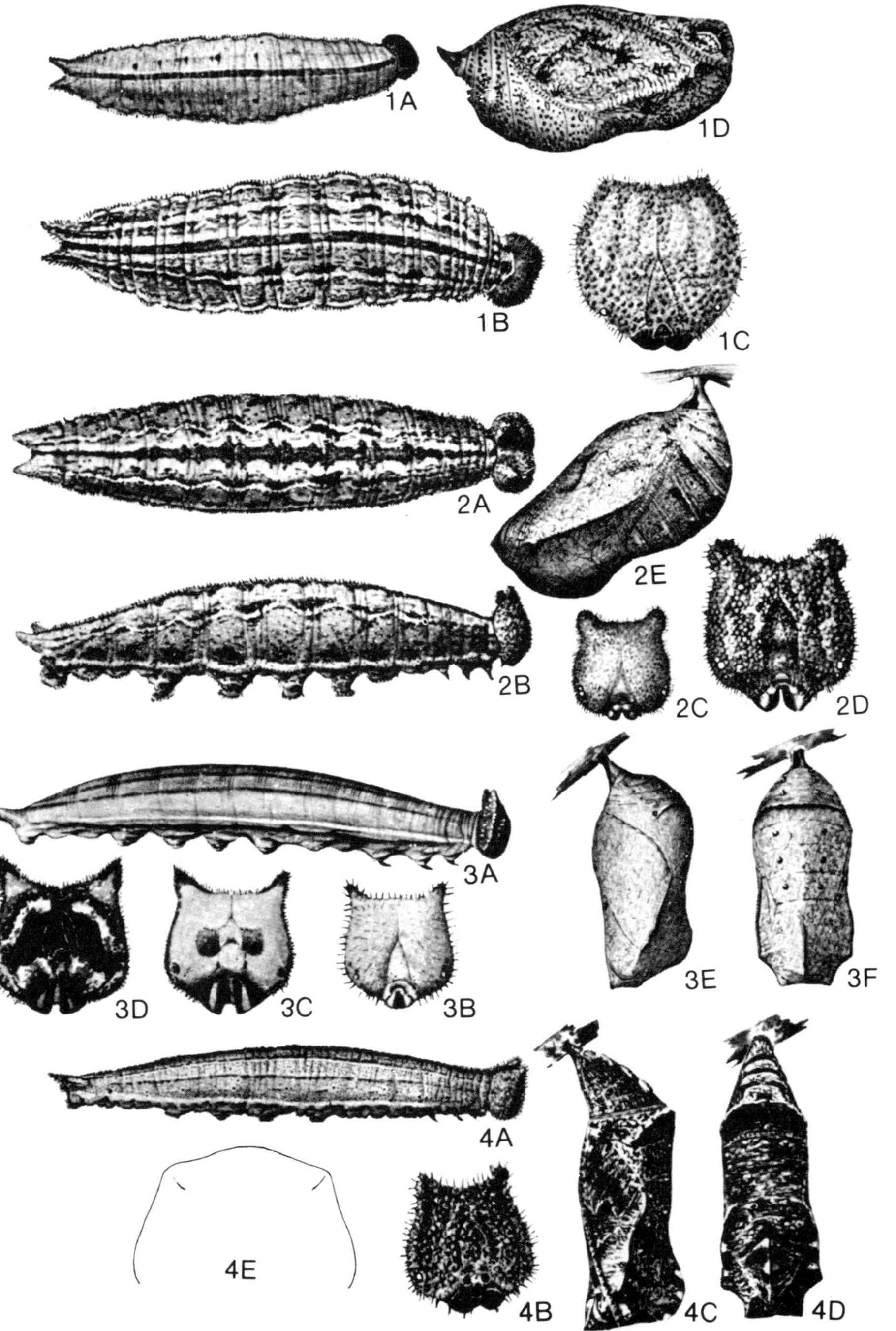

1A

1D

1B

1C

2A

2E

2B

2C 2D

3A

3D 3C 3B

3E 3F

4A

4E

4B 4C 4D

Plate XIV

NYMPHALIDAE

1–5, Larval heads of the five instars of *Polyura pyrrhus sempronius* (p. 202)
6–13, Eggs, larval heads, larva and pupa of *Argynnina cyrila* (p. 170)
(Enlarged)

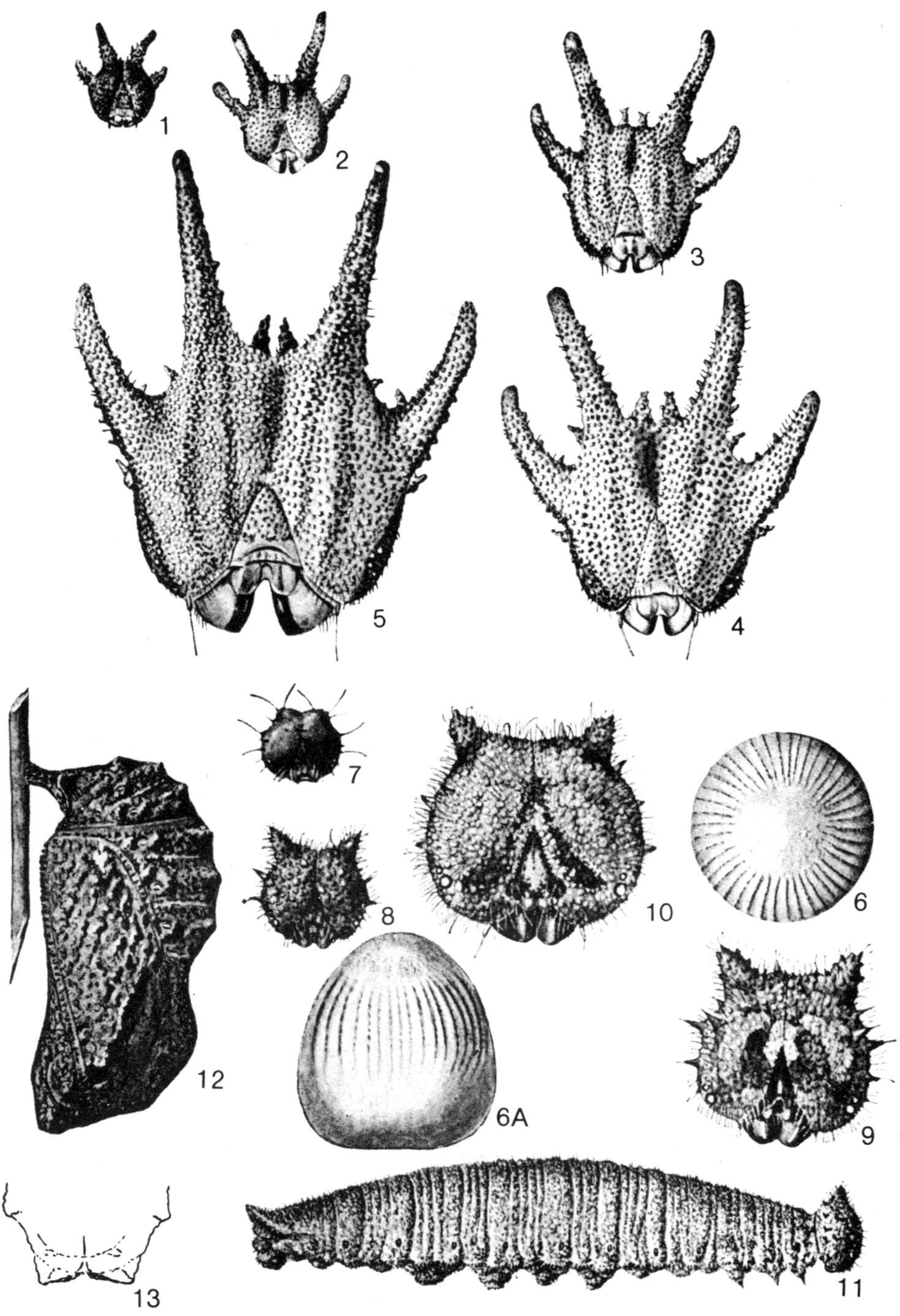

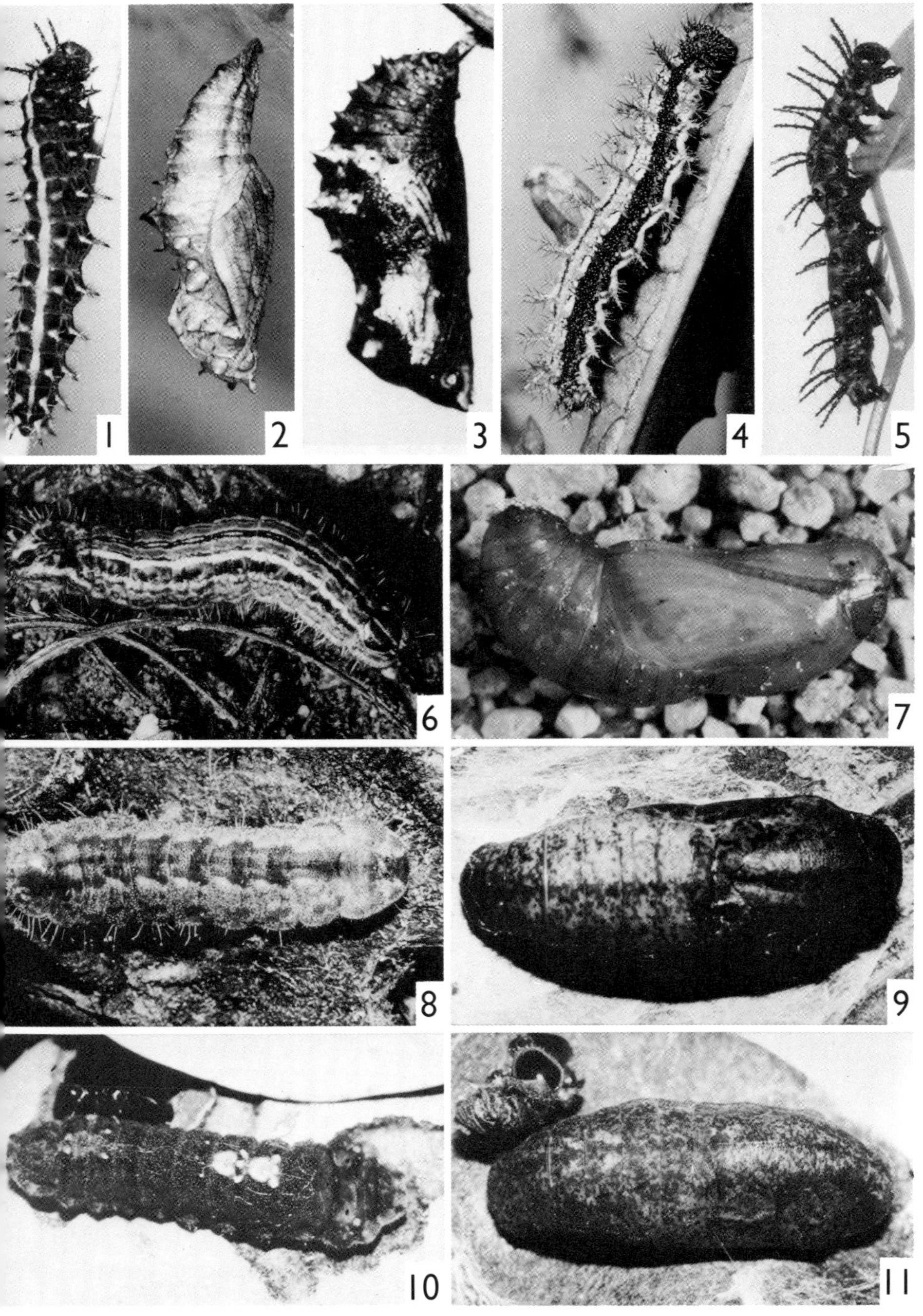

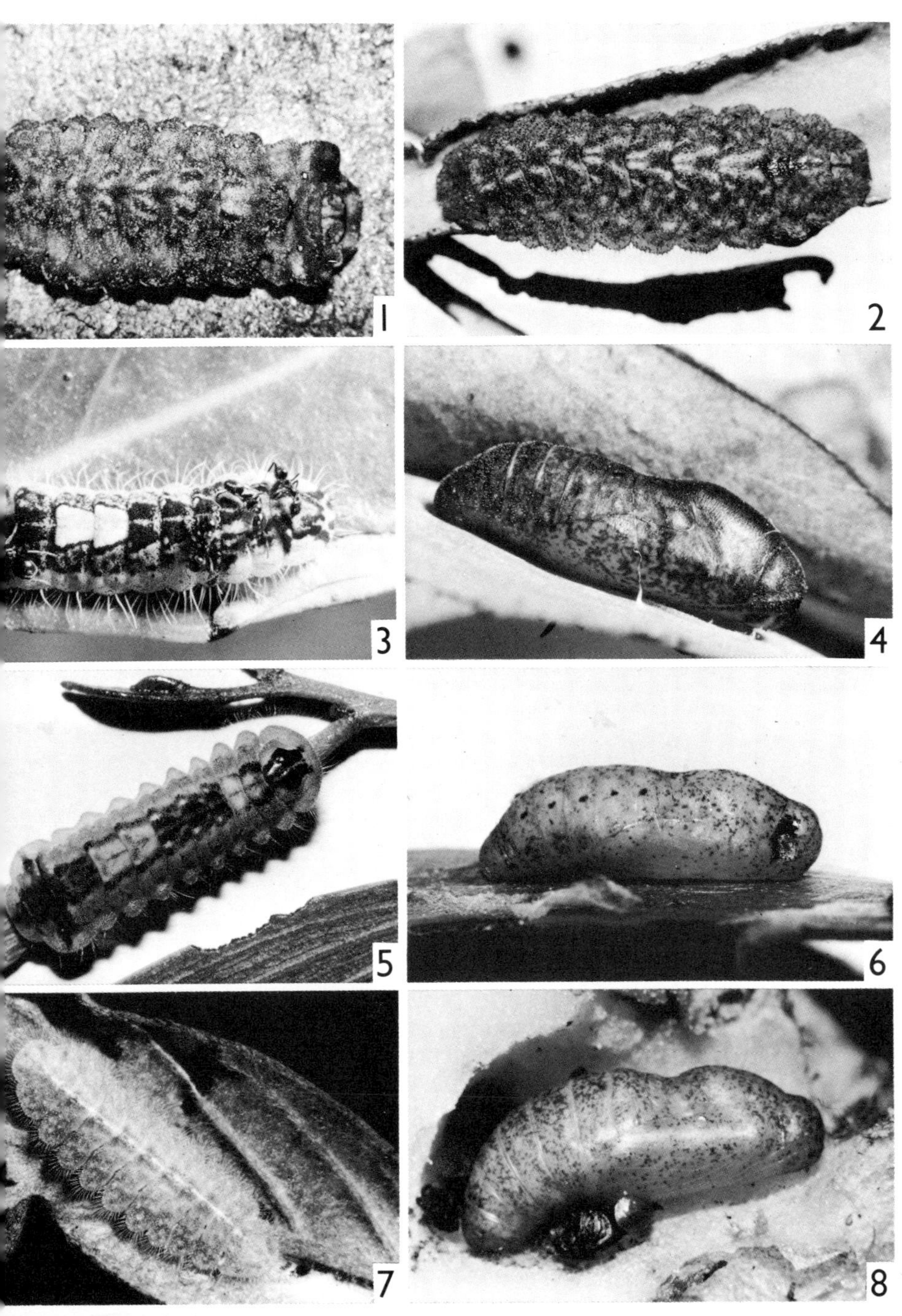

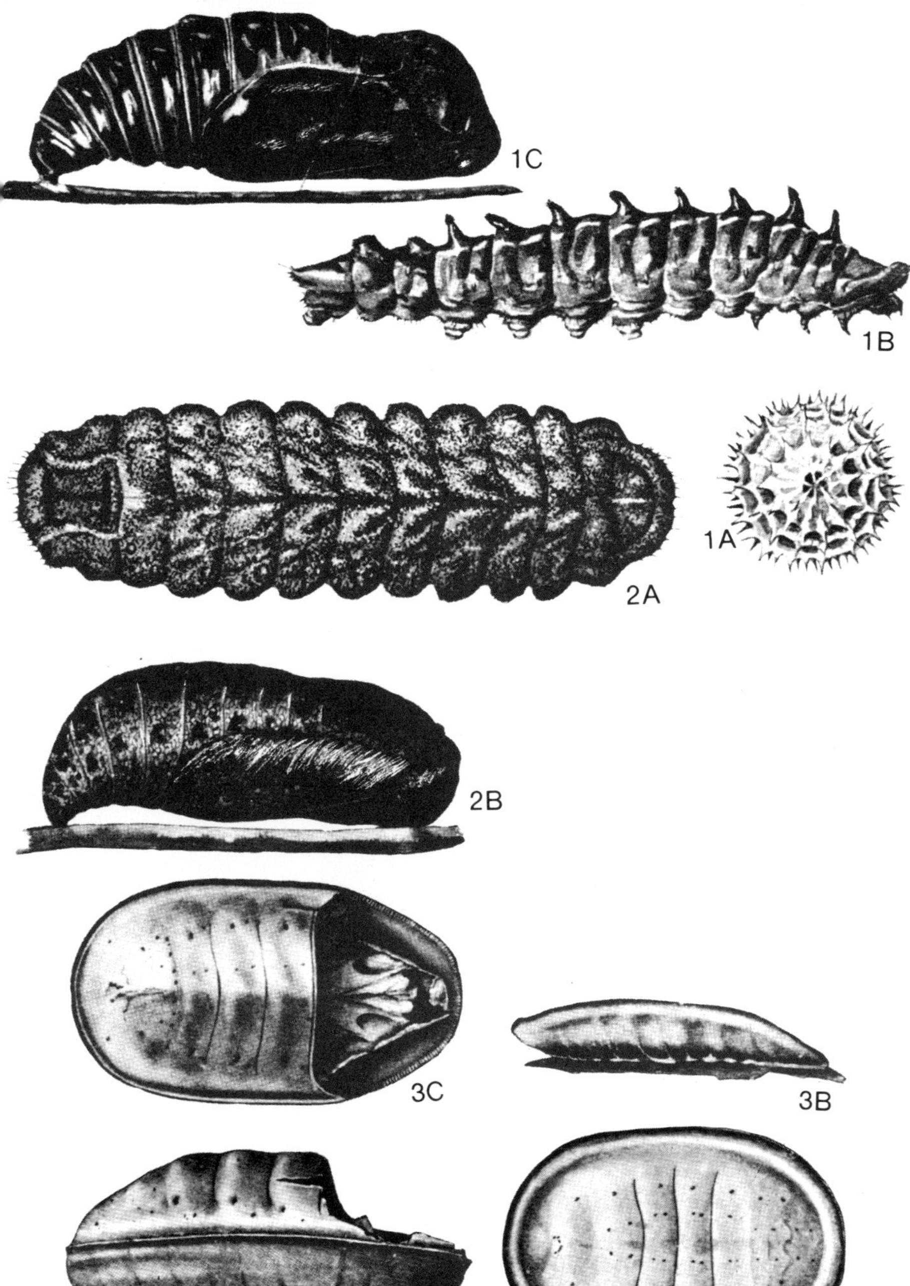

1C
1B
1A
2A
2B
3C
3B
3D
3A

Plate XVIII

LYCAENIDAE

1, 2, Mature larva and newly emerged adult of *Liphyra brassolis major* (p. 230)
3, 4, Mature larva and pupa of *Hypolycaena phorbas phorbas* (p. 295)
5–7, Mature larvae of *Jalmenus ictinus* (p. 284), *J. pseudictinus* (p. 285) and *J. evagoras evagoras* (p. 282)
8, Pupa of *J. icilius* (p. 288)

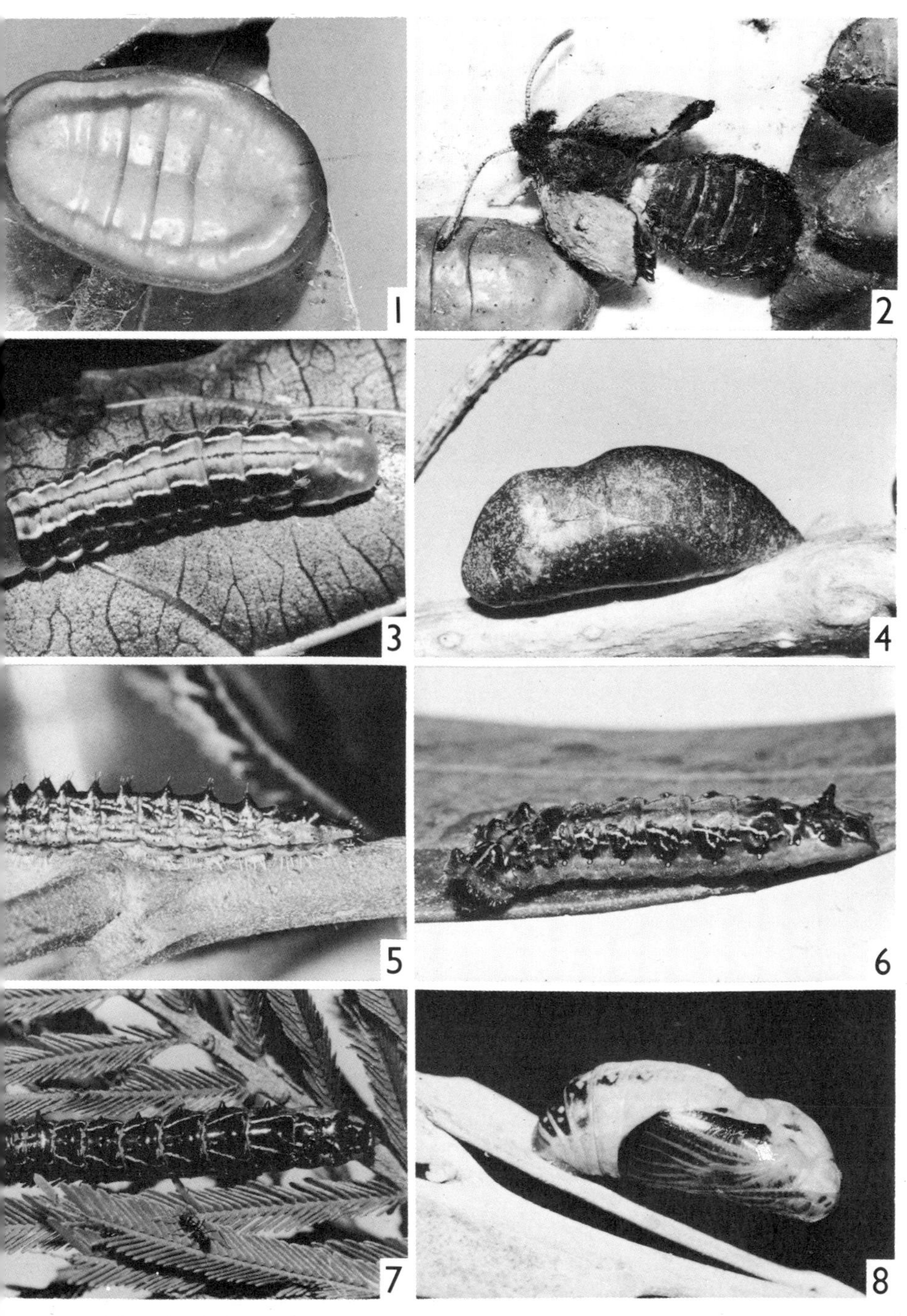

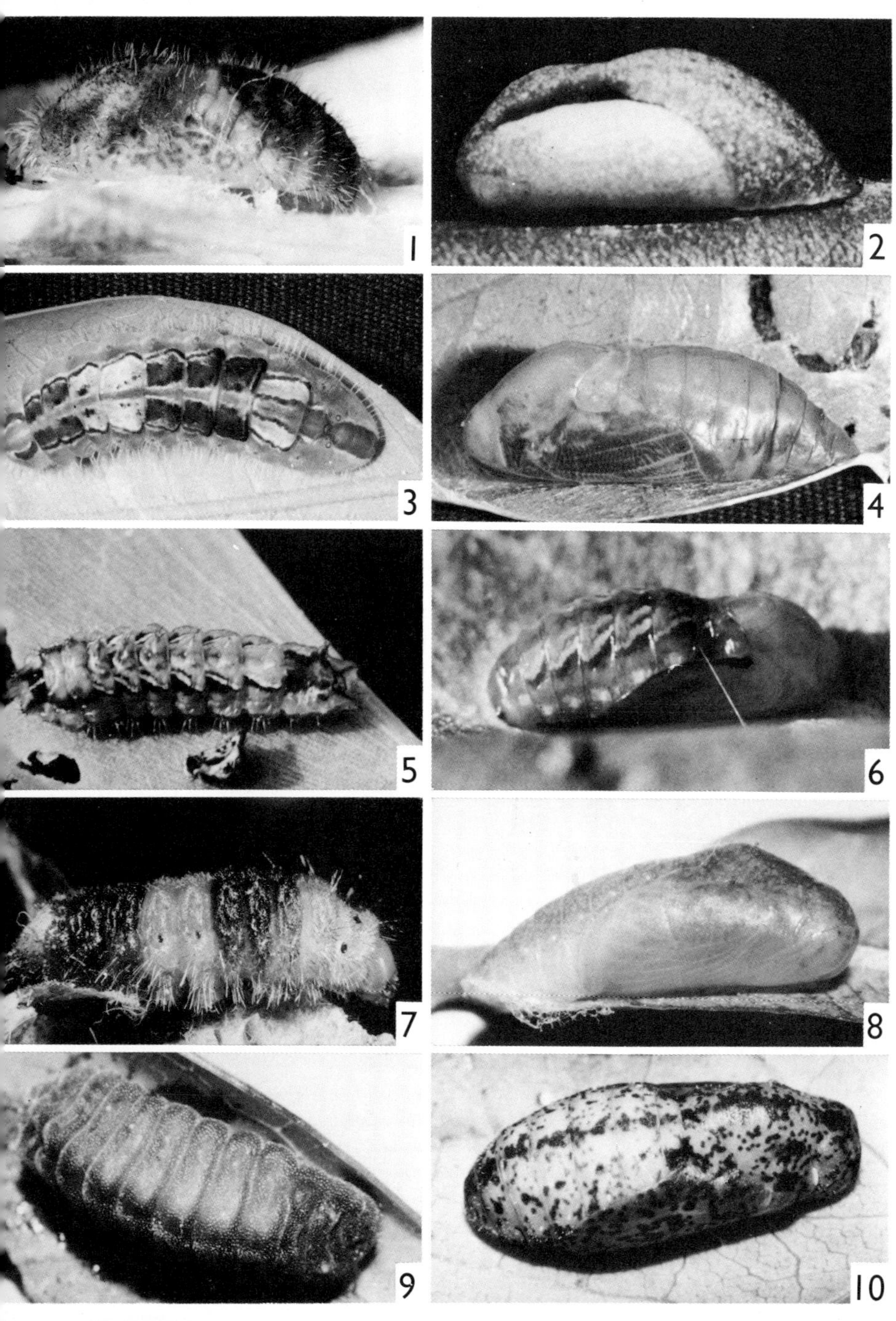

Plate XX

LYCAENIDAE

1A–1E, Larvae and pupae of *Candalides hyacinthinus hyacinthinus* (p. 308)
2A–2C, Larvae and pupa of *Syntarucus plinius pseudocassius* (p. 338)
3A–3C, Larvae and pupa of *Zizina labradus labradus* (p. 340)
4A–4C, Larvae and pupa of *Theclinesthes sulpitius sulpitius* (p. 332)
5A–5C, Larvae and pupa of *Prosotas felderi* (p. 317)
(Enlarged)

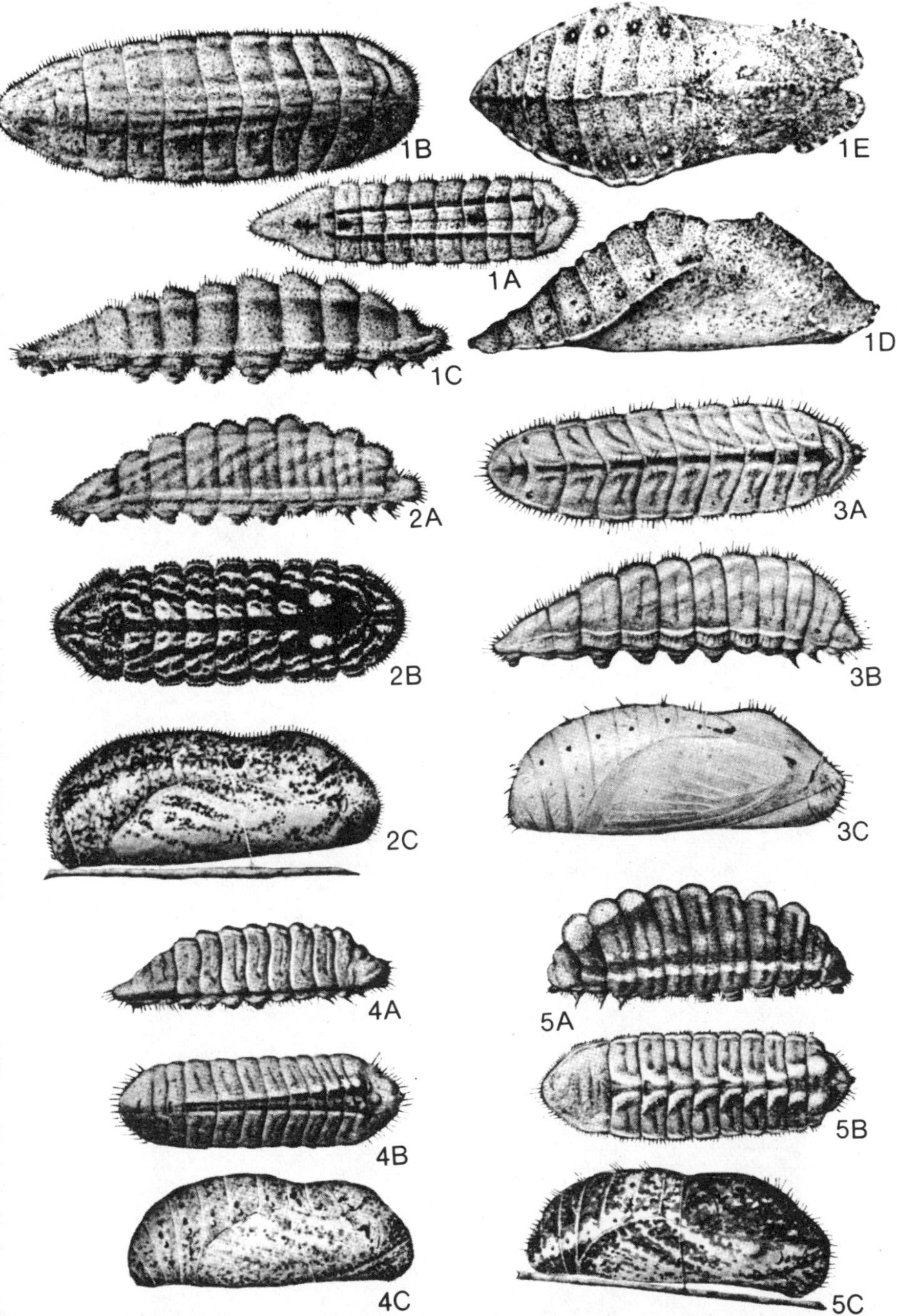

Structure and Life History

EGG

BUTTERFLY eggs are diverse in shape, always with a depression at the top containing the minute aperture or micropyle, through which the sperm from the male enters to fertilize the egg. They may be domed or hemispherical, mandarin- or turban-shaped, spherical, subconical, or spindle-shaped, but always with a more or less flattened base. The egg shell or chorion is sometimes smooth, but more often sculptured in various ways. Frequently there are prominent vertical ribs radiating from the micropylar depression, and sometimes fine horizontal ribs as well.

Butterflies lay their eggs either singly or in groups, usually on a leaf of the larval food plant and often on the underside where they are better protected from extremes of heat and desiccation. They are normally glued to the leaf by a secretion of the female accessory glands. A few species drop their eggs loosely, either during flight or while the female is settled on the food plant or on a nearby object.

LARVA

The butterfly larva or caterpillar (Fig. 1) is elongate and usually roughly cylindrical in shape, but in the Lycaenidae is more or less flattened. It is primarily concerned with the efficient acquisition and utilization of food, resulting in its growth. As its outer wall or cuticle provides for little expansion, it can only grow to its final size by

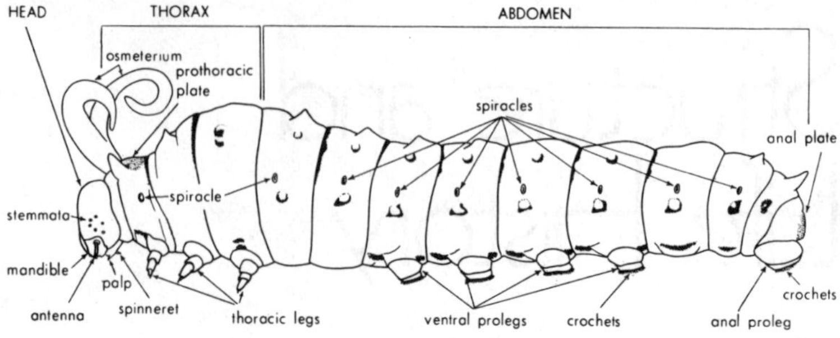

Fig. 1. Structure of mature larva of *Papilio demoleus* (p. 123).

moulting its cuticle several times. Each moult is called an ecdysis and the periods between successive ecdyses are known as instars. Thus a larva in its first instar has hatched from the egg but has not had its first moult. Butterfly larvae usually moult four or five times during their growth; that is, they have four or five instars.

Head   The head is a rounded sclerotized capsule bearing six primitive visual organs on each side, known as stemmata or ocelli, and a pair of small three-segmented antennae furnished with sensilla.

Thorax   The thorax is divided into three segments, each bearing a pair of legs consisting of five short segments and a single apical claw. The first thoracic segment or prothorax usually has a central dorsal plate, the prothoracic plate or shield, and a pair of lateral breathing apertures or spiracles. A fixed pattern of tactile hairs or primary setae is always present in the first instar, though often lost, modified, or supplemented by numerous secondary setae later in larval life. The second and third segments of the thorax, known as the mesothorax and metathorax, are without spiracles, and their pattern of setae differs from that of the other segments.

Abdomen   The abdomen of the larva has ten segments, the first eight of which bear a pair of spiracles each. Segments 3 to 6 each have a

2

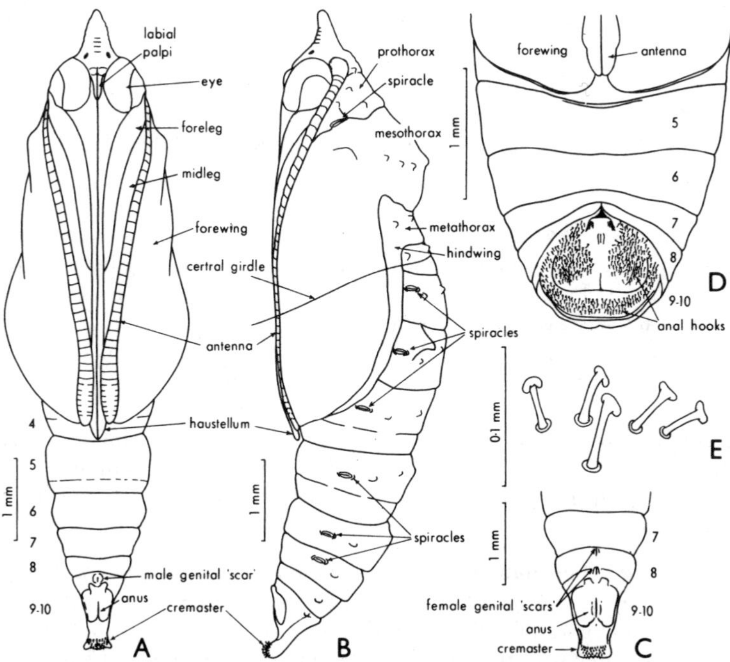

Fig. 2. Structure of pupae: A, B, C, Ventral and lateral views of male, and posterior segments of female abdomen of *Anaphaeis java* (p. 147); D, E, Ventral view of posterior segments of abdomen and anal hooks of *Ogyris amaryllis* (p. 279)

pair of fleshy leg-like organs, the ventral prolegs, and segment 10 has a somewhat similar pair of anal prolegs or anal claspers and a central dorsal plate called the anal plate. The prolegs are used for walking and grasp the twig or other object with a terminal series of tiny claws or hooks, known as crochets.

The larval cuticle is seldom smooth and is generally densely covered with minute spines, called spinules. As in the thorax, each segment of the abdomen carries a characteristic array of primary setae in the first instar. In later instars the number of setae may be reduced or modified in various ways, or supplemented by numerous secondary setae.

PUPA

The pupa or chrysalis is a non-feeding stage in which there is a very active reorganization and transformation of the larval organs to those of the adult. Pupae vary enormously in shape and colour. As they cannot actively avoid enemies they must bluff or warn them or rely on camouflage. Generally speaking, pupae which are formed in an exposed situation are either brightly or aposematically coloured, or match fairly closely the colour of the surrounding objects, and are often green when formed amongst foliage or on green twigs. Not infrequently they are angular or have the skin sculptured in some way. Pupae which are hidden in shelters are generally brown in colour and do not show nearly as much variation in colour or form.

Structure   The main divisions of the pupa (Fig. 2) into head, thorax and abdomen, as well as the appendages, now represent more those of the developing adult rather than the larva. Prominent eyes, antennae and palpi, a central proboscis or haustellum, three pairs of legs, and two pairs of wings may be distinguished, each of which is separately ensheathed and firmly held together within the pupal cuticle.

ADULT

The adult butterfly is responsible chiefly for dispersal and reproduction. It is a familiar object with easily distinguished head, thorax and abdomen. Most of the body and its appendages are covered with a layer of special flattened hairs known as scales.

Head   The head features a pair of large compound eyes, made up of hundreds of facets called ommatidia. They are effective visual organs, enabling the insect to recognize shapes and colours. Sometimes short erect hairs are found between the ommatidia. Ocelli are absent in butterflies but are found commonly in moths, just above the eyes. Arising from between the eyes is a pair of long segmented antennae.

The thick basal segment of each antenna is called the scape, the smaller second segment the pedicel, and the remainder of the antenna the flagellum. In butterflies the flagellum is gradually or abruptly thickened towards its tip to form a club, the slender portion of the flagellum sometimes being called the shaft.

Except for a strongly coiled median proboscis or haustellum and a pair of upturned three-segmented labial palpi, the mouthparts of a butterfly are greatly reduced.

Thorax Three segments, each bearing a pair of legs, are present in the thorax, and the second and third segments, the meso- and meta-thorax, each bear a pair of membranous wings.

The legs (Fig. 3) consist of five main segments, coxa, trochanter,

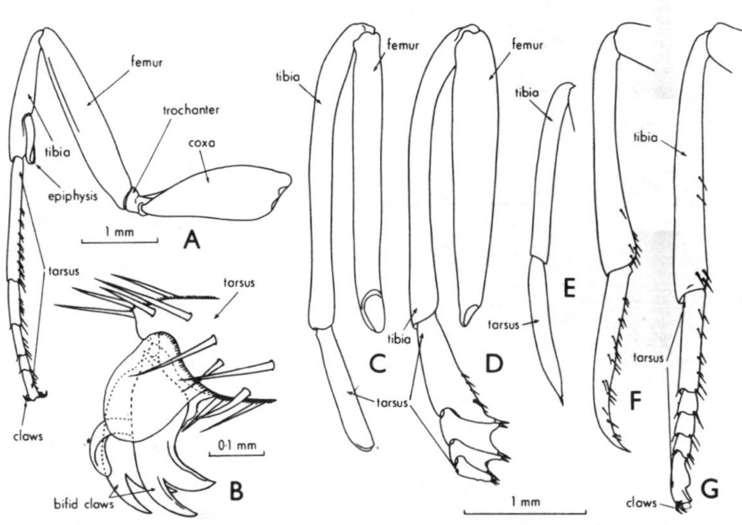

Fig. 3. Forelegs of adult butterflies (below): A, Male of *Toxidia peron* (p. 47); B, Tarsal claws of *Delias nigrina* (p. 146); C, D, Male and female of *Danaus chrysippus* (p. 155); E, Male of *Heteronympha merope* (p. 175); F, G, Male and female of *Jalmenus evagoras* (p. 282).

femur, tibia and tarsus, but the basal two segments are small and the coxa of the mid- and hindlegs is fairly rigidly attached to the thorax. The tarsi are normally five-segmented (Figs. 3A, 3G), with a pair of apical claws, and usually bear chemoreceptors. The forelegs are fully developed for walking in the Hesperiidae, Papilionidae and Pieridae, but are reduced to various degrees, at least in the males, in the Nymphalidae, Libytheidae and Lycaenidae (Figs. 3C–G).

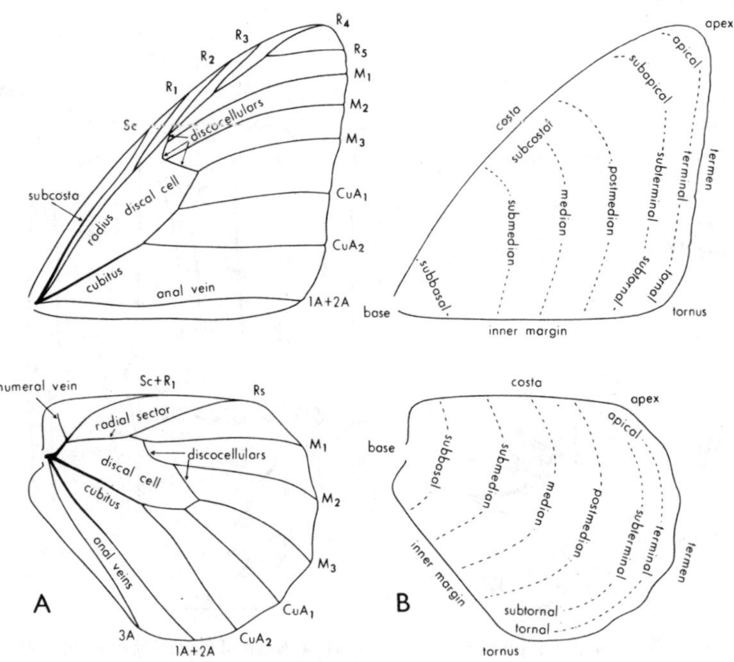

Fig. 4. Structure of wings (p. 7): A, Wing venation, B, Areas of wings, in *Danaus chrysippus* (p. 155).

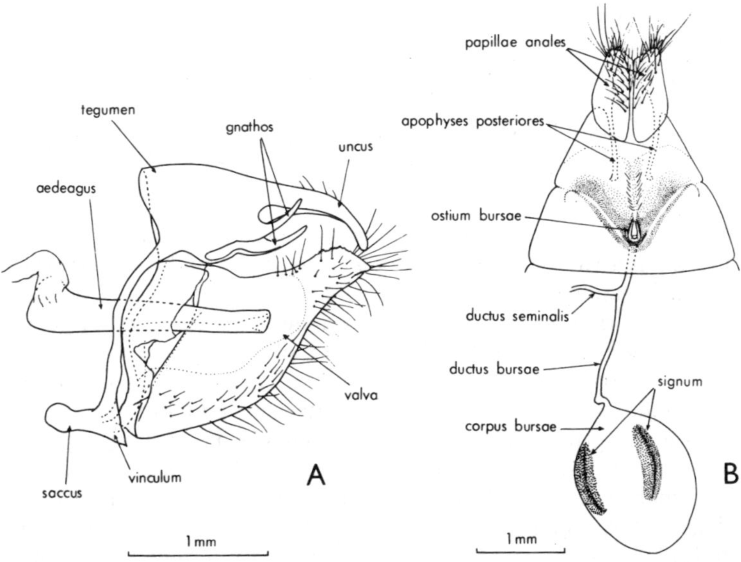

Fig. 5. Structure of genitalia (p. 8): A, Male, lateral view, with left valva removed, of *Argynnina cyrila* (p. 170); B, Female, ventral view, of *Vanessa kershawi* (p. 216).

Adult butterflies always have two pairs of fully developed wings, although wings are sometimes reduced or even absent in a few moths. Each wing is essentially a flattened membranous sac, with the upper and lower membranes pressed closely together and strengthened by a series of hollow tubes called veins (Fig. 4A).

For descriptive purposes each fore- or hindwing is regarded as triangular in shape (Fig. 4B). One corner of the triangle forms the base of the wing, one the apex, and one the tornus. The leading edge of the wing running from the base to the apex is the costa; the outer edge joining the apex to the tornus is the termen; and the trailing edge running from the base to the tornus is the inner margin or dorsum. Names are also given to the main areas of the wings used in technical descriptions.

The upper and lower wing surfaces are densely clothed with overlapping scales arising from minute sockets which, in butterflies, are arranged more or less in transverse rows.

In addition to the normal clothing of wing scales, males of many species carry special scales or androconia associated with pheromone-producing glands, either in characteristic patches called sex-brands or sex-marks, or scattered amongst the normal wing scales.

Abdomen   The abdomen has a total of ten segments, as in the larva, and the last two or three are modified to accommodate the external genital organs.

The external genital organs of the male, and the external genital organs and the internal copulatory ducts of the female, known collectively as the genitalia (Figs. 5A, B), are used widely in the separation of species and even genera of the Lepidoptera including butterflies.

In a few groups of butterflies the males while mating deposit a substance around the female ostium, which soon hardens to form a prominent structure of characteristic shape known as the sphragis (Figs. 6A, 6B). This is believed to form a plug which prevents or delays multiple mating. Amongst the Australian butterflies a prominent sphragis is found only in *Cressida* (Papilionidae) and *Acraea* (Nymphalidae), although much less conspicuous plugs, deposited by the male during copulation, have been reported in many species of butterflies.

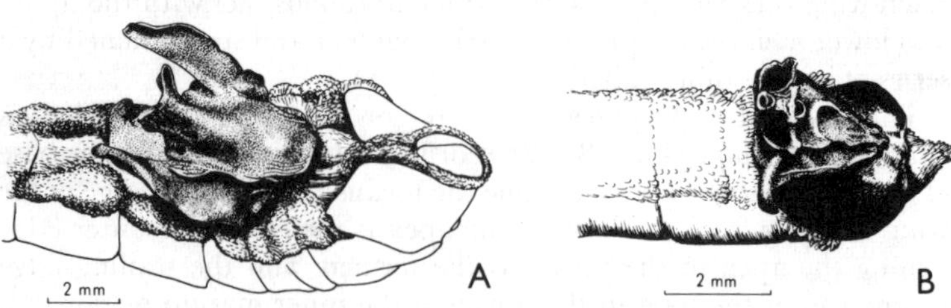

Fig. 6. Sphragis attached beneath female abdomen in A, *Cressida cressida* (p. 125); B, *Acraea andromacha* (p. 224).

8

Geographical Distribution

BUTTERFLIES are found throughout the world except in the polar regions. Some species have an extremely restricted distribution, whereas others range more widely; a few are almost cosmopolitan.

Students of animal distribution or zoogeography have divided the world into six major zoogeographical regions in each of which the fauna is reasonably distinct. These regions are known as the Palaearctic, Ethiopian, Nearctic, Neotropical, Oriental and Australian. The Palaearctic region embraces the whole of Europe, that part of Africa north of the Sahara, and most of Asia north of the Himalayan chain. The Ethiopian region covers Africa south of the Sahara together with Madagascar. The Nearctic and Neotropical regions comprise respectively the continents of North and South America. The Nearctic and Palaearctic have quite closely related faunas and together are referred to as the Holarctic region. The Oriental region covers that part of Asia which lies south of the Himalayas, notably India and south-east Asia, including the Philippines, the Malay Peninsula, and most of the Indonesian islands. Its eastern limit was originally set by A. R. Wallace in 1876 by a line separating Borneo and the Celebes and extending south between the Indonesian islands of Bali and Lombok. The great mass of islands east of this line, together with New Guinea, Australia and New Zealand, he treated as the Australian region. Weber later modified these conclusions, suggesting that a line drawn between the Celebes and the Moluccas and continued south between

Timor and north-western Australia distinguished the Oriental and Australian faunas more effectively. In reality, the broad area between Wallace's and Weber's lines is a transition zone, sometimes called Wallacea, which includes elements from both regions.

The geographical limits of the butterfly fauna included in the present book are roughly the political boundaries of the Commonwealth, including the Australian mainland and its offshore islands, Tasmania,

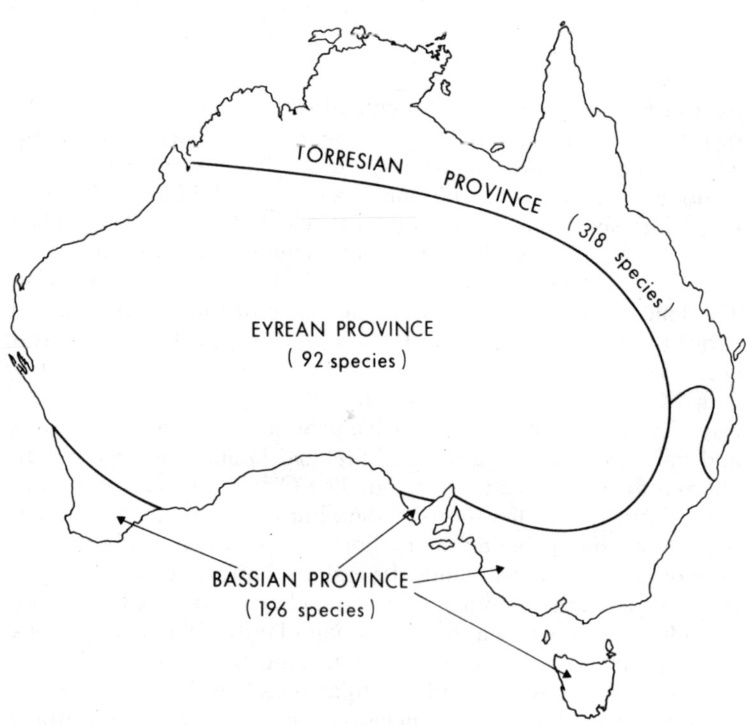

Map 1. Faunal provinces of Australia

Lord Howe Island, and the islands of Torres Strait from Darnley and Murray Islands to Yam, Warraber (Sue), Moa, Badu, Horn, Thursday, Prince of Wales and the associated smaller islands.

The Australian mainland and Tasmania was divided by Spencer late last century into three main faunal provinces (Map 2), *(a)* the Torresian, covering northern Australia, eastern Queensland and north-eastern New South Wales, *(b)* the Bassian, occupying south-eastern New South Wales, Victoria, south-eastern South Australia, south-western Australia, and Tasmania, and *(c)* the Eyrean, comprising the whole of the more arid areas of the continent receiving an annual rainfall of less than about 500 mm annually. The distribution of our butterflies in general supports this subdivision, although in this as in other animal groups, the boundaries must remain reasonably flexible, depending on the group to be studied.

Of the 382 butterfly species recorded from Australia, 174 (46 per cent) are endemic, that is to say, are restricted in their distribution to this continent; and a further 112 (29 per cent) are restricted to the Australian region. The remaining 96 (25 per cent) also have a distribution elsewhere, but 13 of these are predominantly Australian and extend only slightly beyond Weber's line. Another 71 (19 per cent) range widely through the Oriental and Australian regions, 6 of them extending to the Ethiopian and 4 to the Palaearctic. Three or four species have a distribution extending far beyond these limits, notably *Danaus plexippus* which probably reached Australia on the wing more than a century ago, and *Pieris rapae* which is believed to have been introduced accidentally by 1939.

Within Australia the butterflies are primarily a tropical and sub-tropical group. A total of 320 species (84 per cent) has been recorded from the Torresian province compared with only 195 (51 per cent) from the Bassian. They are almost entirely restricted to the less arid areas of the continent, only 92 species (24 per cent) having been recorded from the Eyrean province, several of which range into that area only very slightly or temporarily.

Further additions will certainly be made to the Australian butterfly fauna, either by the discovery of undescribed species or the recognition of species already known elsewhere. The greatest opportunities for collecting new species are in the far north, especially in the Cape York Peninsula, the Northern Territory, and the Kimberley in north-western Australia, where systematic collecting has greatly increased in the last decade. These areas too will yield additional Papuan or Oriental species not so far recorded from Australia. However, more careful collecting and the detailed study of the habits and structure of both adults and immature stages may well reveal additional new species in the more temperate areas.

Classification and Nomenclature

EVERY organism living on the earth today is the result of a long evolutionary history, during the course of which only those species that were best able to cope with their environment at any point in evolutionary time survived.

CLASSIFICATION

All organisms are members of natural populations and the specimens we collect are examples of those populations. A species is a group of interbreeding natural populations that do not normally interbreed with other populations. The species is the unit of animal classification.

The following shows the place occupied by butterflies in the classification of the Animal Kingdom.

Animal Kingdom (all animals)
> Phylum Arthropoda (all animals with jointed legs and an external skeleton or exoskeleton, e.g. crabs, spiders, scorpions, insects etc.)
>> Class Insecta (all true insects, e.g. beetles, wasps, ants, flies, etc.)
>>> Order Lepidoptera (butterflies and moths)
>>>> Superfamilies Hesperioidea (skippers) and Papilionoidea (other butterflies)

From this system it will be seen that the two superfamilies of butterflies (Hesperioidea and Papilionoidea) belong to the great insect order Lepidoptera. This is distinguished from all other orders of insects by a number of characters, including the possession of two pairs of membranous wings clothed with overlapping scales.

There is no single character which will always distinguish butterflies from moths. However, the antennae of butterflies are always clubbed and, except in the male of the Australian genus *Euschemon*, their wings are never fastened together by means of a frenulum and retinaculum. Most moths that have clubbed antennae are day-flying, as are nearly all butterflies, but day-flying moths invariably have a frenulum. At one time the clubbed antennae were thought to separate butterflies from all other Lepidoptera, a conclusion that gave rise to the name Rhopalocera (with clubbed horns) which was applied to butterflies, whereas the name Heterocera (with other horns) was applied to moths. However, these two names have no real scientific significance nowadays.

SCIENTIFIC NAMES

A name provides a convenient and abbreviated means of referring to anything and, if used consistently and unambiguously, always serves to bring to mind the object so named and embraces all of its attributes.

The modern method of naming animals was established by the great Swedish naturalist Carl Linnaeus (or Linné) in the tenth edition of his monumental work *Systema Naturae* published in 1758. Based on the Linnaeus procedure, a binominal system is used in science for naming each species. The scientific name therefore consists of two words, the generic name followed by the specific name.

Species that include two or more subspecies are known as polytypic species. The subspecies that was named first is called the nominotypical or "typical" subspecies, and its subspecific and specific names are identical.

Family Hesperiidae

SKIPPERS
(Plates 1–6, 13, 14, 29, I–VI; Figs. 7–12)

THIS is a family of small to medium-sized butterflies, with thick bodies and relatively short wings. The adults have a rapid and jerky flight, giving rise to their popular name. Wing patterns are usually sombre, often consisting of yellow, white, or hyaline spots on a brownish black background. Few species are brightly coloured. These factors and the frequent similarity of wing pattern, making species identification more difficult than usual, have caused many collectors to neglect the skippers. However, it is now becoming more generally realized that this family includes some of our most interesting butterflies, with a very distinctive structure, and unusual life histories and habits.

The head of a skipper is very broad, with prominent smooth eyes, and antennae widely separated at the base. Towards their tips the antennae are gradually thickened to form a club that is usually strongly curved or hooked (Fig. 7).

The eggs of Hesperiidae are usually hemispherical in general shape, with a micropylar depression above; the surface may be smooth, slightly roughened or, more frequently, with vertical ribs. They are deposited singly on or close to the food plant. The larvae (Plates I–VI) have a prominent head, usually of very characteristic shape, and a body that tapers towards each end. The skin may be smooth or finely spinulose, usually without long hairs, excepting sometimes on the head and anal plate. The larvae feed mainly at night and hide during the day in a shelter of some kind. This is often formed from two or more leaves of the food plant drawn together with silk, or even by folding over part of the margin of a leaf. Sometimes the larva hides in a curled dead leaf near the base of the plant.

The pupae (Plates I–VI) are elongate, more or less cylindrical in shape, or with the abdomen tapering posteriorly. Sometimes there are strongly sclerotized and very distinctively sculptured projections at the anterior end. This area of the pupal skin is usually detached when the adult emerges. Known as the pupal cap, head piece, or operculum, its form is often of great value in identifying the species (Fig. 8). Sometimes the surface of the pupa is partly clothed in short dense erect hairs, and in most species a white waxy powder may cover the pupa and the inside of the shelter. Pupation invariably occurs in a shelter of some kind previously prepared by the larva. This is usually the shelter used by the last larval instar, sometimes firmly lined with silk. Occasionally the mature larva leaves the food plant and pupates in a shelter formed from a curled dead leaf on the ground. The pupa is attached to the silk lining of its shelter by the cremaster, and often by a central silk girdle. The pupal skin is finally cast off within the shelter and the newly emerged adult crawls out before the wings begin to expand. Special care should be taken with hesperiid pupae if normal adults are to emerge in captivity.

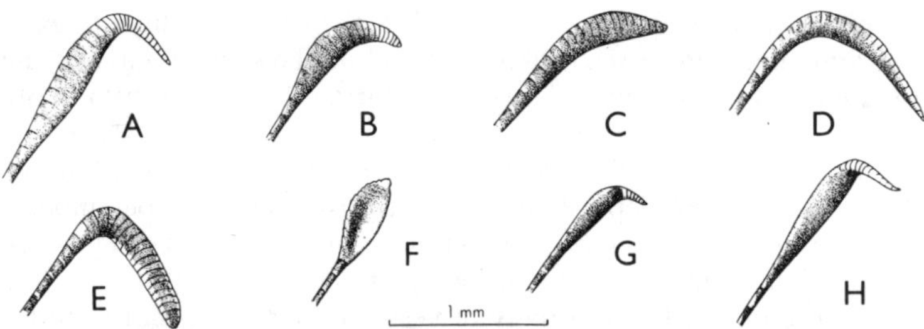

Fig. 7. Antennal clubs in male Hesperiidae: A, *Trapezites maheta* (p. 33); B, *T. sciron* (p. 35); C, *Anisynta dominula* (p. 43); D, *Signeta flammeata* (p. 46); E, *Motasingha dirphia* (p. 74); F, *Taractrocera ina* (p. 88); G, *Ocybadistes walkeri* (p. 90); H, *Telicota ancilla* (p. 102).

16

Most Australian hesperiid larvae feed on monocotyledons, especially grasses (Poaceae) and sedges (Cyperaceae), the Iridaceae and the Xanthorrhoeaceae, but a few feed on dicotyledons, including several families of broad-leafed trees.

Most adult skippers prefer to fly in bright sunshine and many feed freely at flowers, especially during the morning; a few fly only at dusk. Females are most commonly found flying around or near their food plants. Males may also be found near the food plants, often occupying a small flight territory where they patrol up and down and challenge other males that intrude. They tend to settle on exposed twigs or other prominent objects, frequently returning to this perch, even after a serious disturbance. Should a male be captured, another may take over its territory and even adopt the same twig upon which to settle. This behaviour is especially noticeable on hill-tops, where the males of many species are regularly found.

Subfamily COELIADINAE
AWLS
(Plates 1, 14, I–III)

This subfamily contains fairly large skippers with relatively broad wings and short stout bodies. They have a rapid jerky flight and rest with their wings held erect.

The larvae are brightly coloured and feed on dicotyledons.

Genus *ALLORA* Waterhouse and Lyell, 1914

This genus contains only two species restricted to the Australian region, with a range from the Moluccas to New Guinea, Australia and the Solomons. Both species occur in north-eastern Queensland.

1. *Allora doleschallii doleschallii* (Felder), 1860

Plate 1, fig. 10 (male), peacock awl.

DISTRIBUTION.   Moa, Thursday and Prince of Wales Islands, and Cape York, and from Cooktown to Magnetic Island; uncommon.

This species has not yet been taken in the Claudie River area of Queensland. It is slightly smaller than *A. major*, and may be distinguished from it by the extent of the markings beneath, and by the male genitalia. In New Guinea both species usually occur together, but *A. major* is also found at higher altitudes.

The adults fly high along the margins of rain-forest clearings and rarely settle within reach, even of a long-handled net.

2. *Allora major major* (Rothschild), 1916

Plate 14, fig. 19 (male).

DISTRIBUTION.   Claudie River area, Cape York Peninsula.

In Australia this species has been taken so far only near the Claudie River. It is slightly larger than *A. doleschallii*, from which it may be distinguished by the large subtornal spot of the forewing extending inwards to near the middle of the wing, and by the presence of a small spot between veins Rs and M_1 of the hindwing.

Genus *HASORA* Moore, 1881

The species of this genus have dark brown wings without spots in the male, but with hyaline spots on the forewing in the female. Beneath the wings may be suffused with green or purple and, in the Australian species, the hindwing has a cream or white band. The males of *H. chromus chromus* and *H. hurama* have a slender black sex-brand on the forewing; in *H. khoda* it is reduced to four small elliptical patches of raised sex-scales, and it is absent in *H. discolor*.

3. *Hasora discolor mastusia* Fruhstorfer, 1911

Plates 1, fig. 11 (male), II, green awl.

DISTRIBUTION.   Thursday Island, and Cape York to the Clarence River; common locally.

This species is easily identified by the dark blue-green underside with conspicuous cream markings. It often flies very rapidly and energetically, but sometimes fairly close to the ground where it is not difficult to catch. It is found at rain-forest margins or in openings in the rain forest.

The young larva partly cuts out and folds over a small oval section from the edge of a leaflet of the food plant, beneath which it shelters. It feeds on the folded fragment of leaf or on the leaf surface near by, producing series of small irregular holes. The older larva folds or rolls a complete leaflet to form a shelter. The larvae feed on *Mucuna gigantea* (Fabaceae), a common climbing plant of coastal rain forests. *Macaranga tanaria* (Euphorbiaceae) has also been suggested as a possible host, but this has not been confirmed.

4. *Hasora chromus chromus* **(Cramer), 1782**

Plate 1, figs. 14 (male), 14A (female), common banded awl.

DISTRIBUTION. Northern Territory, islands of Torres Strait, and Cape York to Brisbane or rarely Sydney (J. W. C. d'Apice); in northern areas it is on the wing almost throughout the year, but appears to be rare in southern Queensland.

This species is distinguished by the interrupted whitish band beneath the hindwing. The band is broader than in *H. khoda*, the male of which has a reduced sex-brand.

At Townsville and Cairns the larvae feed on *Pongamia pinnata* (Fabaceae), often disfiguring young trees. They live in shelters formed from a folded leaf in which pupation finally takes place.

5. *Hasora hurama hurama* **(Butler), 1870**

Plate 1, fig. 13 (male), broad-banded awl.

DISTRIBUTION. Northern Territory, Moa Island, and Cape York to Mackay; adults are on the wing throughout the year. A worn specimen has been taken at Cooloola, south of Maryborough (M. De Baar).

This species may be recognized by the broad white continuous band beneath the hindwing. The adults have a rapid, jerky flight and settle frequently. It is a rain-forest species.

At Cairns the eggs are laid on the young shoots of the food plant *Derris trifoliata* (Fabaceae). The larva forms a shelter by joining two or more leaflets together with silk and feeds on the surrounding foliage, often causing considerable damage. Pupation occurs in the larval shelter.

6. *Hasora khoda haslia* **Swinhoe, 1899**

Plate 1, fig. 12 (female), large banded awl.

DISTRIBUTION. Fraser Island and Noosa, Queensland, to Sydney, New South Wales; usually uncommon at Sydney but many have been taken in recent years at Bayview by L. C. Haines. It is mainly a coastal species and rarely occurs as far inland as Toowoomba. Adults are on the wing from September to May.

This species has the narrowest band beneath the hindwing; it is interrupted near the tornus as in *H. chromus*, the male of which has a much better developed sex-brand. It is a rain-forest species, but adults are occasionally seen in gardens flying around *Wisteria* or feeding at flowers. L. C. Haines has found that it is active mainly at dusk.

The larvae feed on *Millettia megasperma* (native wisteria, Fabaceae) and are found in shelters formed from a folded leaflet. On one occasion, when the native food plant was not available, larvae accepted cultivated *Wisteria* sparingly. The larvae may occasionally occur on garden *Wisteria*, for J. F. R. Kerr has taken an empty pupal skin on this plant which probably belonged to this species. Pupation occurs in the larval shelter.

Genus *BADAMIA* Moore, 1881

Containing only two species, this genus may be separated from *Hasora* and *Allora* by the elongate narrow forewing, and the hindwing which has a strongly concave termen near the tornus. The hind tibia of the male has a hair tuft which fits into a special pouch on the thorax.

7. *Badamia exclamationis* (Fabricius), 1775

Plates 1, figs. 15 (male), 15A (female), I, III, brown awl.

DISTRIBUTION.   North-western Australia, Northern Territory, the islands of Torres Strait, and Cape York to eastern Victoria; adults are on the wing throughout the year in northern areas, but only during the summer in the south; rare at Sydney and farther south.

Adults vary mainly in the development of the hyaline wing spots, the two extreme forms being figured. There appears to be little geographic variation throughout the enormous range of the species. Geographic uniformity is characteristic of very mobile migrant butterflies: vast flights of *B. exclamationis* have been observed on many occasions in north-eastern Australia and recently by M. S. Moulds near Broome, Western Australia. In Queensland these migratory flights seem to be in a general southerly direction in early to mid-summer, whereas mass movements of the next generation are usually in the reverse direction in late summer. Both sexes migrate in about equal numbers, and breeding may intervene where suitable food plants are available.

In hot weather many of these skippers are sometimes noticed on damp sand at the margins of watercourses sucking up the moisture.

The larvae feed on *Terminalia catappa* (Combretaceae) in northern coastal areas, and on *T. oblongata* (yellow wood) in subcoastal areas of central Queensland. They are sometimes so numerous that their food plants are completely defoliated. The larvae are found in shelters loosely formed from a folded leaf of the food plant. Normally the pupae occur in the larval shelter attached to a pad of silk by the cremaster and a central girdle. When larvae are very abundant and the foliage of the food plant is exhausted, pupation often occurs in various, often exposed situations near by.

Subfamily PYRGINAE
FLATS
(Plates 1, I–III, V)

The skippers in this subfamily are large and robust, with broad wings which are expanded flat when the insect is at rest. They fly rapidly and feed freely at flowers. Some fly only in strong sunshine, whereas others are active only at twilight. The species are mainly of tropical or subtropical distribution. Nine species in five genera are known from within Australian limits.

The larvae are brightly coloured or uniformly whitish in colour, with a large head; they are found in shelters formed by joining or folding leaves of the dicotyledon food plants. The pupae are found in similar shelters.

Genus *EUSCHEMON* Doubleday, 1846

This genus includes only the one Australian species, of considerable interest because wing coupling in the male, effected by means of a frenulum and retinaculum, resembles that found in the majority of moths. Both of these structures are absent in the female.

8. *Euschemon rafflesia* (W. S. Macleay), 1826

Two subspecies are recognized, one in north-eastern Queensland and the other in southern Queensland and northern New South Wales.

a. *Euschemon rafflesia rafflesia* (W. S. Macleay), 1826
Plates 1, fig. 1 (male), I, III, V, regent skipper.

DISTRIBUTION.   Many Peaks and Fraser Island (G. B. Monteith), Queensland to Port Macquarie and Bungwahl, New South Wales. Adults are on the wing mainly from January to May, but an adult has been reared in late September from larvae collected near Gympie (M.

De Baar), and some have been reported on the wing at Burleigh Heads in late October (A. N. Burns) and at Port Macquarie in late November (G. Daniels).

Variation is not marked, but in rare examples the yellow hyaline markings are cream or white. The adults have a rapid jerky flight in bright sunshine near the margins of rain forest, and settle with wings outspread on a leaf or, when feeding, at *Lantana* or other flowers.

The larvae feed on *Wilkiea macrophylla* and *W. huegeliana* (Monimiaceae), shrubs or small trees with large coarse leaves, growing in rain forest. They are found in shelters formed by joining two adjacent leaves with silk, and at night eat irregular areas around the edges of the leaves. Larvae have also been recorded on *Tristania conferta* (Myrtaceae), but G. Sankowsky has found that larvae would not accept this plant. Pupation occurs in the larval shelter, the pupa being attached by the cremaster and a special type of silken girdle. The cremaster is fixed to a vertical silken thread stretched across the inside of the shelter. The girdle is formed of two threads of silk attached at one end to the side of the shelter near the wings, joining over the mesothorax, and then passing as a single thread to the opposite side of the shelter. This girdle is incorrectly shown in Plate I, fig. 4D passing between the second and third abdominal segments.

b. *Euschemon rafflesia alba* Mabille, 1904

Plate 1, fig. 2 (male).

DISTRIBUTION.   Cooktown to Paluma, north Queensland, both on the coast and on the tablelands. Specimens have been taken at Kuranda from March to September.

ADULT.   Similar to typical *rafflesia*, but bright yellow hyaline spots usually larger, and submarginal bands above and beneath usually more yellowish green and more extensive.

The life history is similar to that of typical *rafflesia*. In north Queensland the food plant is said to be *Tetrasynandra* (Monimiaceae), a shrub growing along the margins of rain forest.

Genus *CHAETOCNEME* Felder, 1860

The genus contains twelve species restricted to New Guinea, the Bismarck Archipelago and Australia. Four species occur in north-eastern Australia, one of which also occurs in the Northern Territory. A specimen of this genus has been seen, though not collected, at Yampi Sound, north-western Australia. The adults usually fly at dusk, resting during the day with wings expanded flat on the undersurface of leaves.

9. *Chaetocneme denitza* (Hewitson), 1867

Plate 1, fig. 4 (male), rare red-eye.

DISTRIBUTION. Northern Territory, including Elcho Island (I. Morris) and Groote Eylandt (J. Waddy), and Queensland from Cape York to Mt Tamborine and Canungra (A. Bedford-Russell); uncommon even in the north and apparently rare near Brisbane. It is on the wing from November to April.

Specimens from the Northern Territory are smaller and paler than Queensland specimens.

Larvae are stated to live in shelters formed by folding a leaf, or joining adjacent leaves, of *Tristania conferta* (Myrtaceae).

10. *Chaetocneme beata* (Hewitson), 1867

Plates 1, figs. 3 (male), 3A (female), I, common red-eye.

DISTRIBUTION. Cairns, north Queensland, to Wollongong, New South Wales. It appears to be uncommon nowadays in the Sydney area, but is taken more frequently on the coast and tablelands in southern Queensland; a specimen has been taken at Edungalba, central Queensland (A. F. Atkins), more than 100 km from the coast.

There may be some slight seasonal variation in this species, but there is even more variation amongst individuals taken at the one time of the year. Living adults have conspicuous red eyes which retain their colour fairly well after death. Both sexes are active at dusk and feed at flowers. They are sometimes attracted to artificial lights, especially the females. During the day they may be taken at rest on the undersurface of leaves, but they are easily disturbed and then fly rapidly for a short distance before once more taking shelter amongst dense foliage.

The larvae are found between joined leaves of the introduced *Annona reticulata* (custard apple, Annonaceae) and *Cinnamomum camphora* (camphor laurel, Lauraceae); native food plants include *Tristania conferta* and *Acmena* (both Myrtaceae). They feed only at night and at rest assume a hunched posture (Plate I, fig. 3A). Larvae grow slowly during the winter and pupate in the spring. From these pupae adults emerge in September and October. A second generation of larvae feed during the summer, producing adults in late summer and autumn. Pupation takes place in the larval shelters.

11. *Chaetocneme critomedia sphinterifera* (Fruhstorfer), 1910

Plate 1, fig. 6 (male), banded red-eye.

DISTRIBUTION.   Cape York to the Claudie River. At Cape York specimens have been taken from January to April, and in July and November.

This species and *C. porphyropis* are the only Australian species of the genus with a costal fold in the male. In its normal closed position it is difficult to distinguish, but when open (Plate 1, fig. 6) is paler in colour than the remainder of the wing.

12. *Chaetocneme porphyropis* (Meyrick and Lower), 1902

Plate 1, fig. 5 (male), purple brown-eye.

DISTRIBUTION.   North-eastern Queensland; mainly from Kuranda, where adults have been taken in May, and from September to November. The original specimen came from the Johnstone River. The species is confined to Australia.

The costal fold in the male forewing is narrower than in *C. critomedia*. Adults fly at dusk in rain forest.

Genus *NETROCORYNE* C. and R. Felder, 1867

This small genus contains two species, one from eastern Australia and the other from the Moluccas, New Guinea and the Aru Islands. The adults fly during the day and feed at flowers. They rest with wings extended flat, usually on the underside of a leaf.

13. *Netrocoryne repanda* C. and R. Felder, 1867

Two subspecies are recognized, one ranging from central Queensland to Victoria, and the other confined to north-eastern Queensland.

a. *Netrocoryne repanda repanda* C. and R. Felder, 1867

Plates 1, fig. 8 (male), II, III, eastern flat.

DISTRIBUTION.   The coast and tablelands of eastern Australia, from South Molle Island and Mackay to north-eastern Victoria. It has been taken as far inland as the Carnarvon Range in Queensland, 14 km west of Coonabarabran, N.S.W. (J. R. Turner), and at altitudes up to 1,400 m in the A.C.T. The species has an annual life cycle, and adults are on the wing from September to February.

Adults of this species fly either in rain forest or eucalypt forest. In the latter males are usually seen flying on hill-tops, but may also be taken visiting flowers.

Soon after hatching the young larva constructs a shelter, usually about a centimetre from the egg-shell. It almost completely cuts out a circular disc about one centimetre in diameter from the leaf and folds this over the upper surface of the leaf. Having fastened down the edges with silk, it then cuts a small circular hole at the edge of the disc through which it can pass. The young larva hides in this flattened shelter during the day, emerging at night to feed on the surrounding leaf surface. When it grows too large for this it constructs a much larger shelter by cutting and rolling over an elongate section of the leaf, securely fastening it down with silk and leaving the end nearer the leaf stem open. Except for the midrib, the whole of the leaf between the shelter and the stem is progressively devoured. Before feeding on other nearby leaves, the larva firmly attaches the stem of the original leaf to the branch with silk. It finally pupates within this shelter, the pupa being attached to an elongate pad of silk by the cremaster and a silken girdle.

The most usual larval food plant near Sydney is *Callicoma serratifolia* (Cunoniaceae), but the larvae have been reported from plants belonging to several families. Recorded food plants are *Endiandra sieberi* (corkwood, Lauraceae), *Tristania conferta*, *Acmena* (both Myrtaceae), *Elaeocarpus* (Elaeocarpaceae), *Alectryon subcinereus* (Sapindaceae), *Scolopia braunii* (Flacourtiaceae), *Notelaea longifolia* (Oleaceae) and *Podocarpus elatus* (Podocarpaceae). At Stanthorpe and Sydney the larvae have been taken on *Brachychiton populneum* (kurrajong, Sterculiaceae).

b. *Netrocoryne repanda expansa* Waterhouse, 1932

DISTRIBUTION.   North-eastern Queensland; known from Coen, Cooktown (D. Drake), Mt Molloy, Kuranda, and Herberton. This appears to be much less common than the southern subspecies.

ADULT.   Male: similar to typical *repanda* but darker, forewing with whitish streak between costa and cell spot, hyaline spots joined.

Female: similar to typical *repanda* but darker, forewing with whitish streak between costa and cell spot longer than cell spot.

Genus *TAGIADES* Hübner, 1819

This genus is widely distributed from India and China to Australia and the Solomons; it is also represented in Africa and Madagascar. Twelve species are known, many of which are very variable. Two are recorded from north-eastern Australia, but one of these has been taken only on Darnley Island in Torres Strait. The adults fly very rapidly in bright sunlight, usually at the margins of rain forest, and rest with the wings extended flat.

14. *Tagiades japetus janetta* Butler, 1870

Plate 1, fig. 9 (male), black and white flat.

DISTRIBUTION.   Darnley, Moa and Prince of Wales Islands, and Cape York to Shute Harbour (P. Wilson); collected commonly on the Torres Strait islands and at Cape York from January to June and in October. Specimens were recognized but not captured at Mt Dryander, near Proserpine (G. B. Monteith).

The adults fly in rain forest and usually settle on the underside of a leaf with the wings extended flat. J. F. R. Kerr has suggested that this species may have extended its range farther south quite recently. It is now common as far south as Paluma, but early this century was not taken by F. P. Dodd in the Kuranda or Cairns areas, nor by A. N. Burns when he lived at Meringa between 1925 and 1930.

The life history and early stages have not been recorded from Australia. In the Kai Islands the larva feeds on *Ipomoea batatas* (sweet potato, Convolvulaceae).

15. *Tagiades nestus korela* Mabille, 1891

DISTRIBUTION.   From within Australian limits known only from a single male collected on Darnley Island, Torres Strait.

ADULT.   Above forewing brown-black with white hyaline spots, a pair in cell, another between these and costa, two spots near end of cell, and a curved series of five small subapical spots, traces of a white subterminal band; hindwing white, basal one-third, costa and apex broadly brown-black, continued narrowly along termen, two large black subterminal spots towards apex. Beneath forewing as above, hyaline spots a little larger, whitish subterminal band more distinct; hindwing white, costa, apex and termen broadly brown-black, black subapical spots separate from brown-black margin.

Genus *EXOMETOECA* Meyrick, 1888

This genus contains only the one species from south-western Australia. The adults are smaller than related species, but like them rest on leaves with the wings extended flat.

16. *Exometoeca nycteris* Meyrick, 1888

Plate 1, fig. 7 (female), western flat.

DISTRIBUTION.   South-western Australia, from Perth to Albany. Adults are on the wing from October to December.

Meyrick collected the original specimen at Albany and noticed that, when resting on a twig, its wings were extended downwards beneath the thorax with their undersides touching. More usually the adults rest on leaves with the wings extended flat. They fly for about an hour in the morning before ten o'clock and then again for a similar period after midday. They are much easier to capture in the morning.

The early stages and the larval food plant are still unknown.

Subfamily TRAPEZITINAE

(Plates 1–4, 13, 14, 29, I–VI; Fig. 8)

This characteristically Australian subfamily contains skippers of small to medium size, which have a rapid jerky flight and rest with their wings held erect.

The males of this subfamily frequently find their way to the tops of hills, where each flies in a small area often returning time and again to settle on the same twig. The females are seldom taken in these situations unless the food plant happens to be growing near by.

The eggs are hemispherical in shape, usually with vertical ribs. The larvae invariably feed on monocotyledons. The genera can be divided into three groups according to the structure of the larvae and pupae, and their food preferences. *(a)* In *Trapezites, Anisyntoides, Anisynta,*

Dispar, *Toxidia* and *Signeta* the larvae (Plate I, figs. 2A, 2B; Plate II, figs. 4, 5; Plate III, fig. 4) are stout, usually brownish in colour, with a large roughened black or brown head. They feed on *Lomandra* and *Acanthocarpus* (Xanthorrhoeaceae) and grasses (Poaceae). The pupae (Plate I, fig. 2C; Plate III, fig. 5) are also relatively stout, with the abdomen tapering posteriorly and the anterior end rounded and usually without prominent projections. *(b)* In *Hesperilla*, *Oreisplanus* and *Motasingha* the larvae (Plate II, figs. 6–8; Plate III, figs. 6, 7; Plate IV, figs. 1A, 1B, 2A) are more slender, semi-translucent and greenish in colour, often with longitudinal stripes, and resemble those of the Hesperiinae. The head is large and roughened, greenish or brownish in colour, and usually marked with dark bands. They feed on *Gahnia*, *Carex*, *Lepidosperma*, etc. (all Cyperaceae). The pupae (Plate IV, figs. 1C–1E, 2B) are cylindrical, green, brown or black in colour, with the anterior end produced into processes of characteristic shape (Fig. 8), which provide a useful means of identification. With the exception of *Hesperilla chaostola* and *Motasingha atralba*, the larvae pupate head upwards. *(c)* In *Mesodina* the hairy whitish larvae (Plate I, fig. 1A; Plate III, fig. 9) feed on *Patersonia* (Iridaceae). The pupae (Plate I, figs. 1B, 1C; Plate III, fig. 10) vary in colour from pale brown to dull black, with the anterior end smooth. The larvae pupate head downwards. The second and third groups are not sharply defined and *Proeidosa* (Plate VI, figs. 1, 2) embraces features of both.

Genus *RACHELIA* Hemming, 1964

The genus contains only one species which has only recently been recorded from Australia; it also occurs in New Guinea and the Aru Islands. The black wings, with shining light blue bases, make recognition easy.

17. *Rachelia extrusa* (C. and R. Felder), 1867

Plate 14, fig. 12 (male).

DISTRIBUTION. From within Australian limits known from three specimens collected near the Claudie River, Cape York Peninsula, in May and June, 1973 (A. F. Atkins) and five from the same locality in September, 1977 (D. Binns).

The early stages have not been described.

28

Genus *TRAPEZITES* Hübner, 1819

The genus contains twelve species, all of which are restricted to Australia. Ten occur in Queensland, nine in New South Wales, eight in Victoria, four each in the Australian Capital Territory and South Australia, and only one each in Western Australia and Tasmania. Two species are confined to Queensland.

The eggs are roughly hemispherical in shape and are vertically ribbed. The larvae feed at night on *Lomandra* (Xanthorrhoeaceae), and their presence can often be ascertained by recent feeding at the tips of the leaves which are cut off obliquely. During the day larvae hide deep in the *Lomandra* clumps, either amongst the leaf bases, in debris, in curled dead leaves, or in shelters roughly constructed by the larvae. The pupae, which are clothed with fine short hairs, are found in the same situations. The males of all but *T. symmomus* may be taken on hill-tops.

18. *Trapezites symmomus* Hübner, 1823

This is the largest species, with a wing-span of some 50 mm. It occurs on the tablelands of north Queensland, and from central Queensland to Victoria. Three subspecies are recognized.

a. *Trapezites symmomus symmomus* Hübner, 1823

Plate 1, fig. 17 (female), symmomus skipper.

DISTRIBUTION.   From Springsure, the Expedition Range and Kroombit Tops, west of Gladstone (A. F. Atkins), Queensland, to far eastern Victoria; it occurs both on the tablelands and on the coast in southern Queensland, but in southern New South Wales is found only on the coast. Adults are on the wing as early as October in Queensland, but from November to April farther south.

Adults of both sexes are found flying near the food plant, which grows in fairly damp situations, often close to the shore line. Apparently the males do not collect on hill-tops as do those of all the other species of *Trapezites*. Specimens from Springsure and those collected and reared at the Expedition Range are considered by A. F. Atkins to represent a distinct subspecies.

The larva feeds on *Lomandra longifolia*, usually hiding during the day deep down in the clump, in a shelter formed by drawing two or three leaves together with silk. It pupates in this shelter or in a curled dead

29

leaf at the base of the plant. In central and southern New South Wales the eggs are laid in summer and hatch before winter. The larva feeds intermittently throughout the winter, pupating in late spring or early summer. The pupal duration varies from about twenty-five to thirty days. At Stanthorpe there are two generations annually, with adults emerging in October and March.

b. *Trapezites symmomus soma* **Waterhouse, 1933**

DISTRIBUTION. Victoria; most specimens have been taken along Port Phillip Bay, at Mt Oberon, Wilsons Promontory, in the Dandenong Ranges and the Healesville area. The subspecies is also recorded from the Grampians and a specimen has been sighted near Dartmoor, western Victoria (A. F. Atkins). Adults are on the wing from January to March.

ADULT. Similar to typical *symmomus* but forewing with hyaline spots slightly larger and usually with an additional narrow oblique hyaline spot below longest spot; beneath hindwing with more prominent bluish white spots.

The life history is similar to typical *symmomus*, the larvae having been found on *Lomandra longifolia*.

c. *Trapezites symmomus sombra* **Waterhouse, 1933**

DISTRIBUTION. Mt Spurgeon, north-west of Mossman, Kuranda and the Atherton Tableland to Paluma, in December, January and April. Inland it has been taken as much as 65 km south-south-west of Mt Garnet (W. R. Hindson). Its range may well extend farther southward, and a specimen, presumably of this subspecies, was once recorded from Mackay.

ADULT. Similar to typical *symmomus* but slightly paler above and beneath, spots beneath hindwing much smaller and fewer, slightly bluish or dark brown.

The life history is similar to typical *symmomus*. It has been reared from *Lomandra longifolia* at Cairns (M. W. Mules) and at Herberton (A. F. Atkins).

19. *Trapezites heteromacula* **Meyrick and Lower, 1902**
Plate 2, fig. 11 (female), orange white-spot skipper.

DISTRIBUTION. Islands of Torres Strait, and from Cape York to

Kuranda; rare. Adults have been taken in most months.

Above the colour closely resembles that of *T. eliena*. The specimen illustrated (Plate 2, fig. 11) is larger than usual. The species is found in open eucalypt forest and, between Kuranda and Mareeba, J. F. R. Kerr has found it flying in the same situations as *T. macqueeni*, *T. eliena*, *T. iacchus* and *T. petalia*.

At first the young larvae shelter between leaves at the base of the food plant, *Lomandra longifolia*, but in the second instar make a shelter by drawing together leaves of the food plant with silk, emerging at night to feed. Late-instar larvae shelter in a rolled dead leaf secured to the base of the plant with silk, and pupation occurs in these shelters.

20. *Trapezites macqueeni* Kerr and Sands, 1970

Plate 13, figs. 1 (male), 1A (female).

DISTRIBUTION.   North Queensland, from Musgrave (J. C. Le Souef), Hopevale and Fairview, north and west of Cooktown, to the Paluma Range; adults have been taken mainly on hill-tops, between Kuranda and Mareeba, and west of Paluma, from August to October (A. F. Atkins) and March to May.

The white spots beneath the hindwing are often reduced in size and number, or some may have their white centres filled in with black.

The species flies in open *Eucalyptus* forest. Few females have been taken and the early stages are not known.

21. *Trapezites eliena* (Hewitson), 1868

Plate 2, fig. 1 (male), eliena skipper.

DISTRIBUTION.   Eastern and south-eastern Australia, from the Jardine River, near Cape York, to Cooktown and Kuranda, and to southern New South Wales, the Australian Capital Territory, Victoria and south-eastern South Australia; usually common, but rare in South Australia. The original specimen came from Brisbane. In Queensland it has been taken as far west as Springsure and Mitchell. At Stanthorpe there are two generations annually, and the adults fly in October and March, those of the second generation being more abundant. At Sydney, where there are also two generations annually, adults have been taken from August to April.

The hindwing beneath is very variable in colour and the post-median spots may be reduced to brown dots or be entirely absent.

Specimens from western Victoria and South Australia have usually been treated as subspecies *monocycla* Lower, 1911, because of the ochreous yellow colour beneath the hindwing and the absence of the postmedian spots. However, similar specimens also occur in populations in eastern Victoria, New South Wales and Queensland.

The males fly commonly on hill-tops in open eucalypt forest.

The larvae feed on *Lomandra multiflora*, *L. longifolia*, *L. confertifolia*, and *L. filiformis*. The pupal duration at Sydney in the spring is about twenty-five days, whereas a specimen reared in December from a larva taken on *Lomandra* in the Grampians had a pupal duration of sixteen days.

22. *Trapezites iacchus* (Fabricius), 1775

Plate 2, fig. 2 (male), iacchus skipper.

DISTRIBUTION. Prince of Wales and Horn Islands, and Cape York to Grafton; mainly coastal. Except for November and December, adults have been taken throughout the year in the north. In southern Queensland there are two generations annually, adults being on the wing from September to November and again from February to April.

This species is somewhat similar to *T. eliena*, but may easily be separated by the more pointed forewing, especially in the male, the pale yellow rather than orange-yellow spots above, and the constant dull brown coloration beneath. Adult males are usually taken on hill-tops in open eucalypt forest.

The larvae feed on *Lomandra*, including *L. longifolia* and *L. multiflora*. The pupal duration near Brisbane in late summer was thirteen to fifteen days.

23. *Trapezites iacchoides* Waterhouse, 1903

Plate 2, fig. 3 (male), iacchoides skipper.

DISTRIBUTION. Stanthorpe, Queensland (J. Harslett), the New England National Park, Barrington Tops and the Blue Mountains, and on the coast from Gosford to far eastern Victoria; very local. At Sydney and the lower Blue Mountains adults are on the wing only during September and October, but at higher altitudes from November to January. Females are rarely taken.

This species may be separated from *T. maheta*, *T. phigalioides* and

T. phigalia, all of which have rather similar patterns above, by the more numerous and distinct white spots beneath the hindwing. The males are usually collected on hill-tops in open eucalypt forest.

The life history and early stages have not been reported, but the larvae probably feed on *Lomandra*. At Barrington Tops adults have been collected in places where this plant grows very commonly.

24. *Trapezites maheta* (Hewitson), 1877

This species ranges from Kuranda and the Atherton Tableland to Victoria. The wing pattern of the two sexes differs considerably, especially on the hindwing beneath. The female superficially resembles both sexes of *T. phigalioides* and *T. phigalia*. Two subspecies are recognized. North from Sydney there are two generations annually. The species is found mainly in open eucalypt forest, and males tend to collect on hill-tops.

a. *Trapezites maheta praxedes* (Plötz), 1884

Plate 2, figs. 4 (male), 4A (female), maheta skipper.

DISTRIBUTION. Coast and tablelands of New South Wales to eastern Victoria; fairly common at Sydney and the Blue Mountains. In Victoria specimens have been taken as far west as Wilsons Promontory. Adults are usually on the wing in spring and autumn.

The male may be recognized by the conspicuous silver spots beneath the hindwing; the latter is narrower than in other species. In the female the hindwing beneath resembles that of *T. phigalioides*, but in the forewing the subapical spots form a straight band.

The species is found most commonly in open eucalypt forest and the males collect on hill-tops.

LIFE HISTORY. Mature larva very similar to *T. eliena*. Pupa light pinkish brown, covered with white waxy powder. E. D. Edwards collected a larva on *Lomandra confertifolia*, and reared it successfully on *L. filiformis*. Pupation occurs in the larval shelter at the base of the clump or in a curled dead leaf on the ground.

b. *Trapezites maheta maheta* (Hewitson), 1877

DISTRIBUTION. Queensland; from Kuranda and the Atherton Tableland to the coast and tablelands of the south-east. Adults are on the wing from September to January and again in March and April.

ADULT.   Similar to *praxedes* but ground colour beneath variegated with whitish brown in male, and in female paler than in *praxedes*.

In both north and south Queensland males fly on hill-tops in company with *T. eliena*, *T. petalia* and *T. iacchus*.

25. *Trapezites phigalioides* Waterhouse, 1903

Plate 2, fig. 5 (male), phigalioides skipper.

DISTRIBUTION.   Southern Queensland to Victoria. It occurs on the tablelands and mountains up to 1,500 m in southern Queensland, New South Wales and the Australian Capital Territory, to 600 m in the Grampians, western Victoria, but down to sea level near Sydney. In New South Wales it has been recorded as far west as Mt Kaputar. Most adults have been collected from October to December, but at Ebor have been taken in February, and in the Brindabella Range, A.C.T., in December and January.

Both sexes can always be recognized by the three subapical spots of the forewing, the lowest of which is distinctly nearer the apex than the other two. As in the female of *T. maheta* the hindwing beneath is without silvery white spots. This is a eucalypt forest species.

A larva was collected by E. D. Edwards in March on *Lomandra filiformis* at the Boyd River, N. S. W. At maturity the larva wandered away from the food plant and formed a crude shelter for pupation. At Stanthorpe freshly emerged females have been taken at rest on *Lomandra longifolia,* upon which the larvae no doubt feed.

26. *Trapezites phigalia* (Hewitson), 1868

This species ranges from central Queensland to South Australia, both on the coast and the tablelands. Two subspecies have been distinguished.

a. *Trapezites phigalia phigalia* (Hewitson), 1868

Plate 2, fig. 6 (male), phigalia skipper.

DISTRIBUTION.   New South Wales, as far west as Coonabarabran and Cowra, the Australian Capital Territory up to 1,200 m, Victoria, and south-eastern South Australia, including the Mt Lofty Ranges above about 300 m; a common insect at Sydney and in the Blue Mountains. Adults fly only in the spring, from August to October at

Sydney, but from October to December in the mountains of New South Wales, A.C.T. and Victoria.

From the superficially similar *T. phigalioides*, and the females of *T. maheta*, this species can be separated by the darker central area of the forewing above, the smaller central hyaline spot, the shorter orange band of the hindwing above, the very dark central area of the forewing beneath, and the paler grey ground colour beneath. Unlike *T. phigalioides* the subapical band of small spots in the forewing is straight.

The larvae feed at night on the leaves of *Lomandra filiformis* and, during the day, are found in a shelter at the base of the plant constructed wholly or partly of dead leaves from the food plant. Pupation occurs in this shelter or in a curled dead leaf.

b. *Trapezites phigalia philus* Waterhouse, 1937

DISTRIBUTION. Northern, central and south-eastern Queensland: taken at Cairns and the Expedition Range (A. F. Atkins), at Noosa and Stradbroke Island in September, Lamington National Park in December, at Toowoomba, and commonly at Stanthorpe from September to November. It is apparently rare in coastal areas.

ADULT. Similar to typical *phigalia*, but the ground colour beneath the apex of the forewing and beneath the hindwing is pinkish instead of grey, and the broad orange band above the hindwing is broken by brown veins.

27. *Trapezites sciron* Waterhouse and Lyell, 1914

The known distribution of this species is far western Victoria, south-eastern South Australia, and southern Western Australia, but intervening localities will probably be discovered. It is the only species of the genus found in Western Australia.

a. *Trapezites sciron sciron* Waterhouse and Lyell, 1914

Plate 2, fig. 7 (male), sciron skipper.

DISTRIBUTION. Southern Western Australia, from Rottnest Island and from Perth to the Stirling Range and east to Cocklebiddy; rare. Most adults have been taken from October to December.

The early stages have not been recorded.

b. *Trapezites sciron eremicola* **Burns, 1948**

DISTRIBUTION.   Known so far only from the Little Desert and the Big Desert, western Victoria, where it is very local, and from near Lameroo and the Eyre Peninsula, South Australia (R. H. Fisher). Adults have been taken in October and November.

ADULT.   Similar to typical *sciron* but forewing with larger spots; hindwing beneath greyer with spots usually larger, greyish white edged dark brown.

This subspecies is more distinctly marked than typical *sciron*. Both sexes can be found on sand ridges, where the males settle on the sand and display typical hill-topping behaviour. Females occur in these situations in much lower numbers. *Motasingha dirphia trimaculata* often flies on the same ridges.

In western Victoria the food plant is *Lomandra glauca*, but the life history has not yet been fully worked out. An empty pupal skin and the cast larval head capsule were discovered at the base of one of these plants by W. N. B. Quick in the Little Desert. At maturity the larva forms a tubular shelter against the plant projecting above the surface of the sand, and finally pupates in this shelter.

28. *Trapezites petalia* **(Hewitson), 1868**

Plates 2, fig. 10 (male), I, common white-spot skipper.

DISTRIBUTION.   Kuranda and the Atherton Tableland, Queensland to Bulli, New South Wales; inland as well as near the coast. Inland it has been taken at the Carnarvon Range and Mitchell in southern Queensland. There are two generations annually, adults being recorded from September to May.

Females are taken much less frequently than males. The apiculus of the antenna is short and blunt, but larger than in *T. luteus*. The pattern of the hindwing beneath also separates these two species, the central white spot in *T. petalia* being much more distinct than in *T. luteus*, and the ground colour a duller brownish yellow.

The larvae feed on *Lomandra longifolia* and *L. multiflora* (E. D. Edwards).

29. *Trapezites luteus* **(Tepper), 1882**

This species ranges from southern Queensland to South Australia and Tasmania.

a. *Trapezites luteus leucus* **Waterhouse, 1938**

Plates 2, fig. 8 (male), II, rare white-spot skipper.

DISTRIBUTION.   Southern Queensland, New South Wales, and the Australian Capital Territory. Usually rare on the coast, but more common on the tablelands and western slopes. It is common at times near Brisbane, but much more local than *T. eliena*, *T. iacchus*, *T. maheta* or *T. petalia*.

Larvae have been reared on *Lomandra* at Millmerran, and E. D. Edwards recently collected another on *L. filiformis* at The Summit (1,200 m), near Hampton, N.S.W. An egg collected at Millmerran in September by J. Macqueen produced an adult in early February, whereas eggs laid in late October and early November do not produce adults until the following spring.

b. *Trapezites luteus luteus* **(Tepper), 1882**

DISTRIBUTION.   Central and western Victoria, and the lower Murray Valley, the Mt Lofty Ranges, Yorke Peninsula, Peterborough, and the Eyre Peninsula, South Australia; usually rare. In Victoria adults have been taken in October, November, February and March.

ADULT.   Similar to *leucus* but usually smaller, with central spots on forewing smaller, and yellow central area on hindwing smaller and less distinct.

The larva usually feeds at night, sheltering during the day in a crude shelter formed by joining the bases of the leaves with a few strands of silk, or amongst leaf litter, sometimes more than a metre from the food plant. It pupates in this shelter or in a curled dead leaf on the ground. In South Australia the larvae have been found on *Lomandra dura*.

c. *Trapezites luteus glaucus* **Waterhouse and Lyell, 1914**

Plate 2, fig. 9 (male).

DISTRIBUTION.   Northern and eastern coastal Tasmania, to about 300 m in altitude. Adults are on the wing from November to March, but in the south-east appear to be more common in November and January. The subspecies may have a two-year life cycle.

ADULT.   Similar to *leucus* but slightly larger, above with a greenish tint, forewing with smaller central spots, beneath olive-yellow.

At Kingston the larva feeds on *Lomandra longifolia*, spinning a crude shelter of a few strands of silk on the inner surface of a leaf near its base. At maturity the larva leaves this shelter and wanders for several days before pupating head upwards in a shelter formed by drawing two leaves together with silk.

Genus *ANISYNTOIDES* Waterhouse, 1932

Closely allied to *Trapezites*, this genus includes but a single species.

30. *Anisyntoides argenteoornatus* (Hewitson), 1868

This species is confined to Western Australia, being represented by a mainland subspecies in the south-west and an insular subspecies found on various islands along the coast as far north as the Monte Bello Islands, north of the Tropic of Capricorn.

a. *Anisyntoides argenteoornatus argenteoornatus* (Hewitson), 1868

Plate 2, fig. 14 (male), silver-spotted skipper.

DISTRIBUTION.   Mainland Western Australia, from Geraldton to Bunbury. There is only one generation annually, and adults are on the wing from September to December.

The larvae feed on *Acanthocarpus preissii* (Xanthorrhoeaceae) constructing a shelter similar to that of *insula*, in which pupation occurs.

b. *Anisyntoides argenteoornatus insula* (Waterhouse), 1932

DISTRIBUTION.   Islands off the coast of Western Australia, including the Monte Bello Islands, the Abrolhos Islands and Rottnest Island. The original specimens were taken on Hermite Island of the Monte Bello group in June and July, but farther south adults have been collected from September to December.

ADULT.   Similar to typical *argenteoornatus* but with more distinctly chequered scale fringes, beneath hindwing with central and basal spots more elongate, the postmedian series of spots joined to form a shining white band, extended at both ends towards base.

In flight the adults are said to resemble *Trapezites*, but they frequently settle on objects on the ground. At Rottnest Island they have been observed feeding at various flowers, but especially at *Senecio* (Asteraceae) and *Conostylis radicans* (Amaryllidaceae).

The larva constructs a whitish silken shelter, open at one end, amongst the foliage near the top of the food plant *Acanthocarpus preissii*. The larva is rather similar to *T. luteus*. Pupation takes place in a similar white silken shelter, open at the top, or in a shelter formed by joining with silk several narrow leaves of grass or other plant growing adjacent to the food plant. The pupa is very similar to *Trapezites*.

Genus *ANISYNTA* Lower, 1911

This genus includes six species from southern Australia. One occurs in south-eastern Queensland, five in New South Wales, three in Victoria, two each in the Australian Capital Territory, South Australia and Western Australia, and one in Tasmania.

31. *Anisynta sphenosema* (Meyrick and Lower), 1902

Plate 2, figs. 17 (male), 17A (female), wedge skipper.

DISTRIBUTION. South-western Australia, from Geraldton (N. McFarland) to Albany; common. Adults are on the wing from December to May.

In both sexes the development of the spots on the forewing is very variable; the two illustrations represent extremes of variation. The species is easily distinguished by the large blackish wedge-shaped area beneath the forewing.

The larvae are very similar in shape to those of *Trapezites*, but are not as stout. At Perth they feed on grasses.

32. *Anisynta albovenata* Waterhouse, 1940

This species is known from inland New South Wales, South Australia, and south-western Australia. It can be readily recognized by the white veins of the hindwing beneath. The antennae are strongly ringed with white, and the forewing of the male is without a sex-brand. Three subspecies have been described.

a. *Anisynta albovenata albovenata* Waterhouse, 1940

Plate 13, fig. 7 (female), white-veined skipper.

DISTRIBUTION. South-eastern South Australia. Known only from Yorke Peninsula and the area between the Victorian border and the lower Murray Valley; rare. Adults are on the wing in late September and early October.

The adults fly close to the ground and rest on the bare soil. They also feed at flowers.

The larvae feed on *Stipa semibarbata* (Poaceae), hiding during the day in a tubular shelter, open above, formed from several brown grass stalks bound together with silk. Pupation occurs in this shelter. A. F. Atkins also observed a female at Point Pearce, S.A., depositing an egg on *Danthonia setacea* (Poaceae).

b. *Anisynta albovenata fuscata* Parsons, 1965

DISTRIBUTION.   South-western Australia, between Norseman and Esperance. Adults are on the wing in October.

ADULT.   Male: similar to typical *albovenata* but much larger, above darker with spots larger and a deeper yellow, an additional small subtornal spot above vein 1A + 2A.

Female: similar to typical *albovenata* but much larger, darker above, with spots larger and a deeper yellow.

The average length of the forewing in males and females of *fuscata* is 15.6 and 17.8 mm, compared with 12.7 and 15.1 mm in typical *albovenata*.

c. *Anisynta albovenata weemala* Couchman, 1954

DISTRIBUTION.   Inland New South Wales: the Gunnedah district at about 300 m, and at Mt Kaputar at about 600 m (C. W. Frazier). Adults are on the wing in September and early October.

ADULT.   Similar to typical *albovenata* and of similar or slightly smaller size, but above with spots more prominent, with an additional subtornal spot above vein 1A + 2A in male, and spots a deeper yellow in female; beneath usually with a more distinct grey postmedian band of spots, in female with elongate grey central and basal spots.

The adults have been found mainly on hills of about 300 m elevation. They fly rapidly for short distances, settling on bare patches of earth with their wings expanded.

33. *Anisynta cynone* (Hewitson), 1874

This species is known from inland New South Wales, the Murray River Valley, and South Australia. Four subspecies have been recognized. The male is without a sex-brand.

a. *Anisynta cynone cynone* (Hewitson), 1874

Plate 2, fig. 15 (male), cynone skipper.

DISTRIBUTION.   South Australia, along the sea coast from Victor Harbor to Robe; rare. Adults fly only in March and April.

The number of spots on the forewing above is variable, some males having as few as six, whereas females have as many as eleven.

b. *Anisynta cynone gracilis* (Tepper), 1882

DISTRIBUTION.   In coastal areas of Gulf St Vincent, South Australia, to about 40 km south of Adelaide, and inland north of Gulf St Vincent; usually rare. Adults fly during March and April.
ADULT.   Similar to typical *cynone* but grey-brown above, forewing with spots well developed, hindwing beneath with markings sometimes slightly smaller.

The eggs hatch with the first winter rains and the larvae feed on winter grasses including *Brachypodium distachyon* (Poaceae). Larval shelters are formed by drawing two or more leaf blades together with silk or, when the larva is nearing maturity, by incorporating leaf litter. Finally, in early November, the larva forms a shelter out of a dead leaf on the ground, in which it remains dormant until the following February or March, when it pupates.

c. *Anisynta cynone grisea* Waterhouse, 1932

Plate 2, fig. 16 (male).

DISTRIBUTION.   The Murray Valley, Victoria, and parts of south-western New South Wales. Most specimens have been taken near Kerang and Gunbower, during late March and April, but the species has been recorded to at least 80 km north of Deniliquin (E. D. Edwards).

ADULT.   Similar to typical *cynone* but grey-brown above, spots white, variable in number, beneath apex of forewing and hindwing grey-brown, white spots slightly larger.

The larva and pupa have not been described.

d. *Anisynta cynone gunneda* Couchman, 1954

DISTRIBUTION.   Northern inland New South Wales: adults have been collected in the Gunnedah district at altitudes of about 275 to

335 m, from mid-March to mid-May, and at Mt Kaputar (1,520 m) in mid-February.

ADULT. Similar to typical *cynone* but with spots above forewing sometimes reduced to minute dots, beneath forewing with subapical dots only, beneath hindwing suffused dull yellowish, postmedian spots indistinct, not or only obscurely edged with dark brown on inner side.

The larva and pupa have not been described.

34. *Anisynta tillyardi* Waterhouse and Lyell, 1912

Plate 2, fig. 18 (female), Tillyard's skipper.

DISTRIBUTION. South Queensland, at altitudes of 900 to 1,060 m at Stanthorpe, Tannymorel and the Bunya Mountains (January and February); New South Wales, near Guyra, Ebor, and the New England National Park (1,200 to 1,500 m) from December to February, near Murrurundi in February, and Barrington Tops (above 600 m) in January. It is very common at the Bunya Mts and also fairly common at the New England National Park and Barrington Tops.

The adults feed commonly at the flowers of *Verbena bonariensis* (purple-top), and fly close to the ground in grassy areas.

The larvae feed in the late evening and early morning on *Poa labillardieri* (Poaceae) at the Bunya Mts (A. F. Atkins) and during the day are found in spirally twisted or folded shelters formed from dead leaf litter and attached at the open end with silk to the food plant and nearby debris. In captivity some larvae also formed cylindrical shelters from the leaves or stalks of the food plant. Pupation occurs in leaf debris near the base of the plant.

35. *Anisynta monticolae* (Olliff), 1890

Plate 3, fig. 21 (male), mountain skipper.

DISTRIBUTION. Mountains of south-eastern Australia, above about 600 m. Specimens are known from the Jenolan Caves (900 m), the Tinderry Mountains (1,500 m), Brown Mountain (970 m) and Mt Kosciusko (900 to 1,200 m) in New South Wales; the Brindabella Range (760 to 1,500 m) and Molonglo Gorge (600 m) in the Australian Capital Territory; and Mt St Bernard (1,200 to 1,500 m), Mt Buffalo, Walhalla (900 m), north of Heyfield, near Gisborne, and Mt Rosea (E. D. Edwards), Reids and Baroka Lookouts (J. C. Le Souef) in the Grampians, Victoria. Adults are on the wing from January to March.

The pattern beneath bears a superficial resemblance to that of *A. tillyardi*, but *A. monticolae* may be recognized by its smaller size, the presence of distinct yellow spots on the hindwing above, and the presence of a sex-brand on the forewing of the male.

The adults fly in open eucalypt forest or in subalpine woodland, and are usually taken as they feed at flowers.

Larvae were reared by E. D. Edwards from eggs laid by captive females. They were fed on snow grass (*Poa* sp., Poaceae) and pupated head upwards in the last larval shelter, which was similar to that of *A. dominula*.

36. *Anisynta dominula* (Plötz), 1884

This species occurs in the higher mountains of New South Wales, the Australian Capital Territory and Victoria, and up to about 1,240 m in Tasmania. As in *A. monticolae* the scale fringes of both wings are chequered, and the male forewing bears a dull blackish sex-brand. Five subspecies have been recognized, but those from the mainland are at times not clearly distinguished.

a. *Anisynta dominula drachmophora* (Meyrick), 1885

Plate 3, fig. 20 (male), dominula skipper.

DISTRIBUTION. Mt Kosciusko (900 to 1,500 m), New South Wales, and the mountains of eastern Victoria, including Mt Hotham (1,370 to 1,500 m) and Mt Buffalo (1,370 m). Adults are on the wing from January to March.

The adults are usually collected feeding at daisies or other flowers growing in subalpine woodland.

The larvae probably feed on snow grass (*Poa* sp., Poaceae).

b. *Anisynta dominula draco* Waterhouse, 1938

Plate II.

DISTRIBUTION. Mountains of New South Wales, excluding Mt Kosciusko. Specimens have been taken at Dalmorton, Deer Vale, New England National Park (1,370 to 1,460 m), Ben Lomond, Barrington Tops, Hampton, Kanangra Walls, Mt Tinderry (1,500 m), Black Range and Brown Mountain (1,060 to 1,200 m). Adults fly from December to March, and are often common.

ADULT. Similar to *drachmophora* but slightly larger, scale fringes more distinctly chequered, beneath brown-black with costa and apex of forewing deeper orange-brown, hindwing with subbasal spots, in cell and near inner margin, and postmedian spots shining white, edged with brown-black.

Despite the proximity of Mt Tinderry and Brown Mountain to Mt Kosciusko and the Australian Capital Territory, specimens from these localities appear to be better referred to *draco* than to *drachmophora* or *dyris*.

Adults fly close to the ground in grassy areas amongst open eucalypt forest, and are usually collected as they feed at flowers.

The larva feeds on snow grass (*Poa sieberana*, Poaceae), forming a cylindrical shelter in the middle of a grass tussock by joining a series of the narrow leaves with silk. Pupation occurs in the larval shelter. A larva taken at The Summit, near Hampton, N.S.W., has been reared by E. D. Edwards.

c. *Anisynta dominula dyris* **Waterhouse, 1938**

DISTRIBUTION. Brindabella Range, Australian Capital Territory, at about 1,200 to 1,670 m. Adults are on the wing in January and February.

ADULT. Similar to *drachmophora* but usually slightly larger, forewing with variable subapical spots, hindwing beneath richer orange-brown, subbasal spots in cell and near inner margin, and postmedian spots usually dull ochreous, cell spot sometimes slightly shining.

The adults feed at daisies and other flowers growing in subalpine woodland.

d. *Anisynta dominula dominula* **(Plötz), 1884**

DISTRIBUTION. Northern and eastern coastal areas of Tasmania, at altitudes of up to about 300 m; rare. Adults are known from Billop in February, Baghdad in December, and Cranbrook in March.

ADULT. Similar to *drachmophora* but forewing with three larger pale yellow subapical spots, usually with a distinct pale yellow spot in cell and another beyond lower end of cell, beneath pale orange-brown, hindwing with subbasal and postmedian spots cream or pale yellow, dull or only slightly shining, postmedian series usually joined to form a continuous band.

e. *Anisynta dominula pria* **(Waterhouse), 1932**

DISTRIBUTION. Western and central mountains of Tasmania, at altitudes of 760 to 1,240 m; it occurs on the central plateau and in the buttongrass swamps east of the King River. Adults fly in January and February.

ADULT.   Similar to *drachmophora* but much smaller, above grey-brown with small markings, hindwing beneath yellowish brown, postmedian spots usually forming a shining whitish band, only divided by veins, female occasionally with a pair of small postmedian spots on hindwing above.

Genus *DISPAR* Waterhouse and Lyell, 1914

This genus contains only one species distributed from southern Queensland to Victoria.

37. *Dispar compacta* (Butler), 1882

Plate 3, figs. 22 (male), 22A (female), dispar skipper.

DISTRIBUTION.   The coast and tablelands of south-eastern Australia, from the Bunya Mountains, Stanthorpe and Brisbane, Queensland, to western Victoria. In New South Wales it has been taken inland at Mt Kaputar at about 1,400 m (G. Daniels). Adults are on the wing from January to April.

The hindwing at once distinguishes the male from other species. The purplish grey underside, with obscure darker spots, and the central yellow patch on the hindwing above assist in placing the larger female. Superficially it bears some resemblance to the female of *Toxidia doubledayi*, but in that species there is no yellow patch on the hindwing.

The species occurs in eucalypt forest.

Eggs laid in the late autumn produce larvae which feed intermittently and grow slowly during the winter. In spring and early summer, growth is more rapid, pupation occurring during December and January. The pupal duration is about three weeks in New South Wales, and nearly four weeks in western Victoria.

In New South Wales the larvae have been recorded feeding on common grasses. In the Grampians larvae were pale green and fed on *Poa* spp. (Poaceae), including *P. tenera*. They sheltered at the base of the plant during the day and climbed the leaves and flower stems by about midnight to feed. At maturity the larva forms a tube-like shelter by joining the two edges of a grass blade with a series of equally spaced silken straps, within which it pupates. Larvae have also been recorded on *Lomandra* (Xanthorrhoeaceae).

Genus *PASMA* Waterhouse, 1932

The genus contains only one species, which occurs on the coast and tablelands of south-eastern Australia and Tasmania.

38. *Pasma tasmanica* (Miskin), 1889

Plate 2, fig. 13 (female), tasmanica skipper.

DISTRIBUTION. The tablelands of southern Queensland, and New South Wales, south from Stanthorpe, Q., the Australian Capital Territory, and Victoria as far west as Lorne; less common at lower altitudes in southern New South Wales and Victoria; and from sea level to 900 m in Tasmania. Adults are on the wing from October to April.

Both sexes arc casily recognized by the pair of distinct whitish hyaline spots of the hindwing. Adults fly in eucalypt forest.

The larva and pupa have not been described, but the larva is said to resemble that of *Trapezites* and feeds on *Tetrarrhena juncea* (Poaceae) (A. N. Burns).

Genus *SIGNETA* Waterhouse and Lyell, 1914

This genus includes two closely allied species from southern Queensland, New South Wales, the Australian Capital Territory and Victoria.

39. *Signeta flammeata* (Butler), 1882

Plates 2, figs. 20 (male), 20A (female), III, bright shield skipper.

DISTRIBUTION. Southern Queensland to the Grampians, western Victoria. In Queensland it has been taken at Killarney and Stanthorpe; in north New South Wales it occurs near Ebor, in the New England National Park, and at Mt Kaputar (G. Daniels) and Barrington Tops; in central and southern New South Wales, the Australian Capital Territory and Victoria, it occurs at various altitudes up to about 1,500 m. Adults are usually on the wing from January to April, but at Stanthorpe and in northern N.S.W. and Barrington Tops can be taken in December.

Occasionally there are two or even three small subapical dots in the forewing of the male. The species is found in open eucalypt forest and subalpine woodland.

Females have been observed depositing their eggs on dead *Eucalyptus* leaves lying on the ground. In the Grampians the larvae have been taken feeding at night on *Poa* sp. (Poaceae). At maturity the larva formed a loose shelter of dry grass and the pupa was attached within it by a few strands of silk at the cremaster. The larvae apparently grow slowly during the winter, more rapidly in the early summer, and pupate in January.

40. *Signeta tymbophora* (Meyrick and Lower), 1902

Plate 2, fig. 19 (male), dingy shield skipper.

DISTRIBUTION. The mountains of southern Queensland to coastal southern New South Wales; uncommon and local. Adults have been collected in Queensland very commonly at the Bunya Mountains (1,060 m) in January and February, but much less frequently at Mt Glorious, Beechmont and Lamington National Park (760 m) in February and March; in coastal New South Wales they have been taken at Mt Warning and from Gosford to Mt Dromedary from January to March.

The forewing of the male occasionally has two or even three sub-apical dots instead of one. The adults fly at the margins of, or in clearings in, rain forest. Some have been taken feeding at the flowers of *Leichardtia* (Asclepiadaceae).

Egg with about twelve vertical ribs, cream. First instar larva pale yellowish, with obscure brownish bands towards posterior end, head shining black. The mature larva and pupa have not been described.

Genus *TOXIDIA* Mabille, 1891

This genus is distributed from New Guinea and the Kai and Aru Islands, through eastern Australia from the islands of Torres Strait and Cape York to eastern Victoria. One of the nine species is restricted to the New Guinea mainland, another occurs in New Guinea, the Kai and Aru islands, and Cape York, whereas the remaining seven are restricted to Australia.

41. *Toxidia peron* (Latreille), 1824

Plate 3, fig. 13 (male), large dingy skipper.

DISTRIBUTION. Eastern Australia, from the Archer River and

Coen, Cape York Peninsula, and from Kuranda and the Atherton Tableland to New South Wales, the Australian Capital Territory and eastern Victoria. In northern New South Wales and southern Queensland it occurs commonly on the coast and tablelands, but north of Mackay it is mainly a tableland species, although it has been taken at lower altitudes on Cape York Peninsula. In Victoria it has been taken at Mallacoota, Mt St Bernard and the Wellington River valley.

This species is found in open eucalypt forest; males may be collected feeding at flowers or flying on hill-tops.

The larvae feed on common grasses (Poaceae) and, at Expedition Range, central Queensland, also on *Gahnia sieberana* (Cyperaceae). There mature larvae have also been found feeding on *Dianella caerulea* (Liliaceae), although there was a slight doubt that they had actually developed from the egg on this plant.

42. *Toxidia thyrrhus* Mabille, 1891

Plate 3, fig. 9 (male), thyrrhus skipper.

DISTRIBUTION. Eastern Queensland from Cape York to the Atherton Tableland and south to near Manumbar, 40 km west of Gympie (M. De Baar); usually uncommon.

The forewing of the male in this species and in *T. parvula* has the termen more convex than in the other species. The greatly reduced spots in the forewing of both sexes of *T. thyrrhus*, and the sex-brand in the male, will easily separate the two. *T. thyrrhus* is also rather larger than *T. parvula*.

In central Queensland the larvae feed on *Cenchrus echinatus* (Poaceae) and probably also on other soft grasses (A. F. Atkins).

43. *Toxidia parvula* (Plötz), 1884

Plate 3, fig. 8 (male), parvula skipper.

DISTRIBUTION. Eastern Australia: at Kuranda and from Mackay to central Victoria; common on the coast and tablelands of southern Queensland and northern New South Wales, mainly coastal farther south, but has been taken in the Australian Capital Territory at 600 m. Adults are common near Sydney from September to April.

The termen of the forewing in the male is more convex than in other species except *T. thyrrhus*. The species is found mainly in open eucalypt forest. The males do not show such a strong tendency to collect on hill-tops as does *T. peron*.

The larvae feed on common grasses (Poaceae).

44. *Toxidia doubledayi* (Felder), 1862

Plates 3, fig. 10 (male), III, Doubleday's skipper.

DISTRIBUTION. Atherton Tableland to central Victoria. In north Queensland it has been taken at Herberton and at Paluma (J. Macqueen), but is uncommon. Farther south it has been taken at Edungalba, near Rockhampton, and farther inland at the Expedition and Carnarvon Ranges. It occurs on both the coast and tablelands in southern Queensland, but in southern New South Wales is more common on the coast. Adults are common in Sydney gardens from September to April.

This species is found both in rain forest and *Eucalyptus* forest. Adults are often collected at flowers.

Larvae hide during the day in a shelter made by rolling the edge of a dead leaf on the ground. They emerge at night to feed on nearby grasses (Poaceae). Pupation takes place in the larval shelter.

45. *Toxidia rietmanni* (Semper), 1879

This species occurs in rain forest on the Atherton Tableland and in coastal southern Queensland and New South Wales. Two subspecies are recognized.

a. *Toxidia rietmanni rietmanni* (Semper), 1879

Plate 3, fig. 11 (male), white-brand skipper.

DISTRIBUTION. Rockhampton, Queensland, to Batemans Bay, New South Wales; mainly coastal but it has been recorded from Barrington Tops. There are probably two generations annually, adults flying from September to April.

Occasionally in males the two spots beyond the lower end of the cell are reduced to one and the subapical dots are reduced or absent. The male is readily recognized by the long wavy sex-brand. The female resembles *parasema* (Plate 3, fig. 12) but always has a spot in the cell of the forewing. Unlike the female of *T. doubledayi* it never has the broad lilac-white band beneath the hindwing; the subapical spots of the forewing are seldom in a straight line and usually irregular in size.

The early stages have not been described, but C. N. Smithers and J. V. Peters observed a female near Bellingen, N.S.W., depositing its eggs on dead twigs on the ground.

b. *Toxidia rietmanni parasema* (Lower), 1908

Plate 3, fig. 12 (female).

DISTRIBUTION.   Kuranda, from September to May, and Herberton, northern Queensland.

ADULT.   Male: similar to typical *rietmanni* but forewing with cell spot absent, subapical spots often absent, and only one small spot beyond lower end of cell.

Female: similar to typical *rietmanni* but forewing with cell spot absent and subapical spots sometimes absent.

46. *Toxidia melania* (Waterhouse), 1903

Plate 3, fig. 14 (male), black skipper.

DISTRIBUTION.   Northern Queensland, from Kuranda and Herberton to Tully and Paluma; at Kuranda adults have been taken from December to April.

Males can be distinguished from *T. inornata* by their larger size and the more oblique whitish sex-brand. Both sexes have the slight yellowish suffusion above and the whitish scale fringe to the hindwing. The latter, however, is lost in worn specimens. The species is confined to rain forest.

47. *Toxidia inornata inornata* (Butler), 1883

Plate 13, fig. 6 (male).

DISTRIBUTION.   Captain Billy Creek on Shelburne Bay and the Claudie River, Cape York Peninsula. This subspecies also occurs in the Kai and Aru Islands. Others are known from New Guinea, and from Goodenough and Tagula Islands.

The wings above and beneath are darker than in *T. melania*, and above they are without the yellowish suffusion; the hindwing has a brown-black scale fringe. In the male the termen of the forewing is less convex than in *T. melania*, and the sex-brand is slightly less oblique and uniformly dull black. This species is confined to rain forest.

48. *Toxidia andersoni* (Kirby), 1893

Plate 4, fig. 7 (male), Anderson's skipper.

DISTRIBUTION. Southern Queensland to central Victoria. In Queensland it has been taken at Cunninghams Gap (E. J. Dumigan), Mt Mistake and Tannymorel (S. Johnson), and near Stanthorpe (M. De Baar); in New South Wales it has been taken fairly commonly at New England National Park, and rarely at Wilsons Peak, Murrurundi, Bulli, Mt Kembla, Clyde Mountain and North Durras; and in Victoria at Mallacoota, Healesville and the Dandenong Ranges; not common. Adults have been collected from November to early March.

This species is found in *Eucalyptus* forest.

Adults were observed ovipositing at Kallista, Victoria, by F. E. Wilson in early January (D. F. Crosby), and a larva collected on *Tetrarrhena juncea* (wire grass, Poaceae) at Kallista by B. B. Given was reared by F. E. Wilson. The larval head, pupal skin and reared female adult are now in the National Museum of Victoria.

Genus *NEOHESPERILLA* Waterhouse and Lyell, 1914

This genus contains four closely allied species from the Northern Territory and Queensland, one of which reaches north-eastern New South Wales.

49. *Neohesperilla xiphiphora* (Lower), 1911

Plate 3, fig. 6 (male), xiphiphora skipper.

DISTRIBUTION. Northern Territory, including Groote Eylandt, Moa, Thursday and Prince of Wales Islands, and Cape York to Cairns, the Atherton Tableland, and an area 96 km west of Bowen. It is fairly common at Darwin, where adults have been taken from November to April.

Males may be recognized by the thick black sex-brand. Females are difficult to separate from the next species.

50. *Neohesperilla crocea* (Miskin), 1889

Plate 3, fig. 5 (male), crocea skipper.

DISTRIBUTION. Northern Territory, including Groote Eylandt, Moa, Thursday and Prince of Wales Islands, and Cape York to Townsville. Adults have been taken at Kuranda throughout the year. The species also occurs in southern New Guinea.

Both *N. xiphiphora* and *N. crocea* have a pair of distinct hyaline spots on the hindwing. The male of *N. crocea* has a narrower sex-brand and, in the female, the median spot above vein 1A + 2A is relatively less conspicuous than in *N. xiphiphora*. *N. crocea* is also slightly larger and paler.

Adults fly close to the ground in open eucalypt forest and along the margins of rain forest.

51. *Neohesperilla senta* (Miskin), 1891

Plate 3, fig. 7 (male), senta skipper.

DISTRIBUTION.   Northern Territory and north-eastern Queensland from Cooktown to Kuranda, Herberton and Cardstone (W. R. Hindson). At Kuranda adults have been taken in most months. In the Northern Territory it has been collected at Brocks Creek and Eureka.

Adults fly in open eucalypt forest.

52. *Neohesperilla xanthomera* (Meyrick and Lower), 1902

Plate 3, figs. 4 (male), 4A (female), xanthomera skipper.

DISTRIBUTION.   Northern Territory; and from Cooktown, Kuranda and Herberton, north Queensland, to Grafton, New South Wales; not common. Specimens have been taken at many localities in central and southern Queensland, and at Grafton between September and March. In the Northern Territory the species has been collected at Brocks Creek and Melville Island.

This is the largest of the four species and may be recognized by the absence of spots beneath the hindwing. The adults occur in open eucalypt forest and have been taken settling on grass.

Genus *HESPERILLA* Hewitson, 1868

This is a purely Australian genus of fourteen species widely distributed in eastern and southern areas. Eleven of the species are found in southern Queensland, nine in New South Wales, nine in Victoria, five in Tasmania, four in South Australia, three in Western Australia, and only one each in the Australian Capital Territory and the Northern Territory. The first five species (the *malindeva* group) are mainly northern in distribution (Atkins, 1978).

Hesperilla and allied genera have distinctive larvae and pupae. The larvae are relatively slender and semi-translucent, resembling

52

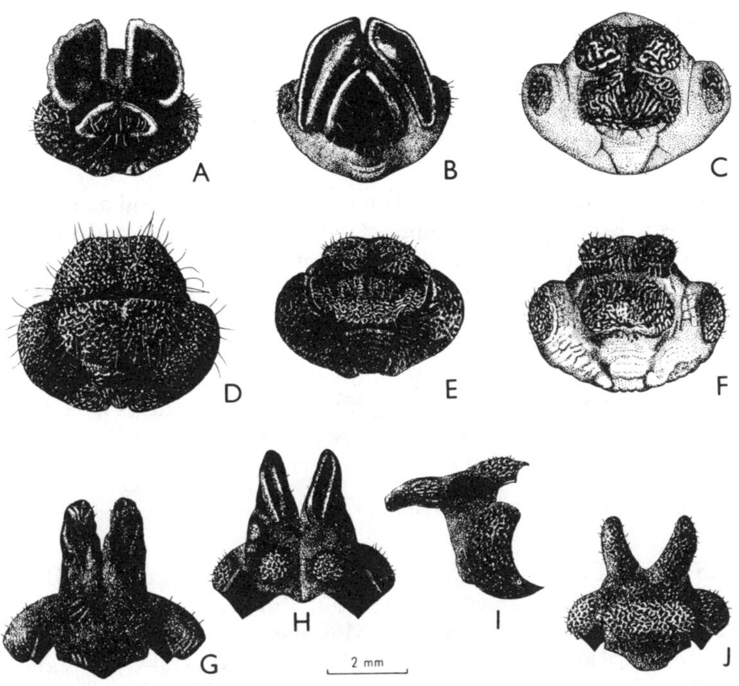

Fig. 8. Pupal caps in *Hesperilla*, Hesperiidae: A, *H. sexguttata* (p. 56);
B, *H. malindeva* (p. 56); C, *H. idothea* (p. 57); D, *H. chrysotricha* (p. 64);
E, *H. chaostola* (p. 67); F, *H. mastersi* (p. 68); G, *H. picta*, dorsal (p. 70);
H, I, *H. ornata*, dorsal and lateral (p. 69); J, *H. crypsargyra*, dorsal (p. 71).

Hesperiinae. They feed on Cyperaceae, usually on *Gahnia* (sword-grass), but also on *Cyperus*, *Carex*, *Scleria* and *Baumea*. The pupae are long and cylindrical in shape, with the anterior end hairy and strongly sculptured. This area of the pupal skin, or pupal cap, is usually detached when the adult emerges; its form (Fig. 8) is characteristic of each species.

The adults lay their eggs singly on the leaves of the food plant. The first instar larva, after hatching, crawls to the centre of a shoot and shelters between two or more leaves that it has joined with silk. It feeds at night on the tips of the young leaves. Later it joins larger leaves together to form a more substantial tubular shelter, and that part of the leaves above the shelter often bends over at a sharp angle, perhaps because of the irregular growth of the joined leaves. The presence of larvae may be detected by noting if the tips of the leaves have been eaten recently and whether any leaves are joined or bent over. Pupation takes place in the larval shelter.

53. *Hesperilla sarnia* Atkins, 1978
Plate 14, fig. 18 (male).

DISTRIBUTION. Mt Stuart (580 m), near Townsville (D. P. Sands and J. F. R. Kerr), Expedition Range, and Gayndah and Palmwoods, southern Queensland. Most specimens have been taken at Expedition Range, at about 900 m elevation in central Queensland.

The pointed forewings, the broad sex-brand in the male, the shining reddish brown colour of the wings above, with greatly reduced spots, and the ochreous suffusion of the wings beneath, distinguish *H. sarnia* from other species in the *malindeva* group. Males have been collected hill-topping on sandstone ridges, as they patrol and vigorously defend sites used annually by successive generations. The flight is rapid and males settle on dead twigs and bushes about one metre above the ground, in small sunny clearings. The only female so far collected, at the Expedition Range, was feeding at the flowers of *Leptospermum flavescens* (Myrtaceae).

The larvae and pupae are not known, but it is thought that the larvae may feed on *Scleria sphacelata* (Cyperaceae).

54. *Hesperilla furva* Sands and Kerr, 1973
Plate 14, fig. 6 (male).

DISTRIBUTION. Central and southern Queensland, from Westwood, Duaringa, Expedition Range, Springsure, Carnarvon Range to Nambour and Stanthorpe. It is common locally in the central highlands, especially on the higher sandstone ridges and ranges.

The male differs from *H. sexguttata* by its more evenly rounded hindwing, by the spots on the forewing which are fewer and pale cream in colour, by the different sex-brand, and by the grey-brown coloration beneath the hindwing. The forewing is not as pointed as in *H. sarnia*, has three or more spots and a narrower sex-brand, and the hindwing is more evenly rounded. The female of *H. furva* differs from *H. sexguttata* and *H. sarnia* by the grey-brown colour beneath the hindwing and the pale cream spots on the forewing.

Adults feed at flowers and the males fly on hill-tops, usually resting on low bushes and rock faces.

Larval food plants include *Baumea* sp., *Scleria sphacelata* and *S. mackaviensis* (all Cyperaceae). In central Queensland the larvae have usually been found on *S. sphacelata*, but sometimes on *S. mackaviensis* if *S. sphacelata* is growing near by.

55. *Hesperilla crypsigramma* (Meyrick and Lower), 1902

Plates 3, fig. 15 (male), III, small dingy skipper.

DISTRIBUTION. Laura, west of Cooktown, and the Atherton Tableland, Queensland to Grafton, New South Wales; fairly common on the western slopes of the northern tablelands of Queensland and on hills in the Brisbane area; specimens have also been taken at several localities in central Queensland as far west as the Carnarvon Range from August to November and February to May, and at Toowoomba, Bunya Mts, Lamington National Park and Springbrook. At Brisbane adults fly from November to April. Females are seldom taken.

This is an open eucalypt forest species. Adults of both sexes feed during the morning at flowers growing in gullies or on the lower slopes of hills; after mid-morning the males are usually found on hill-tops, where they frequently settle on the bare ground or on rocks and bushes close to the ground. Year after year they have been noticed settling in precisely the same small areas.

In central Queensland the larvae feed on *Scleria mackaviensis* (Cyperaceae), a small grass-like sedge, but farther north they are more common on *S. sphacelata*. Both larvae and pupae are found in upright shelters on these plants, to which leaf debris is often attached.

56. *Hesperilla sexguttata* Herrich-Schäffer, 1869

Plates 3, fig. 16 (male), III, six-spot skipper.

DISTRIBUTION. North-western Australia, from Prince Regent River and Lake Argyle (K. L. Dunn); Northern Territory, from Brocks Creek, Darwin, Cobourg Peninsula and the McArthur River (E. D. Edwards), Groote Eylandt and Winchelsea Island; and eastern Queensland, from Moa and Badu Islands, and Cape York to Stanthorpe. In central Queensland it has been taken inland as far as the Expedition Range and Carnarvon Range.

The absence of a spot in the cell of the forewing and of markings on the hindwing will assist in placing this species. Adults of both sexes are found at the edges of swampy areas, along rivers and creeks, and on the slopes of low hills, where they feed at flowers and settle on twigs and low bushes in sunny clearings. Females are usually more plentiful than males. The males do not collect on hill-tops.

The larvae feed on *Cyperus javanicus* (Cyperaceae) in central Queensland, growing either in swamps, on flood plains or behind coastal sand dunes, or on gravel and sand banks in creek beds in inland areas. In north Queensland the larvae feed on this plant, but also on *C. decompositus*. The larvae are much paler in colour than others of the *malindeva* group. The young larvae make tubular shelters amongst the upper leaves of the food plant, but older larvae are found in upright shelters nearer the base of the plant, in which pupation occurs. Larvae are able to withstand temporary flooding of their habitats.

57. *Hesperilla malindeva* Lower, 1911

Plate 4, figs. 15 (male), 15A (female), malindeva skipper.

DISTRIBUTION. Moa Island, Torres Strait, and from Kuranda, Mareeba and Herberton, Queensland, to near Grafton, New South Wales; more common in central and southern Queensland than in the north. It has been taken as far inland as the Burra Range, 135 km south-west of Charters Towers, and at the Expedition and Carnarvon Ranges. Near Grafton it has been taken between September and April.

This species somewhat resembles *Toxidia peron* (Plate 3, fig. 13) but can be distinguished by its larger size, the shorter cell spot in the forewing, and the pair of blackish dots beneath the hindwing. The sex-brand in the male is much narrower than in *T. peron*. The species is much more common in dry inland areas than on the coast.

The larvae feed on *Gahnia aspera* (Cyperaceae) and live in a cylindrical shelter about 15 cm long formed by joining two leaves of the food plant with silk. In their early instars they feed at night on the tips of the leaves forming the shelter, but later leave it to feed on the surrounding leaves. Within the shelter the larva usually rests with its head about 2 cm from the opening. When disturbed it retires to the bottom of the shelter and rotates its head. This behaviour may assist in repelling an intruder. At maturity the larva lines the upper 4 cm with silk, with an opening at the top, in which pupation takes place head upwards.

58. *Hesperilla idothea* (Miskin), 1889

This fine species occurs in the mountains from southern Queensland to Victoria, and in south-eastern South Australia and Tasmania. Two subspecies are recognized.

a. *Hesperilla idothea idothea* (Miskin), 1889

Plate 4, figs. 8 (male), 8A (female), flame skipper.

DISTRIBUTION. Stanthorpe and Springbrook, south-eastern Queensland, mainly the mountains and tablelands of eastern New South Wales, the coast and mountains of eastern Victoria, and at altitudes up to about 300 m in western, southern and eastern Tasmania. Adults are on the wing from November to March.

The female is a large and distinctive skipper not readily confused with any other. Its large size and the pattern beneath the wings will assist in placing the male. The species is found in *Eucalyptus* forest and males may be taken flying on hill-tops.

The larva resembles *H. donnysa* (Plate IV, figs. 1A, 1B) but has a much narrower head. Its most usual food plant is *Gahnia sieberana*, but in Victorian localities where *G. sieberana* is absent, it sometimes feeds on *G. radula* (A. Bishop) or *G. melanocarpa*. In the Tinderry Mts, N.S.W., it feeds on *G. subaequiglumis*.

b. *Hesperilla idothea clara* Waterhouse, 1932

DISTRIBUTION. The Grampians, western Victoria (425 to 760 m), the Mt Lofty Ranges and the far south-east of South Australia; rare in South Australia. Adults fly from November to January.

ADULT. Male: similar to typical *idothea* but larger, hindwing above with well marked elongate pale orange patch, beneath with up to three small postmedian dots.

Female: similar to typical *idothea* but forewing above with larger spots, two small spots above vein 1A + 2A fused, and an additional median spot, hindwing beneath without black dots or with one dot.

In the Grampians males have been taken flying on the mountain tops.

The larvae feed on *Gahnia sieberana* and *G. trifida*, and in the Grampians have been reported on *G. grandis*. Young larvae form shelters near the tip of a leaf by rolling over one edge and attaching it with silk. Older larvae make similar shelters farther down the leaf, or join two or three leaves together with silk, forming a tubular shelter. For pupation the larvae often use the natural tubular base of a leaf, lining it with silk, the pupa being attached by the cremaster to the silk head upwards and with a fine silken girdle.

59. *Hesperilla donnysa* Hewitson, 1868

This is the most widespread species of *Hesperilla*, occurring in Tasmania and in each of the southern mainland States, as far north as Maroochydore in the east and Geraldton in the west. It is found both on the coast and in the mountains up to about 1,500 m. The larvae feed on several species of *Gahnia*, depending on the locality and altitude. Nine subspecies have been distinguished. Some are difficult to separate unless series of specimens are available for comparison.

a. *Hesperilla donnysa donnysa* Hewitson, 1868

Plates 4, fig. 2 (male), IV, donnysa skipper.

DISTRIBUTION. Coastal New South Wales from Lake Macquarie to the Illawarra district; common where the food plant grows. There are two generations annually, adults appearing in spring and autumn. Some specimens taken or reared from Black Range (east of Canberra) (D. Ferguson), the Tinderry Mts (1,500 m), N.S.W., and at Mt Tidbinbilla, (1,500 m) and Smokers Gap (1,240 m), A.C.T. appear to be most closely allied to typical *donnysa*.

This subspecies is mainly restricted to heathland close to the shore line. Although Hewitson's original specimen was stated to be from Moreton Bay, it is now thought to have really come from Sydney. Specimens from the Tinderry Mountains and from Mt Tidbinbilla have a variable number of brown dots beneath the hindwing.

58

The larvae feed on a coarse *Gahnia*, *G. erythrocarpa*, near the coast; there and farther inland they are found on a second species, probably *G. aspera*, and at the Tinderry Mts and Mt Tidbinbilla on *G. subaequiglumis*. The young larva at first lives in the centre of the shoot, but the older instars make a shelter amongst the larger leaves.

b. *Hesperilla donnysa icaria* **Waterhouse, 1941**

DISTRIBUTION. Coastal southern Queensland and northern New South Wales. Known localities are Maroochydore, Caloundra, Burleigh Heads, Stradbroke Island and the Richmond River. Three males have been taken at Toowoomba (J. Macqueen). Specimens from Stanthorpe in October (J. Harslett), and larvae found on *Gahnia* in the Pilliga Scrub, northern N.S.W., may belong to this subspecies. *H. donnysa* probably occurs in many localities between the Richmond River and Lake Macquarie, but until specimens are available for study their relationship to typical *donnysa* or to *icaria* cannot be foreseen.

ADULT.   Male: similar to typical *donnysa* but larger, above darker brown, spots on both wings smaller and paler, subapical spots in a straight band, sometimes with a minute spot below them, usually no yellow spot above vein $1A + 2A$; beneath with apex of forewing and hindwing pinkish grey, hindwing with an additional dark brown dot above cell.

Female: similar to typical *donnysa* but larger and differing in colour and markings in same way as male; hindwing beneath usually pinkish grey, but sometimes a deeper pink, dark brown dots absent between veins M_1 and M_3.

The life history is similar to typical *donnysa*. The food plant at Maroochydore and Caloundra is *Gahnia sieberana*. At Burleigh Heads the larvae feed on *G. erythrocarpa* growing in swampy areas, and at Stanthorpe on a *Gahnia* doubtfully identified as *G. melanocarpa*. Larvae appear to be more common on isolated clumps of *G. erythrocarpa*, away from all shelter and shade.

c. *Hesperilla donnysa samos* **Waterhouse, 1941**

DISTRIBUTION.   Mainly in the Blue Mountains, New South Wales, above 600 m; common. Specimens have also been taken at Mittagong and inland west of Cowra and Coolah. In the Blue Mountains adults are on the wing from November to the beginning of

January, and there is a single generation annually; near Coolah they are also on the wing in April (J. V. Peters), and there are almost certainly two generations each year.

ADULT.   Male: similar to typical *donnysa* but smaller and darker, spots smaller and paler; forewing with subapicals smaller, the middle one the smallest, sometimes reduced to one subapical spot, rarely a small spot near inner margin; beneath grey, forewing with yellow in cell restricted, cell spot distinct, hindwing often with an additional dark brown dot above cell, sometimes brown dots represented by small brown rings.

Female: similar to typical *donnysa* but smaller and differing in colour and markings in same way as male.

The life history is similar to typical *donnysa*. In the Blue Mountains the larvae feed on *Gahnia filifolia*, *G. microstachya* and *G. sieberana*, approximately in decreasing order of preference. Near Coolah the food plant is *G. aspera*.

d. *Hesperilla donnysa patmos* Waterhouse, 1941

DISTRIBUTION.   Eastern and central Victoria, from Mallacoota to Gisborne, including Wilsons Promontory, the Dandenong Ranges, and the eastern shores of Port Phillip Bay; common. There are two generations annually, adults being on the wing during November and early December, and again in February and March.

ADULT.   Male: similar to *samos* but paler brown above and spots larger, often with two orange spots near inner margin, hindwing with larger yellow patch; beneath variable in colour, hindwing with some of the dark brown dots represented by brown rings.

Female: similar to *samos* but differing in colour and markings in same way as male.

Specimens from near Ararat, V., may belong to this subspecies, but also show some characteristics of *delos*. Near Melbourne the adults begin to fly about three weeks later than those of *H. flavescens*.

The life history is similar to typical *donnysa*. The larvae feed mainly on *Gahnia radula*, but occasionally on *G. sieberana* and *G. grandis*.

e. *Hesperilla donnysa aurantia* Waterhouse, 1927

DISTRIBUTION.   Throughout Tasmania, from sea level to about

1,060 m. Adults are on the wing from December to February. At the higher altitudes there is only a single generation annually, but on the coast there may be two.

ADULT.   Male: similar to typical *donnysa* but forewing usually with larger cell spot and sometimes with two small yellow spots below vein CuA_1, occasionally with additional spots; hindwing with central patch larger and bright orange in colour; hindwing beneath with larger but more obscure spots, the one in cell with a white centre.

Female: similar to typical *donnysa* but differing in colour and markings in same way as male, usually with two yellow spots below vein CuA_1, which are sometimes joined, occasionally two small spots below subapicals, forewing then with a continuous postmedian band of spots from costa to near inner margin.

The life history is similar to typical *donnysa*; the larvae feed on *Gahnia grandis* and other species of *Gahnia*.

f. *Hesperilla donnysa delos* Waterhouse, 1941

DISTRIBUTION.   South-western Victoria, from Ararat and the Grampians south to the coast between Port Campbell and Nelson; also south-eastern South Australia, in restricted localities in the Mt Lofty Ranges up to 400 m, the south-east corner south from Beachport, and on Kangaroo Island. Adults are on the wing mainly in November and there is only one generation annually.

ADULT.   Male: similar to typical *donnysa* but often much larger, above darker brown, forewing with cell spot larger, a small spot above vein 1A + 2A; beneath pale brown, hindwing usually with brown dots, but cell spot often represented by a brown ring with pale centre.

Female: similar to typical *donnysa* but much larger with larger spots; hindwing often with additional spots below central patch; hindwing beneath as in male.

The larvae feed on *Gahnia sieberana*, *G. trifida*, *G. radula* and *G. filum*. When mature they form tubular shelters by joining several leaves of the food plant, one of which is usually strongly looped. Pupation occurs in this shelter.

g. *Hesperilla donnysa diluta* Waterhouse, 1932

DISTRIBUTION.   South-eastern South Australia, near Goolwa

and Port Elliot, in restricted localities between the lower Murray Valley and the Victorian border, and on Yorke and Eyre Peninsulas. Adults are on the wing mainly in September and October and in March and April, but occasionally from December to February. The life cycle extends over twelve months.

ADULT.   Male: similar to typical *donnysa* but more strongly suffused with yellow, above forewing with cell spot larger, two spots below vein CuA_1, subapicals paler than other spots, sex-brand slightly broader and duller; hindwing with central patch dull yellow; beneath hindwing suffused with pink.

Female: similar to typical *donnysa* but above forewing usually with pale dots between veins M_1 and M_3, and two almost confluent spots below vein CuA_1, the lower one large; beneath hindwing as in male.

The larvae feed on *Gahnia ancistrophylla* and *G. deusta*.

h. *Hesperilla donnysa albina* Waterhouse, 1932
Plate 4, fig. 4 (male).

DISTRIBUTION.   South-western Australia, from Rottnest Island and Fremantle to the Stirling Range and Salmon Gums, north of Esperance (C. G. Miller); rare. Adults fly from October to December and from late February to April.

The females of this subspecies superficially resemble *H. chrysotricha* on the upperside, but may be distinguished by the underside. The male of *albina* has a more pointed forewing and a narrower sex-brand than *H. chrysotricha*, and the two may be readily separated by the underside.

Specimens from north of Esperance are much smaller than those from near Perth, with smaller hyaline spots on the forewing and smaller and duller orange patches on the hindwing (C. G. Miller).

Pupae have been collected from characteristic shelters on a *Gahnia* near Fremantle.

i. *Hesperilla donnysa galena* Waterhouse, 1927

DISTRIBUTION.   Geraldton district, Western Australia; known only from the original series collected as pupae in September, emerging in September and Octber.

ADULT.   Male: similar to typical *donnysa* but above with basal half heavily suffused with yellowish, forewing with cell spot larger and followed by a broad dark brown streak, usually four subapicals, spots beyond lower end of cell larger, sex-brand more irregular, grey; hindwing with yellow central patch crossed by dark veins M_3 and CuA_1; beneath greyish brown, forewing with central area dark brown, cell largely yellow, spots larger than above, hindwing with cell spot and a series of six postmedian spots silvery white ringed brown-black.

Female: similar to male but termen of both wings more rounded, forewing with spots larger, a large yellow spot near inner margin and a smaller one above it, hindwing with central patch larger and brighter.

This is a very distinct yellowish subspecies, and may eventually prove to be a separate species. It somewhat resembles *H. flavescens*, but is smaller and not as yellow. The sex-brand in the male is more irregular than in other forms of *H. donnysa*.

The original pupae were taken on *Gahnia trifida*.

60. *Hesperilla flavescens* **Waterhouse, 1927**

This species occurs in the western half of Victoria and near Adelaide in South Australia. It was originally thought to be a subspecies of *H. donnysa*, which it resembles in shape but the male genitalia, the early stages, and the larval food preferences suggest they are distinct. The adults are rather variable, but both wings are usually strongly suffused with yellow. They appear about three weeks earlier in the season than *H. donnysa delos*, the distribution of which apparently overlaps that of *H. flavescens* in western Victoria. Two subspecies have been recognized, one in Victoria and the other in South Australia. As in *H. donnysa diluta*, adults are on the wing in spring and autumn, although the life cycle of each species occupies a full twelve months. Apparently a few adults emerge during the summer, ensuring that some interbreeding occurs between earlier and later populations.

a. *Hesperilla flavescens flavescens* **Waterhouse, 1927**

Plate 4, fig. 3 (male), yellowish skipper.

DISTRIBUTION. Central and western Victoria, at Altona, Bellarine Peninsula, near Lake Connewarre, and near Ararat and Natimuk, near Horsham. Adults are usually taken in October and November and from the end of February to early April.

The larva and pupa are very similar to *H. donnysa* (Plate IV, figs. 1A–1E) but in the latter the larva is yellowish green with a darker finely granulose head and the sculpturing of the head piece in the pupa is slightly more angular. At Altona and near Lake Connewarre the larvae apparently always feed on *Gahnia filum*, which grows in clumps behind the shore-line. Near Natimuk adults have also been reared from larvae feeding on *G. filum*, whereas at Ararat the larvae were on *G. radula*.

b. *Hesperilla flavescens flavia* **Waterhouse, 1941**

DISTRIBUTION. Eastern shore of Gulf St Vincent, South Australia; very local. Formerly specimens were taken at Henley Beach, West Beach and St Kilda, near Adelaide, but urban development has probably eliminated the species from these localities. A few specimens which may belong to this subspecies have been collected on the Eyre and Yorke Peninsulas, and from near the mouth of the Murray River and the Coorong.

ADULT. Male: very similar to typical *flavescens*, above with strong yellowish suffusion and similar variable spots.

Female: very similar to typical *flavescens*, with yellowish suffusion often stronger than in male, spots similarly variable.

The life history and early stages are almost identical to those of typical *flavescens* and the larvae likewise feed on *Gahnia filum*.

61. *Hesperilla chrysotricha* **(Meyrick and Lower), 1902**

This species is restricted to the most southern States, Victoria, Tasmania, South Australia and south-western Australia. The forewing of the male is less pointed than in *H. donnysa* with a more distinct sex-brand. Six subspecies are recognized. Of these, typical *chrysotricha*, *naua* and *lunawanna* are rather smaller and darker, especially beneath the hindwing, than the other three subspecies, and the postmedian brown spots beneath the hindwing are smaller and less distinctly marked with silvery white.

a. *Hesperilla chrysotricha chrysotricha* **(Meyrick and Lower), 1902**

Plate 4, figs. 5 (male), 5A (female), chrysotricha skipper.

DISTRIBUTION. South-western Australia, from Rottnest Island and Fremantle to the Stirling Range and Esperance. Adults are on the wing in October and November and again in February, and there are apparently two generations annually.

The red-brown colour and the spots beneath the hindwing easily distinguish the females from the superficially similar *H. donnysa*.

The larvae feed on *Gahnia trifida* and, in the Stirling Range, a large tough *Gahnia*.

b. *Hesperilla chrysotricha naua* Couchman, 1949

DISTRIBUTION. Eyre Peninsula, South Australia, from near Port Lincoln to Coffin Bay. Many specimens have been taken at Wanilla Fountain, about 19 km north-west of Port Lincoln, during October. Specimens taken on Kangaroo Island may be nearer this subspecies or typical *chrysotricha* than *leucosia*.

ADULT. Similar to typical *chrysotricha* but hindwing above with central patch narrower, distinct, rectangular, beneath with cell spot and four of the postmedian series of six brown spots with distinct silvery white centres; in female one of these two small spots reduced to a brown dot, the other an elongate brown ring.

The larvae are similar to *leucosia* and feed on *Gahnia trifida*. The pupal shelters are similar to those produced by *leucosia*, *cyclospila* and other subspecies.

c. *Hesperilla chrysotricha leucosia* Waterhouse, 1938

DISTRIBUTION. The lower Glenelg River, south-western Victoria, and south-eastern South Australia; rare. In South Australia it occurs in the south-east corner, the southern Mt Lofty Ranges up to about 300 m, and on the Yorke Peninsula. Adults are on the wing from October to December, and occasionally as late as January.

This subspecies is similar in size to *cyclospila*, but usually the central orange patch of the hindwing is more clearly defined, the underside is more reddish brown, and the spots beneath the hindwing are much smaller. However, some specimens from Robe suggest that these two subspecies may merge into one another.

The head of the larva is flatter and narrower than that of *H. donnysa*. The larvae form a slightly twisted shelter from several leaves of the food plant, the upper end of which is closed with a silken pad before pupation. At Dartmoor, V., adults have been reared from pupae on *Gahnia trifida*, but in South Australia the early stages also occur on *G. sieberana* and *G. filum*.

d. *Hesperilla chrysotricha cyclospila* (Meyrick and Lower), 1902

DISTRIBUTION.   Southern central and western Victoria, from Westernport Bay to the Grampians. It was once taken at Inverloch and Fernshaw, near Healesville. Adults fly in November and December.

ADULT.   Male: similar to typical *chrysotricha* but larger, above forewing with spots not as bright, hindwing with central patch longer and narrower, more diffuse, beneath dull brown, postmedian spots above veins M_1, CuA_1 and CuA_2 elongate, cell spot and four of six postmedian spots with silvery white centres, those between M_1 and M_3 brown rings.

Female: similar to male but termen of both wings more rounded, above forewing usually with an additional dot above spot near tornus, beneath hindwing with an additional postmedian spot above Rs with clear silvery white centre.

The larvae feed on *Gahnia*, living in a shelter formed by joining several leaves in a special manner with an opening at the top. The shelter has a characteristic twisted appearance. At maturity the larva spins a silken pad across the shelter some distance from the entrance and pupates head upwards beneath it. This pad, which can be seen through the opening of the shelter, is never found in shelters spun by the larvae of *H. donnysa*. In many areas the larvae feed on *G. filum*, but east of Port Phillip Bay also on *G. sieberana*, *G. radula* and *G. trifida* (W. N. B. Quick). South of Halls Gap in the Grampians pupae have been collected by R. C. Manskie on *G. microstachya*.

e. *Hesperilla chrysotricha plebeia* Waterhouse, 1927

Plate 4, fig. 6 (male), plebeia skipper.

DISTRIBUTION.   Coastal and river shores of north, eastern and south-eastern Tasmania; uncommon and local. Adults are on the wing from late December to late February.

The first three larval instars live in slender straight shelters formed by joining the tips of two or three leaves of the food plant. The older larva forms a spiral shelter lower down in the plant from four or five leaves spun together with silk, one of the leaves being looped over in a characteristic way. The larva emerges at night to feed at the tips of the leaves above the shelter, gradually eating them back towards the shelter. The larvae feed on *Gahnia trifida* and *G. filum*.

At maturity the larva lines its shelter with silk, spins a silken pad between itself and the opening above, and pupates usually but not always head upwards. However, pupal shelters spun on *G. filum* have a much thinner lining than those on *G. trifida*, and the enclosed pupa has much more freedom of movement.

f. *Hesperilla chrysotricha lunawanna* Couchman, 1949

DISTRIBUTION.   South Bruny Island, D'Entrecasteaux Channel, south-eastern Tasmania; very local. Adults are on the wing in January.

ADULT.   Male: similar to *plebeia* but smaller, above forewing with subapical spots white, hyaline, sex-brand grey-black, interrupted, forming a series of crescents; hindwing with central patch elongate, orange, diffuse below, crossed by dark vein M_3; beneath hindwing grey-brown, postmedian brown spots variable in size but that above vein M_1 always much larger than others, and like cell spot with silvery white centre.

Female: similar to male but larger, termen of both wings more rounded, above forewing with two spots beyond lower end of cell of about same size, an additional smaller spot above vein $1A + 2A$, beneath hindwing with cell spot and postmedian spots above veins M_1 and CuA_1 with silvery white centres, that above CuA_2 faintly white-centred.

The life history and larvae and pupae are closely similar to *plebeia*. The larvae feed on *Gahnia trifida* and construct characteristic twisted shelters similar to those of other subspecies.

62. *Hesperilla chaostola* (Meyrick), 1888

This species occurs rarely in the mountains of central New South Wales, but is slightly more common in Victoria. It is a rare and very local insect in Tasmania. Three subspecies are known. The apical segment of the labial palpi is longer than in other species of *Hesperilla*, and there is always a pair of small subterminal spots on the forewing below the subapicals. Both *chaostola* and *chares* are known to have a two-year life cycle.

a. *Hesperilla chaostola chares* Waterhouse, 1933

Plates 4, fig. 1 (male), II, chaostola skipper.

DISTRIBUTION. Victoria; known from Erica, Moe, the Dandenong Ranges, the eastern suburbs of Melbourne, and the Grampians; not common. The type specimens were collected at Beaconsfield in October and November.

The most usual larval food plant is *Gahnia radula*, but in the Grampians larvae have been found on *G. sieberana* (R. C. Manskie) and *G. microstachya* (E. D. Edwards).

b. *Hesperilla chaostola chaostola* (Meyrick), 1888

DISTRIBUTION. Blue Mountains, New South Wales; very rare.

ADULT. Similar to *chares* but darker, forewing with spots slightly smaller, beneath dull dark purplish.

The larvae feed at night on *Gahnia filifolia*. They hide during the day in a shelter formed by joining the edges and the tips of several *Gahnia* leaves, providing an entrance at the bottom. Unlike other *Hesperilla* larvae it rests in this shelter with its head directed downwards, and at maturity pupates head downwards attached by the cremaster to a silken web, but without a pad of silk inside the entrance.

From eggs laid in October and November larvae develop slowly during the summer and following winter, reaching their final instar about December. They remain quiescent in this instar until the end of the second winter, pupating in the early spring and emerging as adults in October and November, two years from the time the eggs were laid.

c. *Hesperilla chaostola leucophaea* Couchman, 1946

DISTRIBUTION. Eastern coastal areas of Tasmania; very local. Most specimens have been collected near Kingston.

ADULT. Similar to typical *chaostola* but beneath apex of forewing and hindwing brownish grey, with very obscure markings.

The larva feeds on *Gahnia radula* and hides during the day in a shelter similar to that of typical *chaostola*. On the western bank of the Derwent River, the species appears to favour sandy hillsides with a northerly aspect.

63. *Hesperilla mastersi* Waterhouse, 1900

This fine species has mainly a coastal distribution in southern Queensland, New South Wales, eastern Victoria and north-eastern

Tasmania. With its bright upperside and distinctive underside it cannot be confused with any other species. Two subspecies have been recognized.

a. *Hesperilla mastersi mastersi* **Waterhouse, 1900**

Plates 4, fig. 9 (male), II, Masters' skipper.

DISTRIBUTION. Mt Glorious-Mt Nebo, Mt Tamborine and Cunninghams Gap, south-eastern Queensland, eastern New South Wales and eastern Victoria west to the Mitchell River, from sea level to more than 900 m at Mt Warning, Barrington Tops and the Blue Mts; fairly common but local in New South Wales, rare in Victoria. Adults have been taken from November to March. There may be two generations in northern areas, but only a single generation annually in Victoria, where adults fly from December to February.

The larvae feed on *Gahnia melanocarpa* growing amongst thick undergrowth in deep shade, usually producing slightly twisted shelters.

b. *Hesperilla mastersi marakupa* **Couchman, 1965**

DISTRIBUTION. Known from a single female reared from a pupa collected near Bridport, north-eastern Tasmania, in December.

ADULT. Female: similar to typical *mastersi* but three subapical and two postmedian spots separated by dark veins, the latter slightly smaller than in typical *mastersi*, beneath hindwing with cream markings more restricted.

The larval food plant was *Gahnia melanocarpa*, growing at the edge of a swampy heathland. The larva constructs a characteristic shelter, as in typical *mastersi*, in which the joined leaves have a slight spiral twist, somewhat similar to that of *H. chrysotricha* but not as strongly twisted.

64. *Hesperilla ornata* (**Leach**), **1814**

This species has a wide distribution in eastern Australia from the Claudie River and Cooktown, Queensland, to Victoria. The forewing of the male is more pointed than in other species. The pattern beneath the wings bears some resemblance to that found in *Oreisplanus perornatus* (Plate 4, fig. 16), but in that species the wings are more rounded and the male is without a sex-brand. This and the next two species, *H. picta* and *H. crypsargyra*, form a natural group with very similar larvae and pupae. Two subspecies of *H. ornata* are recognized.

a. *Hesperilla ornata ornata* (Leach), 1814

Plate 4, fig. 13 (male), spotted skipper.

DISTRIBUTION.   From Mackay, Queensland, to the Grampians, western Victoria; rare at Mackay and on the tablelands and coast of south-eastern Queensland, common in coastal New South Wales and in the Blue Mountains, fairly common in the Dandenong Ranges, Victoria, and in the outer eastern suburbs of Melbourne (W. N. B. Quick). At Sydney adults may be taken from October to April.

Adults of both sexes from southern Queensland have much smaller spots on the forewing than in those from southern New South Wales and Victoria, but never as small as in *monotherma*. The yellow band of the hindwing is also much smaller.

Both larva and pupa are very similar to *H. picta*. The larvae usually feed on *Gahnia erythrocarpa*, *G. aspera* and *G. sieberana* but also on *G. melanocarpa*, *G. radula*, *Carex appressa* and *C. brunnea*, forming tubular shelters by joining together several leaves with silk. In Victoria eggs laid towards the end of November hatched in twenty days.

b. *Hesperilla ornata monotherma* (Lower), 1907

Plate 4, fig. 14 (female).

DISTRIBUTION.   North-eastern Queensland, from the Claudie River and from Cooktown to the Atherton Tableland and the Paluma Range; fairly common locally. At Kuranda adults have been taken in September and November and again from February to April.

65. *Hesperilla picta* (Leach), 1814

Plate 4, fig. 10 (male), painted skipper.

DISTRIBUTION.   Coastal areas from Cooroy, Caloundra and Beerwah, Queensland, to Wilsons Promontory and Cape Liptrap, Victoria, and in the Blue Mountains; common but local in coastal New South Wales, usually rare in southern Queensland but fairly common near Caloundra. In Victoria it has been taken at several localities between Mallacoota and Lake Tyers from late January to early March, and on Wilsons Promontory in February. At Sydney adults have been taken from September to April, but especially in spring and autumn.

Specimens from southern Queensland have somewhat smaller spots on the forewing than those from near Sydney. Adults are found in swampy areas and damp gullies where the food plant grows.

The larvae and pupae are very similar to *H. ornata*, but the projections of the pupal cap are perhaps a little thicker and longer than in *H. ornata*. The larvae feed at night on *Gahnia clarkei* growing in swampy gullies, usually near estuaries. They have also been recorded on *G. melanocarpa* but this requires confirmation. The larva forms a shelter by joining two or three of the terminal leaves with silk, lining it with white silk before pupation.

66. *Hesperilla crypsargyra* (Meyrick), 1888

This is another handsome species from the mountains of southern Queensland, New South Wales, and eastern and western Victoria. Three subspecies are distinguished.

a. *Hesperilla crypsargyra crypsargyra* (Meyrick), 1888

Plate 4, fig. 11 (male), silvered skipper.

DISTRIBUTION.   Central New South Wales; it occurs in the Blue Mountains (above 600 m), at Carrington Falls, near Robertson (517 m), Berowra (one specimen) and Ku-ring-gai Chase (one specimen). Adults are common in the Blue Mountains from November to February, but are apparently much less so in the other localities.

Both larvae and pupae resemble *H. picta* but are smaller, and the anterior projections of the pupal cap are much more divergent. The larva feeds on *Gahnia microstachya*, a smaller species than those chosen by *H. picta* and *H. ornata*.

b. *Hesperilla crypsargyra hopsoni* Waterhouse, 1927

Plate 4, fig. 12 (female).

DISTRIBUTION.   Near Stanthorpe, Queensland (J. Harslett), and Deer Vale near Dorrigo, Ebor, New England National Park, and Barrington Tops, New South Wales. Adults are on the wing from November to February.

This subspecies is richer in colour than typical *crypsargyra* and in size approaches *H. picta*.

At New England National Park the larvae feed on *Gahnia sieberana* (R. P. Field, E. D. Edwards).

c. *Hesperilla crypsargyra lesouefi* Tindale, 1953

DISTRIBUTION.   Victoria; near Briagolong (60–120 m) and near

Licola (300 m), eastern Victoria, and on Mt William, from 600 to 900 m, in the Grampians, western Victoria; very local. Adults are on the wing from November to February.

ADULT.   Similar to typical *crypsargyra* but smaller, above darker, forewing with spots paler yellow, hindwing with central band orange-yellow, rather broader and uninterrupted at veins; beneath forewing darker, hindwing with larger silvery white spots.

The larva feeds on *Gahnia microstachya* which grows on steep slopes with a southerly aspect. The larvae are found mainly on clumps of *Gahnia* growing in relatively open areas amongst the mountain gums, but larvae and pupae have also been collected from clumps growing in dense undergrowth. The larva constructs a shelter by drawing together three upright leaves and joining them with silk, one of the three leaves often with a loop. Pupation occurs in the larval shelter. The pupal duration is about twenty-one days.

Genus *OREISPLANUS* Waterhouse and Lyell, 1914

This small genus of two species is restricted to the mountains of south-eastern mainland Australia, and coastal areas of north-western Tasmania. Superficially the adults resemble *Hesperilla* but the club of the antenna is evenly curved in the middle and the apiculus is not as pointed as in *Hesperilla*. The forewing in the male is without a sex-brand. The larvae feed on *Carex*, *Scirpus* and *Gahnia* (Cyperaceae).

67. *Oreisplanus munionga* (Olliff), 1890

This handsome species occurs in the mountains of the south-east, from Barrington Tops to Victoria, and near the coast in north-western Tasmania. Two subspecies have been distinguished.

a. *Oreisplanus munionga munionga* (Olliff), 1890

Plate 4, fig. 17 (male), alpine skipper.

DISTRIBUTION.   The mountains of central New South Wales, above about 1,100 m, and southern New South Wales, the Australian Capital Territory and eastern Victoria, above about 900 m. In New South Wales it occurs at Barrington Tops (J. W. C. d'Apice), Jenolan Caves, Kanangra Walls, and Brown Mountain, near Nimmitabel, and is at times common in the Mt Kosciusko area. It is locally common between Omeo and Mt Hotham in the Victorian Alps from January to March (W. N. B. Quick) and also occurs north of Heyfield.

The adults feed freely at the yellow flowers of everlasting and other daisies, and at rest their colours harmonize well.

The pupa may be easily separated from *O. perornatus* by the much longer anterior projections, which more closely resemble those of *Hesperilla picta* and *H. ornata*. The larva feeds on *Carex appressa* growing in swampy creeks. They construct shelters by drawing together several leaves of the food plant and joining them with silk. Pupation takes place with head directed upwards in the larval shelter. At Kanangra Walls E. D. Edwards found larvae on *Scirpus polystachyus*. On one occasion, in the absence of the normal food plant, larvae accepted *Gahnia*, but the resulting adults were smaller than normal.

b. *Oreisplanus munionga larana* Couchman, 1962

DISTRIBUTION. Marrawah district and near Stanley (M. Pickett), north-western Tasmania, near sea level; very local. Adults were reared at the end of January and in early February.

ADULT.   Male: similar to typical *munionga* but above forewing with cell spot narrower and tapering obliquely downwards, postmedian spot between vein M_3 and CuA_1 connected to spot above $1A + 2A$ by a narrow bar; beneath forewing with spots as above, but that above $1A + 2A$ not connected to lower postmedian spot, apex and hindwing with yellow areas smaller and paler.

Female: similar to male but larger, termen of both wings more rounded, above forewing with spots larger, submedian spot above $1A + 2A$ distinct, hindwing with orange central band broader and extending along veins M_2 to CuA_2 towards termen; beneath as in male, but yellow areas slightly larger.

Larvae and pupae have been taken in characteristic shelters on a coarse tufted *Carex*, possibly *C. appressa*, during January, and adults emerged from the pupae in from fourteen to eighteen days.

68. *Oreisplanus perornatus* (Kirby), 1893

Plates 4, fig. 16 (male), IV, mountain spotted skipper.

DISTRIBUTION. Blue Mountains to Victoria; usually at lower altitudes than *O. munionga*, but the two species occur together at about 1,060 m at Brown Mountain, N. S. W. In Victoria it occurs in the Wonnangatta River valley, the Dandenong Ranges, the Kinglake National Park, Mt Macedon, and in the Grampians (425 to 600 m).

Adults are on the wing from October to December, but mainly from early to mid-November.

The females have a relatively slow fluttering flight around the food plant.

The larvae feed on coarse *Gahnia*, forming a shelter by joining two or three leaves of the food plant with silk, in which pupation occurs head upwards. At Brown Mountain and at Kinglake the larvae feed on *G. sieberana*. In Victoria eggs laid in late November hatched in sixteen days.

Genus *MOTASINGHA* Watson, 1893

Including only two species, this genus is widely distributed in southern Australia, south from Brisbane in the east and Carnarvon in the west. It has not been recorded from Tasmania. The club of the antenna (Fig. 7E) is strongly bent before the middle and the apiculus is blunt. A sex-brand is present in the forewing of the male.

69. *Motasingha dirphia* (Hewitson), 1868

This species occurs widely in southern mainland Australia, from Brisbane around to Carnarvon. On the forewing above, the two sub-terminal spots found in *M. atralba* between veins M_1 and M_3 are absent, and there is a large silvery white cell spot beneath the hindwing. The female usually has a round cell spot in the hindwing above. Four subspecies are known.

a. *Motasingha dirphia dilata* Waterhouse, 1932

Plate 3, figs. 19 (male), 19A (female), dirphia skipper.

DISTRIBUTION.   Southern Queensland and New South Wales. It is apparently rare in Queensland, but has been taken recently at Stanthorpe (J. Harslett); it is known principally from along the coast in the Sydney area, adults flying from October to December. Specimens from 45 km north of Braidwood, N. S. W., may belong to this subspecies.

The general colour above is more yellowish than in the other subspecies; the outer spots of the forewing are less regular in shape, and are usually slightly produced outwards along the veins. The white spots beneath the hindwing are usually very distinct, and also more numerous than in the other subspecies.

The larva feeds on *Lepidosperma concavum* (Cyperaceae), hiding during the day in a shelter formed by drawing leaves together with silk and lined with silk. Recent feeding at the tips of the leaves indicates the presence of a larva. The projections of the pupal cap protrude rather more than those of *M. atralba*.

b. *Motasingha dirphia dea* Waterhouse, 1933

Plate II.

DISTRIBUTION.   Blue Mountains, New South Wales. Adults are on the wing from November to February.

ADULT.   Similar to *dilata* but slightly smaller, above forewing with smaller spots; beneath hindwing with less than three white post-median spots in males, and usually less than three in females.

Specimens from Killara and Berowra are perhaps nearer this sub-species than *dilata*.

The life history and early stages are similar to those of *dilata*, and the larvae (Plate II, fig. 8) also feed on *Lepidosperma concavum*.

c. *Motasingha dirphia trimaculata* (Tepper), 1882

DISTRIBUTION.   Dimboola, Little Desert and Big Desert, western Victoria, and South Australia from the Victorian border to the Mt Lofty Ranges, up to about 400 m, the Yorke and Eyre Peninsulas, Kangaroo Island and the Flinders Ranges. Adults are on the wing in October and November in western Victoria. The species is rare in South Australia, where most adults have been collected in November and December.

ADULT.   Similar to *dilata* but smaller, above forewing with spots usually smaller; beneath hindwing reddish brown, distinct white spot in cell, usually three small white postmedian spots.

The life history and early stages are similar to those of *dilata*. The larvae feed on *Lepidosperma carphoides* (Cyperaceae) in western Victoria and south-east South Australia, and on *L. viscidum* in western Victoria and in the Flinders Ranges.

d. *Motasingha dirphia dirphia* (Hewitson), 1868

DISTRIBUTION.   South-western Australia, from Carnarvon to Twilight Cove and Madura (J. C. Le Souef). Adults have been taken from October to December.

ADULT. Similar to *dilata* but much greyer, forewing with spots smaller and paler, sex-brand in male broader, more distinct, interrupted at veins; beneath hindwing variable but usually much darker, and with less than three small postmedian white spots.

The larvae feed on *Lepidosperma angustatum* (Cyperaceae), a small flat-leaved sedge growing on sand dunes.

70. *Motasingha atralba* (Tepper), 1882

This species is restricted to north-western Victoria and South and Western Australia, more usually near the sea coast. The forewing above has a pair of small subterminal spots between veins M_1 and M_3. Five subspecies have been recognized.

a. *Motasingha atralba atralba* (Tepper), 1882

Plate 3, fig. 17 (male), black and white skipper.

DISTRIBUTION. Lake Hattah area and Big Desert, north-western Victoria, to the Mt Lofty Ranges, and Yorke and Eyre Peninsulas, South Australia. Populations in north-western Victoria may represent a separate subspecies. Adults are on the wing from September to December and in March and April.

In both sexes the termen of the forewing is more convex and the spots are much larger and whiter than in the four western subspecies. The adults settle on the ground.

The larva usually feeds on *Gahnia lanigera* but occasionally on *G. ancistrophylla* (Cyperaceae). During the day it is found head downwards in a shelter it has formed by joining the tips and edges of several leaves with silk. The opening is at the lower end. Pupation occurs within the shelter and the pupa is fastened by the cremaster with the head directed downwards. No silken pad is spun within the entrance below the pupa. The rate of development of the larva is very variable and the life cycle of many individuals may occupy a full twelve months.

b. *Motasingha atralba anaces* Waterhouse, 1937

Plate 3, fig. 18 (male).

DISTRIBUTION. National Park, Lesmurdie, Waroona and Hamel, south-western Australia. Adults have been taken in October.

This is the largest of the five subspecies, and the spots beneath the hindwing are the least conspicuous. As in the other western subspecies the male sex-brand is prominent.

c. *Motasingha atralba anapus* **Waterhouse, 1937**

DISTRIBUTION.   Western Australia from Katanning to Twilight Cove and Madura, including the Stirling Range, in October and November. Specimens have been taken at Katanning and Cape Riche by K. R. Norris, near Ravensthorpe, north of Esperance, and Twilight Cove by M. S. Upton, and between Esperance and Madura by J. C. Le Souef.

ADULT.   Male: similar in shape to *anaces* but much smaller, and similar in size to *dactyliota* and *nila*, above forewing with spots larger than in *anaces*, but smaller than in *dactyliota*, with postmedian spot near lower end of cell round; beneath forewing with apex grey, and additional spot above vein 1A + 2A, hindwing grey, spots much more distinct than in *anaces*.

   Female: similar to *anaces* but smaller, beneath as in male.
The larvae feed on *Gahnia lanigera* (Cyperaceae).

d. *Motasingha atralba dactyliota* **(Meyrick), 1888**

DISTRIBUTION.   Geraldton, Western Australia. Known only from the original type specimens, two males and a female, taken at the end of October.

ADULT.   Similar to *anaces* but much smaller, and similar in size to *anapus* and *nila*, above forewing with spots larger than in *anaces*.
Further specimens are needed to confirm the status of this form.

e. *Motasingha atralba nila* **Waterhouse, 1932**

DISTRIBUTION.   Known only from Dirk Hartog Island, south-west of Carnarvon, Western Australia. Adults were taken in August.

ADULT.   Male: similar in shape to *anaces* but much smaller, with spots of forewing smaller, sex-brand a little narrower, hindwing yellowish brown; beneath darker than in other subspecies, apex of forewing and hindwing yellowish brown with very obscure spots.

   Female: similar to male but much larger, sex-brand absent, forewing with spots larger.

Genus *MESODINA* **Meyrick, 1901**

This genus is represented in the Northern Territory, in south-eastern mainland Australia, south from Bowen, and in south-western

Australia. The club of the antenna is strongly bent before the middle and the apiculus is very blunt. There is no sex-brand in the forewing of the male. As in *Croitana*, but unlike other Trapezitinae, the posterior tibiae are without median spurs. The uncus of the male genitalia is long and slender. Two species are recognized. The larvae of both feed on *Patersonia* (Iridaceae) and pupate with the head directed downwards.

71. *Mesodina halyzia* (Hewitson), 1868

This distinctive species occurs in the Northern Territory, eastern Australia from Bowen, Queensland, to eastern Victoria, and south-western Australia. It has not been found in western Victoria or South Australia. Two subspecies are recognized, but specimens from the Northern Territory may represent a third.

a. *Mesodina halyzia halyzia* (Hewitson), 1868

Plates 3, fig. 1 (male), III, halyzia skipper.

DISTRIBUTION. Northern Territory, and from north Queensland to eastern and central Victoria. In the Northern Territory a few specimens have been taken at Melville Island, Maningrida, the Cadell and Blyth River area, and at Nabarlek. In Queensland the species occurs at Bowen, Byfield, Expedition Range and Isla Gorge, and south from Cooloolah (M. De Baar, G. B. Monteith) and Noosa Heads, and at Stanthorpe and Millmerran. In eastern New South Wales it is common at Sydney and in the Blue Mountains, and in Victoria ranges as far west as Bacchus Marsh. Near Sydney adults fly from August to May, and in the Blue Mountains from October to March.

One or two subapical spots are occasionally present in the forewing of males from eastern Australia, and occur in the few known from the Northern Territory. Specimens from the latter locality are also smaller and duller than adults from eastern localities and probably represent a separate subspecies. Adult females are usually collected much less frequently than males. In coastal northern New South Wales and in southern Queensland this species appears to be largely restricted to sandy areas.

Both larva and pupa are very similar to *M. aeluropis*. The larva feeds during the day on *Patersonia*, including *P. glabrata* and *P. fragilis* near

Sydney, *P. sericea* in the Blue Mountains, and *P. fragilis* near Stanthorpe. E. D. Edwards found larvae at Nabarlek, N. T., on a *Patersonia* thought to be *P. macrantha*. It constructs a vertical tubular shelter by joining three leaves of the food plant with silk, in which it rests with head directed downwards. This shelter is densely lined with silk and has the entrance at the bottom. Before pupating the larva spins a horizontal pad of silk within the entrance; the head of the pupa is also directed downwards.

b. *Mesodina halyzia cyanophracta* **Lower, 1911**

DISTRIBUTION.   South-western Australia, from Geraldton to the Stirling Range. Adults have been taken from November to March.

ADULT.   Similar to typical *halyzia* but above grey-brown, forewing in male with costa grey and with three subapical dots, rarely one or more dots absent, beneath bluish grey.

The early stages are similar to those of typical *halyzia* and the larva also feeds on *Patersonia*.

72. *Mesodina aeluropis* **Meyrick, 1901**

Plates 3, fig. 2 (male), I, V, aeluropis skipper.

DISTRIBUTION.   Ebor, the Blue Mountains from about 820 m to 1,200 m, and southern New South Wales. Adults have been taken from October to February in the Blue Mountains, and in December at Ebor. Recently J. W. C. d'Apice collected adults near the Tantangara Dam, south of the A. C. T., in February. The species has also been reported from Mt Kembla, N. S. W.

The seasonal history of this species is not well understood for, in November, all the immature stages and the adults may be collected at the same time.

The larva feeds on a pale-leafed form of *Patersonia sericea*, forming a shelter similar to that of *M. halyzia*, in which it rests head downwards. Pupation finally occurs in this shelter, the head of the pupa also being directed downwards.

Genus *PROEIDOSA* **Atkins, 1973**

This genus includes only one species which was for many years associated with *Pasma tasmanica* in *Pasma*, from which it differs in the venation and the male genitalia. A. F. Atkins has shown that it is more closely related to *Mesodina*, *Motasingha* and *Croitana*.

73. *Proeidosa polysema* (Lower), 1908

Plates 2, fig. 12 (male), VI, polysema skipper.

DISTRIBUTION. North-western Australia, Northern Territory, including Groote Eylandt and the Alice Springs area (E. D. Edwards), and northern and central Queensland; local and usually scarce. Specimens have been taken at Exmouth, W. A. (S. Wallace), at Daly River, Darwin, McArthur River (E. D. Edwards) and Groote Eylandt, N.T., on Stanley and Flinders Islands, the Claudie River, Petford, near Chillagoe, the Belyando River, Expedition Range, Theodore, Moura, Canoona, Glen Geddes and Isla Gorge in Queensland. Adults are more common in late spring and in autumn.

In central Queensland the larvae feed on *Triodia mitchellii* and *T. hostilis* (Poaceae), and in north Queensland on another species of *Triodia*, possibly *T. pungens*. Near the South Alligator River, N.T., they have been found on *T. microstachya* (E. D. Edwards). Young larvae form a shelter by joining two or more leaf-tips with silk, and later a shelter of six or seven leaves. The larva feeds at night, resting during the day in a head-downward position in the shelter. The shelter of a mature larva includes eight or nine leaves, and within it pupation takes place head downwards.

Genus *CROITANA* Waterhouse, 1932

This genus contains three small species from Western Australia and the southern part of the Northern Territory (E. D. Edwards). The club of the antenna is strongly bent before the middle and the apiculus is very blunt. As in *Mesodina* the hind tibiae lack median spurs and there is no sex-brand in the male. The uncus of the male genitalia is short and broad with a pair of apical projections.

74. *Croitana croites* (Hewitson), 1874

Plate 3, fig. 3 (male), croites skipper.

DISTRIBUTION. Western Australia, from Carnarvon to Bunbury. Mainly coastal but occurring up to 80 km inland at Pindar. Most specimens have been taken near Bunbury from September to November, where it is said to be very local amongst sand dunes.

The yellow or orange markings above vary considerably and the range includes two specimens from Pindar which were separated as subspecies *pindar* Waterhouse, 1932.

Females are much less frequently caught than males, and at Bunbury have been observed resting on the flower-heads of *Conostylis* (Amaryllidaceae), the underside of the wings blending well with the colour of the flower-heads.

The food plant and the larger larvae and pupae are still unknown. A. F. Atkins, who obtained a first instar larva from an egg dissected from a captured female, considers that the young larva is very similar to that of *Proeidosa polysema*.

75. *Croitana arenaria* Edwards, 1979

Plate 29, figs. 1 (male), 6 (female).

DISTRIBUTION.   Southern areas of the Northern Territory; specimens have been taken near Hermannsburg, at several localities within 60 km of Alice Springs, and two localities up to 245 km east-north-east of Alice Springs. Adults have been collected in September and October.

The termen of the forewings in both sexes is rather more convex than in *C. croites*, and the postmedian spots beneath the hindwing form a continuous band, whereas in *C. croites* and *C. aestiva* this band is broken between $Sc + R_1$ and between Rs and M_1.

C. arenaria occurs in the ranges and in hilly and sandplain country. Males patrol territories and settle with wings held erect on small stones or on the sand, where they are difficult to see. Females have been taken feeding at daisies (Asteraceae). Nothing is known of the early stages.

76. *Croitana aestiva* Edwards, 1979

Plate 29, fig. 10 (male).

DISTRIBUTION.   Known from an area 25 to 40 km west of Alice Springs, Northern Territory. Adults have been taken in February.

The wings in both sexes are narrower than in *C. croites* and *C. arenaria*, and the submedian patch in the forewing and the central patch in the hindwing above are both yellow in *C. aestiva*, whereas in *C. arenaria* the forewing patch is pale yellow and the hindwing patch orange. The series of spots beneath the hindwing is broken in *C. aestiva* and *C. croites*, but confluent in *C. arenaria*. The wing margins are only faintly chequered in *C. aestiva*, but very distinctly so in the other two species.

Subfamily HESPERIINAE
DARTERS AND SWIFTS
(Plates 1, 4–6, 14, IV, V; Figs. 9–12)

This is an extensive tropical subfamily which has extended into northern Australia, either through New Guinea to Cape York and the eastern coast, or through Timor to north-western Australia. All the genera and most of the thirty-six species found in Australia also occur in New Guinea or south-east Asia. The number of genera or species in the Australian fauna therefore decreases markedly from north to south.

The species are of small to medium size; the smallest Australian skippers are included here. At rest the adults hold their wings erect, but when feeding or settled in sunshine, they may depress their hindwings. The forewing of the male is without a costal fold, but often bears a sex-brand.

The eggs are relatively large and hemispherical, with a smooth surface. The larvae are usually pale green in colour, with inconspicuous hairs, and feed exclusively on monocotyledons, especially grasses and palms. The pupae are usually pale green or brown, sometimes covered with a white waxy powder.

Genus *NOTOCRYPTA* de Nicéville, 1889

This genus ranges from Sri Lanka and India to China and the Philippines, and through south-east Asia to New Guinea, the Bismarck Archipelago, the Kai and Aru Islands, and north-eastern Australia.

77. *Notocrypta waigensis proserpina* (Butler), 1883
Plate 1, fig. 16 (female), banded demon.

DISTRIBUTION. The islands of Torres Strait, and Cape York to Paluma. Adults have been taken at Kuranda during most months.

Adults fly in clearings or at the margins of rain forest.

Near Cairns the eggs are laid on young plants of *Alpinia caerulea* (wild ginger, Zingiberaceae) growing in rain forest. The larva forms a tubular shelter by curling in the edges of a leaf, and pupation finally occurs in this shelter. At the Claudie River larvae have also been found on *Hornstedtia scottiana* (Zingiberaceae) (G. B. Monteith).

Genus *TARACTROCERA* Butler, 1870

This genus includes fourteen very small and superficially similar species, distributed from Sri Lanka and India to China, and through south-east Asia to New Guinea and Australia. The genus is easily recognized by the curious flattened spoon-shaped club of the antenna (Fig. 7F), with the apiculus nearly or completely absent. The uncus of the male genitalia is undivided and usually tapers to a sharp down-wardly curved point. The five Australian species are often difficult to distinguish, especially the females. The forewing in the males of some species carries a continuous or interrupted sex-brand (Fig. 9), the shape of which assists in their identification. However, for some species the male genitalia (Figs. 10A–10E) provide the most reliable means of recognition.

Only one species, *T. papyria*, is really well represented in collections, the others probably being overlooked because of their superficial resemblance to the much more common species of *Ocybadistes*. However, in Queensland at least the species of *Taractrocera* are much more common than the number of specimens in many collections would suggest. The adults fly in grassy areas and are at times numerous in the coastal Wallum districts of southern Queensland and in dry open forest country inland. J. F. R. Kerr has noted that in some places, such as near Mareeba, he has found four species (*T. papyria*, *T. anisomorpha*, *T. dolon* and *T. ina)* flying together.

78. *Taractrocera papyria* (Boisduval), 1832

This is the best known and most widely distributed species, ranging down the eastern coast from Kuranda, Queensland, to Victoria, Tasmania, South and Western Australia. In the male genitalia (Fig. 10A), the uncus is truncate and slightly indented at the tip and, unlike the other Australian species, the valva is not divided apically. Two very distinct subspecies are known.

a. *Taractrocera papyria papyria* (Boisduval), 1832

Plate 4, figs. 18 (male), 18A (female), white grassdart.

DISTRIBUTION.   Eastern Australia, from Kuranda, Queensland,

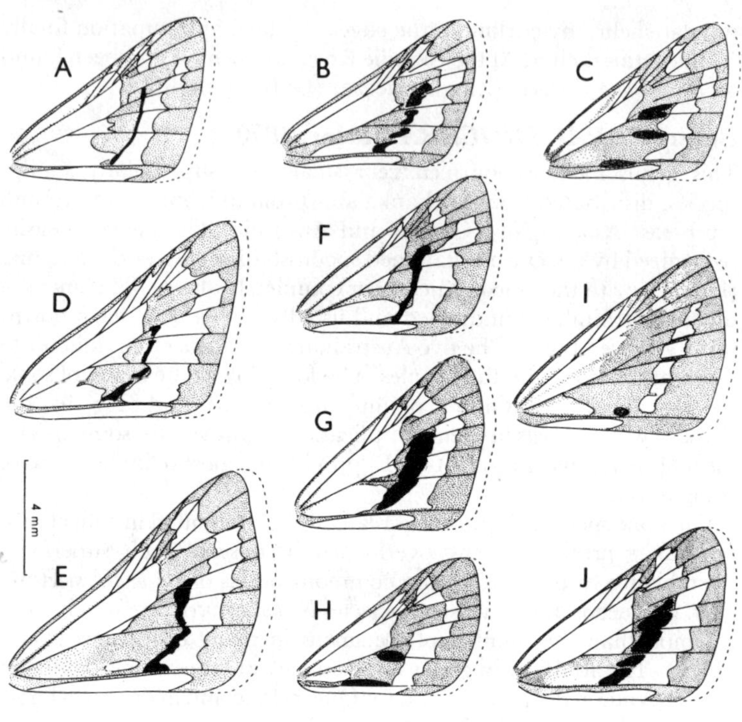

Fig. 9. Sex-brands in male Hesperiidae: A, *Taractrocera papyria* (p. 83);
B, *T. ilia* (p. 87); C, *T. dolon* (p. 87); D, *T. anisomorpha* (p. 86);
E, *Ocybadistes hypomeloma* (p. 92); F, *O. flavovittatus* (p. 89); G, *O. walkeri*
(p. 90); H, *O. ardea* (p. 92); I, *Suniana lascivia* (p. 93); J, *S. sunias* (p. 95).

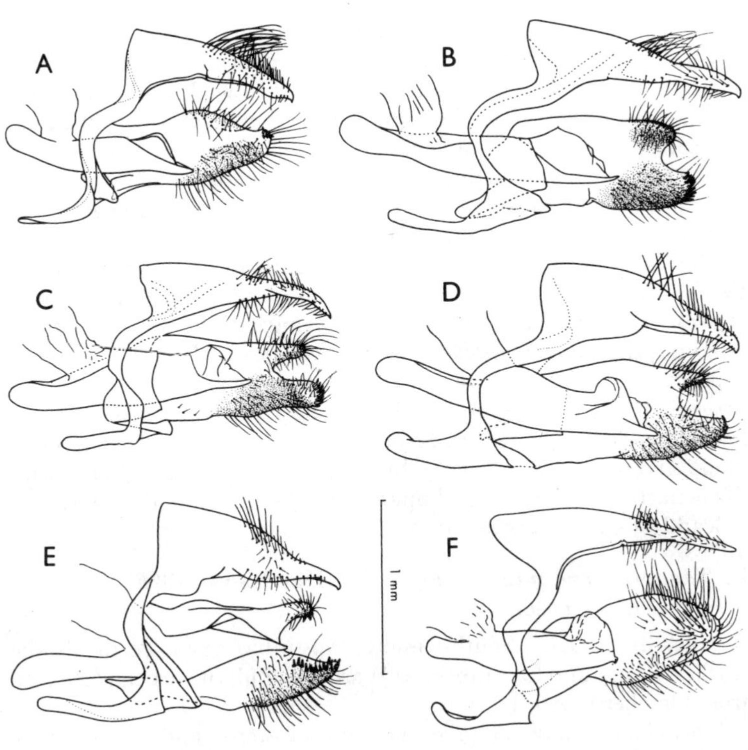

Fig. 10. Male genitalia, lateral view with left valva removed: A, *Taractrocera papyria* (p. 83); B, *T. dolon* (p. 87); C, *T. ilia* (p. 87); D, *T. anisomorpha* (p. 86); E, *T. ina* (p. 88); F, *Ocybadistes flavovittatus* (p. 89).

to Victoria, on the coast, tablelands and western slopes, south-eastern South Australia as far north and west as Wallaroo on Spencer Gulf, and coastal Tasmania up to 600 m; also Lord Howe Island. Adults are common on the east coast throughout the warmer months, and on the tablelands and near Adelaide mainly from February to April.

The small size and distinct white band beneath the hindwing separate this from all the other small skippers. Newly emerged specimens may have a greenish tint beneath.

The larva and pupa resemble those of *Ocybadistes hypomeloma* (Plate IV, figs. 3A–3C) but are smaller. The larva feeds mainly on common grasses (Poaceae), including the following: *Pennisetum clandestinum* (kikuyu), *Echinopogon caespitosus*, *Microlaena stipoides*, *Paspalum dilatatum*, a perennial *Poa* sp., *Danthonia* sp., *Oryza sativa* (rice), *Imperata* and *Cynodon dactylon* (couch). At the Boyd River, N.S.W., it feeds on *Carex gaudichaudiana* (Cyperaceae). It forms a shelter by folding over a grass blade on broad-leafed grasses or by joining several leaves together with silk in narrow-leafed species. Pupation occurs in the larval shelter, or in a dead leaf-base near the ground.

b. *Taractrocera papyria agraulia* (Hewitson), 1868

Plate 4, fig. 19 (male), western grassdart.

DISTRIBUTION. South-western Australia, from Perth to the Stirling Range and Esperance. At Waroona adults have been collected from October to April.

This is much more orange than typical *papyria* and has no white markings beneath the hindwing. It is the smallest skipper to be found in south-western Australia.

The life history and early stages are said to resemble those of typical *papyria*.

79. *Taractrocera anisomorpha* (Lower), 1911

Plate 5, figs. 4 (male), 4A (female), orange grassdart.

DISTRIBUTION. North-western Australia, Northern Territory, the islands of Torres Strait, and Cape York to Burleigh Heads (A. N. Burns) and Millmerran; inland it has been taken at the Fortescue River in Western Australia, the Alice Springs area (E. D. Edwards), Tennant Creek, Pine Creek and Brocks Creek in the Northern Territory, and Mitchell in southern Queensland; usually not common, but

86

has been taken in fair numbers by J. F. R. Kerr in March between Kuranda and Mareeba, north Queensland. At Burleigh it flies in October and November and from March to May.

This species is larger than *papyria*, and the male has more pointed forewings than other species. The irregular, slightly interrupted black sex-brand (Fig. 9D) and the valva of the genitalia (Fig. 10D) will place the male; the uncus has a sharp point. The waved inner margin of the postmedian band in the forewing of both sexes is also a useful guide. As in *T. ilia* and *T. ina* the hindwing is without a separate small post-median spot below vein Rs. However, the orange markings above are more extensive than in *T. ina*, which is about the same size as *T. anisomorpha* and with which it flies. Occasionally the hindwing beneath is without markings and may have a slightly greenish tint.

The species is found in open eucalypt forest.

The life history and early stages have not been described, but A. F. Atkins noticed a female, probably of this species, laying eggs on *Cenchrus ciliaris* (buffel grass, Poaceae) in central Queensland.

80. *Taractrocera ilia ilia* Waterhouse, 1932

Plate 5, fig. 1 (male), northern grassdart.

DISTRIBUTION.   Northern Territory; adults have been collected at the King River, Darwin, Alligator Rivers and Melville Island; rare.

The forewing is broader than in other species, with squarer tornus, and both edges of the postmedian band of the forewing are fairly straight. The sex-brand (Fig. 9B) in the male is brown and more or less interrupted at the veins and midway between CuA_2 and $1A + 2A$. The postmedian band of the hindwing is continued to Rs. The valva (Fig. 10C) of the male genitalia is more deeply divided apically than in *T. anisomorpha* and *T. dolon*, but not as deeply as in *T. ina*.

81. *Taractrocera dolon* (Plötz), 1884

This small species is known from the Northern Territory, coastal Queensland and northern New South Wales, but also from the Loui-siade Archipelago. The uncus of the male genitalia ends in a fine point, and the valva (Fig. 10B) is moderately divided apically.

a. *Taractrocera dolon dolon* (Plötz), 1884

Plate 5, figs. 2 (male), 2A (female), dingy grassdart.

DISTRIBUTION.   Eastern Australia, from the Hann River, 85 km north-west of Laura (M. De Baar), Kuranda and Atherton to Port Macquarie; fairly common.

In the male the interrupted sex-brand (Fig. 9C), of the same colour as the dark brown wing, as well as the genitalia (Fig. 10B) will place this small, inconspicuous species. The postmedian bands on both fore- and hindwing tend to be divided by darker veins into separate spots. There is usually a small spot above vein M_1 on the hindwing and, when present, enables *dolon* to be distinguished from the other Australian species except *papyria*.

The species is found in open eucalypt forest.

b. *Taractrocera dolon diomedes* Waterhouse, 1933

DISTRIBUTION.   Northern Territory; specimens have been taken at Darwin, Brocks Creek, Daly River Crossing and Alligator Rivers.

ADULT.   Similar to typical *dolon*, and with similar sex-brand in male, but larger, markings above broader, with both wings beneath much brighter orange.

This subspecies may be separated from the larger *T. ina*, which occurs in the same areas, by the presence of the sex-brand in the male, the confluent subapical spots in the forewing, and the small variable spot above M_1 of the hindwing, when present.

82. *Taractrocera ina* Waterhouse, 1932

Plate 5, fig. 3 (male), ina grassdart.

DISTRIBUTION.   Northern Territory including Alice Springs and the Plenty River (E. D. Edwards), Yam Island, Torres Strait (G. B. Monteith), and from Cape York to southern and western Queensland and western New South Wales; often common, for example, on hill-tops near Brisbane and behind the beach south of Noosa Heads.

The absence of the sex-brand immediately distinguishes the male from the other species in the genus. The male can usually be recognized by the shape of the wings and, if still in doubt, by examining the genitalia. The uncus of the male genitalia is pointed and the valva (Fig. 10E) deeply divided apically. The separation of the female from neighbouring species is more difficult, but the absence of the small spot above vein M_1 in the hindwing, and in the forewing the irregular postmedian band and the separate subapical spots, should help to place it.

The species probably has a continuous distribution across northern Australia and through Queensland and inland New South Wales. It is found in open eucalypt forest.

The larva makes a tubular shelter at the tip of a grass blade by cutting into the midrib from each edge above and beneath. As the midrib is progressively weakened by continued feeding, it bends over and the larval shelter is probably thus gradually lowered to the ground. At Mackay eggs laid in October hatched in six days, the larvae reached maturity and pupated in about four weeks, and the adult emerged in eleven days. Near Darwin the species has been reared by C. S. Li from *Oryza sativa* (rice) and by N. Forrester from *Cymbopogon citratus* (lemon grass). In Queensland the larvae have been recorded on *Brachiaria decumbens*, *Paspalum urvillei*, *P. dilatatum* and *P. conjugatum* (all Poaceae).

Genus *OCYBADISTES* Heron, 1894

Distributed from Sumba to Timor, Tanimbar, the Kai and Aru Islands, and from New Guinea to Australia, this genus contains only five species. Four of these are represented in Australia, but only one is confined to this country. Superficially the genus resembles *Taractrocera*, but adults can be distinguished readily by the antennal club (Fig. 7G), with a short slightly bent apiculus. The uncus of the male genitalia (Fig. 10F) is produced to a long narrow point and, as in *Taractrocera*, is undivided. However, the valva in the various species does not differ very much in form. The male forewing usually bears a sex-brand (Figs. 9E–9H), the shape and position of which, taken in conjunction with the markings, is a good guide to the species. Females are more difficult to identify.

83. *Ocybadistes flavovittatus* (Latreille), 1824

This is a common species occurring in the Northern Territory and in eastern Australia from the islands of Torres Strait, and from the Claudie River to southern New South Wales. Four subspecies are recognized, three from Australia and the other from New Guinea. The species is found in grassy areas in eucalypt forest, and often in gardens.

a. *Ocybadistes flavovittatus flavovittatus* (Latreille), 1824

Plate 5, figs. 11 (male), 11A (female), common dart.

DISTRIBUTION. The coast and tablelands of south-eastern

Queensland to coastal southern New South Wales; common.

The sex-brand (Fig. 9F) of the male could only be confused with that of *O. hypomeloma*, but *O. flavovittatus* is much smaller and is without the very distinct whitish streak along the inner margin of the hindwing found in *O. hypomeloma*. The female is rather similar to *O. walkeri sothis* but the hindwing beneath is not greenish yellow in colour and there is no spot above vein Rs.

b. *Ocybadistes flavovittatus vesta* (Waterhouse), 1932

DISTRIBUTION.   North-western Australia and Northern Territory. Specimens have been taken at the Drysdale River, W.A., and at Daly River, Alligator Rivers, McArthur River, Adelaide River, and commonly at Brocks Creek from January to April and in June.

ADULT.   Similar to typical *flavovittatus* but smaller and usually with broader orange markings.

c. *Ocybadistes flavovittatus ceres* Waterhouse, 1933

DISTRIBUTION.   Warraber (Sue) Island (E. D. Edwards) and Horn Island (G. B. Monteith), Torres Strait, and eastern Queensland, from the Claudie River, Cooktown and the Atherton Tableland to Yeppoon.

ADULT.   Similar to typical *flavovittatus* but smaller, relatively more specimens with broader orange markings above than in *flavovittatus* and less than in *vesta*. It is usually larger than *vesta*.

84. *Ocybadistes walkeri* Heron, 1894

This is the most widely distributed species, ranging from Sumba, Flores and Timor, through Tanimbar, the Kai and Aru Islands and New Guinea to northern and eastern Australia, South Australia and Tasmania. Six subspecies are recognized, four of which are found in Australia. The species is found in grassy areas near and in eucalypt forest, and is sometimes common in suburban gardens.

a. *Ocybadistes walkeri sothis* Waterhouse, 1933
Plate 5, figs. 9 (male), 9A (female), yellow-banded dart.

DISTRIBUTION.   From Rockhampton, Springsure and the Expedition Range, Queensland, to southern New South Wales, the Australian Capital Territory, central and eastern Victoria, and northern and eastern coastal areas of Tasmania. Inland it is widely distributed from Mitchell, Q., to Deniliquin, N.S.W., and Kerang (A. F. Atkins, A. D. Bishop), Gunbower and Rutherglen, Victoria. It has recently been recorded at Black Rock and East Brighton, near Melbourne. At Sydney adults are on the wing from August to April.

This subspecies can be separated from *O. flavovittatus flavovittatus*, which often occurs in the same areas, by the shape of the sex-brand (Fig. 9G) in the male and the additional spots beneath the hindwing, the ground colour of which is greenish yellow rather than brownish orange. However, worn specimens of *sothis* tend to lose the greenish yellow scales beneath and appear browner than usual.

The larva and pupa resemble those of *O. hypomeloma* (Plate IV, figs. 3A–3C). The larva feeds on *Cynodon dactylon* (couch grass) and no doubt other common grasses (Poaceae). An adult has been reared from a larva found feeding on *Dianella* (Liliaceae) at Stanthorpe, Queensland, in early January. The pupal duration was eight days, but from eggs laid at Sydney in April, adults did not appear until October, the pupal duration being five weeks.

b. *Ocybadistes walkeri hypochlorus* Lower, 1911

Plate 5, fig. 10 (male), southern dart.

DISTRIBUTION.   South-eastern South Australia; near Adelaide adults are on the wing from October to May.

The young larva forms a short tubular shelter by rolling over the edges of a leaf of the food plant and fastening them together with silk. Older larvae construct a larger shelter, joining several leaves. Pupation may occur in this shelter, but more often in a shelter spun in a dried leaf sheath close to the ground. The larva feeds on *Brachypodium distachyon*, *Cynodon dactylon* (couch grass) and *Pennisetum clandestinum* (kikuyu) (all Poaceae).

c. *Ocybadistes walkeri sonia* Waterhouse, 1933

DISTRIBUTION.   North-eastern Queensland from Weipa and the Dalhunty and Claudie Rivers to Mackay. Adults have been taken at Kuranda in June, July, September and October.

ADULT.   Similar to *sothis* but smaller, beneath not tinged greenish.

The larvae have been recorded feeding on *Thuarea involuta* (Poaceae) on Hoskyn Island, Queensland.

d. *Ocybadistes walkeri olivia* **Waterhouse, 1933**

DISTRIBUTION. Northern Territory. Adults have been collected at Darwin, Brocks Creek, Pine Creek, Daly River Crossing and Alligator Rivers.

ADULT. Similar to *sothis* but smaller, with much broader orange markings.

85. *Ocybadistes hypomeloma* **Lower, 1911**

This is the only species of the genus restricted to Australia. It occurs in north-western Australia, the Northern Territory, the islands of Torres Strait, and eastern Australia from west of Cooktown to Sydney. Two subspecies are known.

a. *Ocybadistes hypomeloma hypomeloma* **Lower, 1911**

Plates 5, figs. 8 (male), 8A (female), IV, pale orange dart.

DISTRIBUTION. From 45 km west of Laura, Cairns and Kuranda, Queensland, to Jervis Bay, New South Wales. Adults are common at Sydney, where two generations are completed annually.

This is the largest species of *Ocybadistes*. It can be identified by the white streak beneath the hindwing in both sexes and the sex-brand of the male (Fig. 9E).

The larvae feed on common grasses (Poaceae).

b. *Ocybadistes hypomeloma vaga* **(Waterhouse), 1932**

DISTRIBUTION. North-western Australia, Northern Territory and Horn (G. B. Monteith) and Prince of Wales Islands, Torres Strait. In Western Australia it has been taken at the Prince Regent River, in the Northern Territory at Brocks Creek, Darwin (J. F. R. Kerr) and at the Cobourg Peninsula and the McArthur River (E. D. Edwards).

ADULT. Male: similar to typical *hypomeloma* but above with markings broader and a deeper orange, forewing with subapicals elongate and joining or nearly joining cell spot and postmedian band, hindwing with spot above M_1 touching or joining postmedian band.

Female: similar to that of typical *hypomeloma*, but hindwing with spot above M_1 joining postmedian band.

86. *Ocybadistes ardea heterobathra* **(Lower), 1908**

Plate 5, fig. 7 (male), dark orange dart.

DISTRIBUTION. Islands of Torres Strait, and eastern

Queensland from Cape York to Burleigh Heads (A. N. Burns). At Burleigh it flies from September to November and again from March to May. It is common at Cape York and Kuranda.

This is a small, very distinctly marked species, with a modified sex-brand (Fig. 9H) in the male. It is found mainly in wet coastal districts.

Genus *SUNIANA* Evans, 1934

This genus, with three small species, ranges from Timor and the Moluccas to New Guinea, the Bismarck Archipelago, the Solomons and Australia. Two species are represented in Australia. The antennal club is slightly flattened, with a short curved apiculus, and the forewing of the male has a sex-brand. Unlike *Taractrocera* and *Ocybadistes*, the uncus of the male genitalia (Fig. 11A) is deeply divided, each half being long, pointed and curved downwards.

87. *Suniana lascivia* (Rosenstock), 1885

This species occurs in Timor, New Guinea, and northern and eastern Australia as far south as Victoria. Six subspecies are recognized, four of which are known from Australia, one from New Guinea and one from Timor.

a. *Suniana lascivia lascivia* (Rosenstock), 1885

Plates 5, fig. 14 (male), V, dingy dart.

DISTRIBUTION.   Eastern Australia, mainly coastal, from Mackay to Melbourne, but also on the tablelands of southern Queensland, including the Carnarvon Range, and northern New South Wales. At Sydney there are two generations annually and adults are on the wing from October to April.

Both sexes can be recognized by their dull brown coloration and reduced spots and bands.

The larva feeds on *Imperata cylindrica* (blady grass, Poaceae), constructing a vertical cylindrical shelter by bending round the edges of a section of leaf and joining them with silk. At maturity it leaves the food plant and pupates in a shelter formed from several dead leaves at the base of the plant.

b. *Suniana lascivia neocles* (Mabille), 1891

Plate 5, fig. 15 (male), northern dingy dart.

DISTRIBUTION.   North-eastern Queensland, from Horn Island (G. B. Monteith) and from Cape York to Ingham, including the Atherton Tableland.

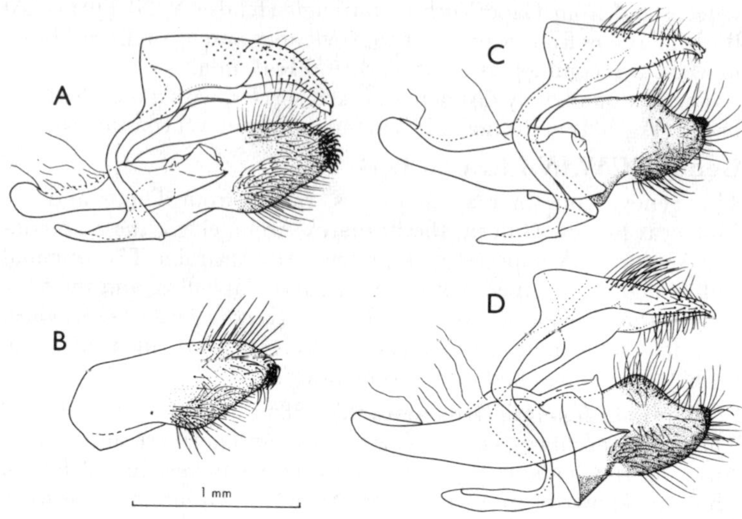

Fig. 11. Male genitalia, lateral view with left valva removed: A, *Suniana sunias* (p. 95); B, *S. lascivia*, right valva only (p. 93); C, *Arrhenes marnas* (p. 97); D, *A. dschilus* (p. 97).

c. *Suniana lascivia lasus* Waterhouse, 1937

DISTRIBUTION. Bathurst Island, Northern Territory, in October.

ADULT. Similar to typical *lascivia*, but much smaller, above with markings bright orange, clearly defined, hindwing with postmedian band broader, beneath forewing with three subapicals, hindwing with band well defined.

The small size and relatively bright, clearly defined markings distinguish this subspecies. It may be simply a form of the New Guinea subspecies *pola* Evans, 1934, but further specimens are needed to determine the relationship of these two and *larrakia*.

d. *Suniana lascivia larrakia* **Couchman, 1951**

DISTRIBUTION.   North-western Australia and Northern Territory, including Groote Eylandt; fairly common at Darwin, N.T. (J. F. R. Kerr).

ADULT.   Male: similar to typical *lascivia* but markings more distinct, orange, above forewing with well-defined patch covering cell and costa to beyond end of cell, three small subapicals, postmedian band of five spots broader, straight and of fairly uniform width from veins M_1 to $1A + 2A$, sex-brand a small slightly raised patch of scales on middle of vein $1A + 2A$, hindwing with postmedian band of four spots broader, a minute spot above M_1, beneath forewing with apex suffused orange, hindwing brownish orange, with postmedian band faintly edged brown-black.

Female: similar to male but markings orange-yellow, spots more distinctly separated from one another, postmedian bands crossed by dark veins, sex-brand absent.

This was originally thought to represent a separate species, but it now seems almost certain that it is a geographical form of *lascivia*. Adults are almost as large as typical *lascivia* but the markings are more distinct in both sexes. The sex-brand in the allotype male from north-western Australia, and a paratype male from Groote Eylandt, is of the same colour as the dark background, but is slightly raised and can be distinguished by the specialized nature of the scales. The genitalia of both sexes match those of typical *lascivia*. This subspecies is larger than *lasus* with markings not as bright.

88. *Suniana sunias* **(Felder), 1860**

This species ranges from Sumba and the Moluccas to New Guinea, the Bismarck Archipelago, the Solomons and Australia. Eight subspecies are recognized, three of which are found in Australia. The forewing of the male bears a broad black sex-brand (Fig. 9J). The species is found mainly in wet coastal districts, especially in grassy clearings in rain forest.

a. *Suniana sunias rectivitta* **(Mabille), 1878**

Plate 5, fig. 12 (male), orange dart.

DISTRIBUTION.   Islands of Torres Strait, and from Cape York to Rockhampton; common.

Larvae have been found at Paluma, Queensland, on *Paspalum urvillei* (Poaceae) (J. F. R. Kerr).

b. *Suniana sunias nola* **(Waterhouse), 1932**

Plate 5, figs. 13 (male), 13A (female), southern orange dart.

DISTRIBUTION.   Southern Queensland and northern New South Wales, from Noosa to Taree.

This subspecies is not as bright as *rectivitta*.

E. D. Edwards has reared the larva on common grasses (Poaceae). It constructs a shelter by joining several leaves together with silk, in which pupation takes place head upwards. In Queensland the larvae have been found near Brisbane on *Panicum maximum* and near Caboolture on *Leersia hexandra* (both Poaceae).

c. *Suniana sunias sauda* **Waterhouse, 1937**

DISTRIBUTION.   Northern Territory, including Groote Eylandt.

ADULT.   Similar to *rectivitta* but orange markings above paler and more extensive, beneath with ground colour and markings paler.

Genus *ORIENS* Evans, 1932

This widespread genus is distributed from India to the Philippines and Celebes, and also occurs in Fiji and Samoa. One species has been recorded from near Darwin and Townsville, but additional specimens are needed to confirm the occurrence of this species in Australia.

89. *Oriens augustulus augustulus* **(Herrich-Schäffer), 1869**

DISTRIBUTION.   Townsville, Queensland (one male in South Australian Museum), and Darwin, Northern Territory (one male in British Museum). This subspecies is fairly common in Fiji. If it in fact occurs naturally in Australia, it has been overlooked by collectors.

ADULT.   Male: above dark brown, markings orange; forewing with costa and base suffused orange, a divided spot in cell, three subapical spots, a broad postmedian band from veins M_2 to $1A + 2A$, paler and semihyaline between veins CuA_1 and CuA_2, a streak along inner margin; hindwing with a spot in cell, a postmedian band with central paler semihyaline area. Beneath forewing dark brown, costa and apex broadly orange-brown, markings as above; hindwing orange-brown, markings as above, but postmedian band interrupted at vein CuA_2.

This species is similar in size and general appearance to *Arrhenes marnas* (Plate 5, fig. 6), but the male is without a sex-brand, and the postmedian band of both wings has a paler central section.

Genus *ARRHENES* Mabille, 1904

The genus ranges from Flores and the Moluccas to New Guinea, the Bismarck Archipelago and north-eastern Australia. Seven species are known, two of which occur in Australia.

90. *Arrhenes dschilus iris* (Waterhouse), 1932

Plate 5, fig. 5 (male), iris skipper.

DISTRIBUTION.   North-eastern Queensland, from Cape York to Mackay; common. At Kuranda adults have been taken nearly throughout the year.

This is the larger of the two Australian species, with the subterminal spot between veins M_1 and M_2 of the forewing absent or reduced, and the postmedian band of the hindwing beneath crossed by dark veins. In the male genitalia (Fig. 11D) the tip of the uncus is pointed or very slightly indented. The species is found mainly in wet coastal districts.

The larva feeds on *Imperata* (blady grass) and also attacks *Saccharum officinarum* (sugar cane) (both Poaceae).

91. *Arrhenes marnas affinis* (Waterhouse and Lyell), 1912

Plate 5, fig. 6 (male), affinis skipper.

DISTRIBUTION.   Queensland, from Cairns to Burleigh Heads; usually uncommon. Adults are much more local than *A. dschilus* and have been taken at Kuranda from February to June, and at Burleigh Heads in March; they are sometimes common near Ingham.

This species is much smaller than *A. dschilus*, the subapical spots of the forewing are elongate and confluent, and the postmedian band in both wings is longer, the spots between veins M_1 and M_3 being broader and joined to the broader section between M_3 and $1A + 2A$. In the male genitalia (Fig. 11C) the tip of the uncus is broader than in *A. dschilus* and more deeply excavated. The species is found mainly in wet coastal districts.

Near Caboolture, Queensland, larvae have been found feeding on *Leersia hexandra* (Poaceae) growing in a swampy area. In March the pupal duration was eight or nine days.

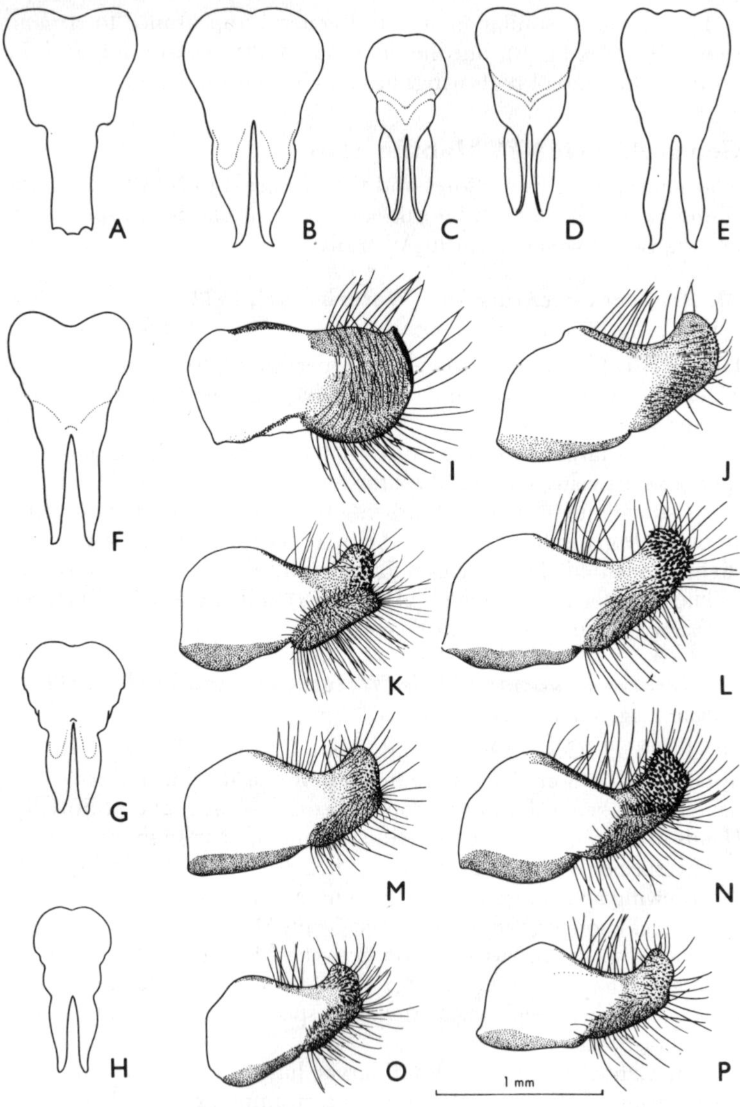

Genus *TELICOTA* Moore, 1881

This is a large and difficult genus containing twenty-two species widely distributed in the Oriental and Australian regions, from India and Sri Lanka to China, through south-east Asia to New Guinea, northern and eastern Australia and the Solomons. The adults are much larger than in *Taractrocera*, *Ocybadistes* and *Suniana*, and superficially resemble *Oriens* and *Arrhenes*.

Several of our species are very similar and their correct identification depends on a careful examination of wing pattern and colour, the form and position of the sex-brand in the male, and the structure of the male genitalia (Figs. 12A–12P). Nevertheless the females are easily confused and their positive identification will ultimately require the study of their genitalia.

92. *Telicota eurotas* (Felder), 1860

This species occurs in the Moluccas, New Guinea, the Aru Islands, and eastern Australia.

a. *Telicota eurotas eurychlora* Lower, 1908

Plate 6, figs. 8 (male), 8A (female), dingy darter.

DISTRIBUTION. Burleigh Heads, south Queensland, to Nowra (E. D. Edwards), New South Wales; usually uncommon, but sometimes common locally.

The very indistinct markings beneath the hindwing distinguish both sexes. The undivided uncus (Fig. 12A) at once separates the male.

The species is apparently found in swampy areas where its food plant probably grows. On one occasion J. F. R. Kerr took adults commonly in a swamp near the Hastings River.

b. *Telicota eurotas laconia* Waterhouse, 1937

DISTRIBUTION. North-eastern Queensland, from the Claudie River, Kuranda and Cairns to Ingham.

ADULT. Male: similar to *eurychlora* but above with markings a

Fig. 12. Male genitalia of *Telicota*, dorsal view of uncus and inner side of right valva: A, I, *T. eurotas* (above); B, J, *T. colon* (p. 100); C, L, *T. augias* (p. 100); D, N, *T. anisodesma* (p. 101); E, K, *T. ohara* (p. 102); F, M, *T. ancilla* (p. 102); G, P, *T. mesoptis* (p. 103); H, O, *T. brachydesma* (p. 104).

darker orange; forewing with subapical spots less clearly joined to costal streak, spots between veins M_1 and M_3 smaller, postmedian band narrower, with straight edges; hindwing with cell spot smaller and postmedian band extending beyond vein M_1; beneath more orange and markings more distinct.

Female: similar to male but forewing with subapicals quite separate from costal streak, without sex-brand.

D. Binns reared an adult from a pupa he found in May 1977 at the Claudie River in a shelter resembling that of a *Hesperilla* on the sedge *Scleria polycarpa* (Cyperaceae). The sedge was growing in a swampy area. At Mission Beach J. F. R. Kerr noticed adults flying around another sedge *Rhynchospora corymbosa* (Cyperaceae).

93. *Telicota colon argeus* (Plötz), 1883

Plate 6, figs. 3 (male), 3A (female), pale darter.

DISTRIBUTION. North-western Australia, Northern Territory, Moa Island and Cape York to Sydney; common north of the Richmond River, rarely taken at Sydney. Specimens have recently been taken on the Fortescue River, W.A.

The uncus is deeply divided, and the two arms are flattened and broad at base, with tapering divergent tips (Fig. 12B); the distal end of the valva (Fig. 12J) is smoothly rounded, with only minute spines. Other characters which help to place the male are the broad greyish sex-brand, the pale rather pointed forewings, and the extension of the orange-yellow postmedian band and the subapicals along the veins to the margin. The markings above are a paler orange-yellow than in any other Australian species. Females are much more difficult to identify, especially if they are worn.

The species is found both in open eucalypt forest and in grassy clearings in rain forest.

94. *Telicota augias* (Linnaeus), 1763

This species ranges from Burma to the Philippines, and south through Indonesia to the Northern Territory and north-eastern Queensland. Seven subspecies are recognized, two of which occur in Australia. The uncus (Fig. 12C) is deeply divided and the two arms form more or less parallel and vertical blades.

a. *Telicota augias krefftii* **(W. J. Macleay), 1866**

Plate 6, fig. 4 (male), Krefft's darter.

DISTRIBUTION.   Islands of Torres Strait, and Cape York to Mackay; common.

This subspecies has the markings above more reddish orange than either *T. colon* or *T. ancilla*. It is about the same size as *T. colon*, but has a narrower and darker sex-brand; it is usually smaller than *T. ancilla*. Males may be separated from *T. mesoptis*, with which they often fly, by the more oblique sex-brand which reaches vein 1A + 2A before rather than beyond its mid-point. Beneath the ground colour is much brighter orange than in allied species and the markings are sometimes obscure or even absent; in the hindwing the postmedian band is not outlined with black.

Adults fly in open eucalypt forest as well as in rain forest.

b. *Telicota augias argilus* **Waterhouse, 1933**

DISTRIBUTION.   North-western Australia and Northern Territory. In the north-west it has been taken at Baudin Island, the Prince Regent and Drysdale Rivers, and Kununurra. At Darwin adults have been collected in May, November and December.

ADULT.   Similar to *krefftii* but slightly smaller, with more rounded termen, markings above broader and brighter, beneath hindwing with postmedian band well developed.

Adults have been reared from cultivated rice in the Northern Territory (C. S. Li), and E. D. Edwards has noticed that adults are found where *Flagellaria indica* (Flagellariaceae) grows.

95. *Telicota anisodesma* **Lower, 1911**

Plates 5, figs. 18 (male), 18A (female), V, large darter.

DISTRIBUTION.   Cooloola, north of Noosa Heads (M. De Baar) and Montville, Queensland, to Gosford, New South Wales; rare. The original specimens came from Ballina. Adults have been taken from October to April. The species has also been recorded from Townsville, Cairns and Mackay, but these localities need confirmation.

This species is usually much larger than others in the genus except *T. ohara*, nearing *Cephrenes augiades* in size. From the latter it may be distinguished by the presence of the sex-brand in the male and by the straight postmedian band of the forewing and the underside of the female. In recent years *T. anisodesma* has been treated as a subspecies of *T. augias*, and the genitalia (Figs. 12D, 12N) show the two are certainly closely related despite the great difference in size and markings.

The species is probably confined to rain forest.

Near Burleigh Heads the larva feeds on *Flagellaria indica* (Flagellariaceae), a tall climbing rain-forest plant with grass-like leaves each of which is tipped with a coiled tendril. The larvae roll the tips of the leaves into cylindrical shelters with the entrance directed towards the main stem. They finally leave the food plant to pupate in curled dried leaves on the ground. They are said to be difficult to rear in captivity, but can be enclosed in a netting "sleeve" on the food plant, some dead leaves being provided for pupation. However, E. D. Edwards has found that the last instar larvae can be easily reared in plastic bags, in which the food plant will remain fresh for up to two or three weeks. At Cairns larvae, said to be *T. anisodesma*, have been reported on a grass, *Leptaspis banksii* (Poaceae).

96. *Telicota ohara ohara* (Plötz), 1883

Plate 6, fig. 2 (male), dark darter.

DISTRIBUTION. Thursday Island, and Cape York to Mackay; uncommon.

The orange markings above are often overlaid with scattered dark scales. This species may be as large as *T. anisodesma*, but in the male the underside is much darker, the two arms of the uncus (Fig. 12E) are much narrower and flatter, and the valva (Fig. 12K) is very different in shape. The female of *T. ohara* has more prominent markings above than most specimens of *T. anisodesma* and beneath is much darker and similar to the male of *T. ohara*.

The species is probably confined to rain forest.

97. *Telicota ancilla* (Herrich-Schäffer), 1869

This is another widely distributed species, occurring from Sri Lanka and India to China and south through the Philippines and south-east Asia to the Moluccas, New Guinea and Australia. Eleven subspecies are recognized, two of which occur in Australia.

a. *Telicota ancilla ancilla* (Herrich-Schäffer), 1869

Plates 6, figs. 5 (male), 5A (female), IV, greenish darter.

DISTRIBUTION. Kuranda, north-eastern Queensland to far eastern Victoria; common north of Batemans Bay, N.S.W.

Males most closely resemble *T. colon*, but the orange-yellow markings are usually a little richer in colour, the sex-brand is narrower and a little darker, and the wings often have a slightly greenish tinge beneath. This is more noticeable in females, and worn specimens of either sex may be dull yellowish brown beneath. In the male the two arms of the uncus (Fig. 12F) are long, narrow and only slightly divergent, and the distal end of the valva (Fig. 12M) is slightly indented with few very small spines.

The species is found in grassy places, usually in eucalypt forest or coastal heath.

The flattened anal segment of the larva serves to distinguish it from *Suniana lascivia* found on the same food plant. The larva feeds on Poaceae such as *Imperata cylindrica* (blady grass), *Paspalum urvillei* and *Sorghum halepense* (Johnson grass) (T. H. Guthrie). The early instars live in a cylindrical shelter formed by rolling over part of a leaf and joining the edges with silk, but later two or more leaves may be joined to form the shelter. At maturity the larva becomes reddish brown in colour and leaves the grass to pupate in a shelter formed from one or more curled dead leaves on the ground.

b. *Telicota ancilla baudina* Evans, 1949

DISTRIBUTION.   North-western Australia and Northern Territory. Specimens are known from Queen, Baudin and Cassini Islands and Parry Harbour in the Admiralty Gulf area of Western Australia, and at the Daly River and Darwin.

ADULT.   Similar to typical *ancilla*, but rather smaller, forewing with squarer tornus, markings broad, hindwing beneath bright orange with distinct postmedian band.

This subspecies resembles *T. augias argilus*.

98. *Telicota mesoptis mesoptis* Lower, 1911

Plate 6, figs. 6 (male), 6A (female), Lower's darter.

DISTRIBUTION.   Northern Territory, and Cape York to Mackay; common at the Claudie River and near Cairns.

This species is similar to *T. augias*, with which it flies, but the sex-brand is rather less oblique, reaching vein 1A + 2A farther from the base than in either *augias* or *ancilla*. The orange markings above are usually a little paler than in *augias* and the hindwing beneath is more yellowish and tinged with green. Adults are smaller than *ancilla*, the arms of the uncus (Fig. 12G) are shorter and broader, and the valva (Fig. 12P) is rather narrower and more curved, but with the distal end also slightly indented.

Adults fly at the margins of rain forest where *Imperata* (blady grass, Poaceae) grows.

99. *Telicota brachydesma* Lower, 1908

Plate 6, figs. 7 (male), 7A (female), small darter.

DISTRIBUTION. Claudie River to Mackay, Queensland; rare. Specimens have been taken at the Claudie River, Cooktown, Kuranda, South Johnstone River and Mackay.

This is the smallest species in the genus, with dark wings and very pointed forewing in the male. In colour above and in general appearance the male might be taken for a small specimen of *T. augias krefftii*, but is much darker beneath, with the postmedian band of the hindwing very obscure. The uncus (Fig. 12H) and valva (Fig. 12O) are rather similar to those of *T. mesoptis* but the valva is shorter and broader. In the female the forewing has a small or obscure orange spot in the cell, and the postmedian bands of both fore- and hindwing are narrower than in *T. mesoptis*; beneath the wings are darker.

The species appears to be confined to rain forest, and adults fly around rain-forest trees.

The larvae feed on *Leptaspis banksii* (Poaceae) growing in rain forest.

Genus *CEPHRENES* Waterhouse and Lyell, 1914

This genus contains five species distributed from India to the Philippines and through south-east Asia to the Moluccas, New Guinea, Australia and the Solomons. The two Australian species superficially resemble *Telicota*.

100. *Cephrenes trichopepla* (Lower), 1908

Plates 5, fig. 16 (male), V, yellow palmdart.

DISTRIBUTION. North-western Australia, north from Broome (M. S. Moulds), Northern Territory, including the Alice Springs area, the islands of Torres Strait, and Cape York to Brisbane (J. F. R. Kerr); common near the food plant in the north.

The larvae feed on *Cocos nucifera* (coconut palm), *Livistona,* and other palms (Arecaceae).

101. *Cephrenes augiades sperthias* (Felder), 1862

Plate 5, figs. 17 (male), 17A, 17B (females), orange palmdart.

DISTRIBUTION. Islands of Torres Strait, and Cape York to the Illawarra district of New South Wales; common near the food plant.

In the male the tip of the uncus is produced into two sharp teeth. Extremes of variation in the female are shown in Plate 5, figs. 17A and 17B. There seems to be a greater tendency for the markings to be reduced in specimens from Queensland.

The larvae feed on the fronds of the native palms *Archontophoenix cunninghamiana* (Bangalow palm) and *Livistona australis* (cabbage tree palm) (both Arecaceae), but also damage palms grown as ornamentals, especially young plants. They shelter between two adjacent sections of the palm frond joined together with many strands of silk, and later pupate in the same situations. The inside of the shelter is often liberally coated with a white waxy powder.

Genus *SABERA* Swinhoe, 1908

This genus occurs in New Guinea and its associated islands, in the Bismarck Archipelago, north-eastern Australia and Fiji. There are eleven species including three from Australia.

102. *Sabera caesina albifascia* (Miskin), 1889

Plate 6, fig. 15 (male), black and white swift.

DISTRIBUTION. Moa and Prince of Wales Islands, and Cape York to Ingham; sometimes common.

This species flies at rain-forest margins or in clearings in rain forest. The antennae are longer than in the other two Australian species, and the white clubs are very noticeable when the insect is settled on a leaf.

103. *Sabera fuliginosa fuliginosa* (Miskin), 1889

Plate 6, fig. 16 (male), white-fringed swift.

DISTRIBUTION. From 30 km north-west of Cooktown, the Daintree River, Rumula, Kuranda, Malanda Falls, and Cairns to Paluma, northern Queensland; less common than *S. caesina.*

This is also a rain-forest species. The antennae are long, though not as long as in *S. caesina*, from which it is easily separated by the absence of markings above, and the form of the sex-brand in the male.

104. *Sabera dobboe autoleon* (Miskin), 1889

Plate 6, fig. 1 (male), Miskin's swift.

DISTRIBUTION. Thursday Island, and Cape York to Mackay, north Queensland; common at times near Cairns and at Shute Harbour, near Proserpine (D. P. Sands).

The antennae are similar in length to those of *S. fuliginosa*, and much longer than in the species of *Telicota* and *Cephrenes* which this species resembles superficially. Like the other two Australian species, it is found in rain forest.

Near Cairns the larva feeds on the leaves of *Cordyline terminalis* (Agavaceae), which grows in the rain forest. It forms a tubular shelter by curling in the edge of a leaf, in which pupation later occurs.

Genus *MIMENE* Joicey and Talbot, 1917

This genus has a distribution through New Guinea and its associated islands, the Bismarck Archipelago, New Georgia in the Solomons, and far northern Queensland. Fifteen species are recognised, including one New Guinea species which has only recently been recorded from Australia (C. G. Miller).

105. *Mimene atropatene* (Fruhstorfer), 1911

Plate 14, fig. 7 (male) (New Guinea).

DISTRIBUTION. Within Australian limits known so far from only two female specimens taken at the Claudie River, Queensland, in May and August.

This species bears some resemblance to *Sabera fuliginosa* and like that species has long antennae. However, the tornus of the hindwing in the male of *M. atropatene* is much more produced and the sex-brand is different. The dark greenish blue coloration beneath the wings immediately distinguishes *M. atropatene*. The life history is not known.

Genus *PARNARA* Moore, 1881

This is a widespread genus which ranges from Africa and Madagascar, through India to China and south-east Asia, the Moluccas, New Guinea and Australia. Four species are known, two of which are found in Australia.

Parnara, Borbo and *Pelopidas* form a compact group of grey-brown species, often suffused with dull yellow or green, with whitish or yellowish hyaline spots, especially on the forewing. The antennae are always short, and the apiculus strongly bent with its tip upturned. The pupae where known are elongate, pale green in colour, with the anterior end produced into a long fine point.

106. *Parnara naso sida* (Waterhouse), 1934

Plate 6, fig. 12 (male), straight swift.

DISTRIBUTION.   Cooktown, Mareeba, Kuranda and Herberton, Queensland, to Port Macquarie, New South Wales; usually not common.

Near Grafton, N.S.W., the males have been observed hill-topping.

The life history and early stages of *sida* have not been recorded, but it has been reared from *Oryza sativa* (rice) at Mareeba. In Asia the larvae of *bada* feed on rice, sugar cane, maize, *Bambusa* (all Poaceae), and *Colocasia* (Araceae).

107. *Parnara amalia* (Semper), 1879

Plate 6, fig. 9 (male), hyaline swift.

DISTRIBUTION. Northern Territory, and from Cairns, Queensland, to Lismore and Maclean, New South Wales.

The adults may be distinguished from others in the *Pelopidas* group of genera by the rich orange suffusion of the wings above.

In northern Australia adults have been reared from larvae feeding on wild rice (*Oryza* sp.) in the Northern Territory (C. S. Li) and from cultivated rice in Queensland. Near Caboolture, Q., larvae were found in March feeding on *Leersia hexandra* (Poaceae) growing in a swampy area.

Genus *BORBO* Evans, 1949

This genus contains eighteen species in Africa and four distributed from Sri Lanka and India to China and through south-east Asia to Australia and the Solomons. Three are known from far northern Australia.

As in *Parnara* the male is without a sex-brand. The hindwing beneath never has a spot in the cell, whereas it is sometimes present in *Pelopidas*. Some of the species are pests of rice and sugar cane (Poaceae).

108. *Borbo impar lavinia* (Waterhouse), 1932

Plate 6, fig. 11 (male), yellow swift.

DISTRIBUTION. Northern Territory and Moa Island, Torres Strait. At Darwin adults have been taken from February to May and from September to December.

109. *Borbo cinnara* (Wallace), 1866

Plate 6, fig. 13 (male), rice swift.

DISTRIBUTION. Northern Territory, the islands of Torres Strait and the Claudie River (G. Daniels); rare.

The larva lives in a shelter formed by folding over a grass blade. Pupation occurs in this shelter. Before pupating the larva lines the shelter with silk to which the pupa is attached by the cremaster and a silken girdle. The larvae feed on grasses, including rice (Poaceae).

110. *Borbo bevani* (Moore), 1878

Plate 6, fig. 14 (male), Bevan's swift.

DISTRIBUTION. Northern Territory. As only three specimens have been taken at Darwin, it is not known if the species is really established in Australia.

This species is very similar to *B. cinnara* but is smaller, of a more decided yellowish brown coloration above, with the tornus of the hindwing not produced. In the forewing the hyaline spots are more distinct, the postmedian spots are not equidistant, and the yellow postmedian spot above 1A + 2A, present in *B. cinnara*, is obscure or absent.

Some authors consider that this species should be placed in a separate genus *Pseudoborbo*.

Overseas the life history and early stages are said to resemble those of *B. cinnara*, and the larvae feed on rice, sugar cane and *Paspalum* (Poaceae).

Genus *PELOPIDAS* Walker, 1870

This genus ranges from Africa and Turkey to Australia and the southwest Pacific. Nine species are known, two of which occur in Australia.

Males usually carry a linear sex-brand on the forewing, and there is sometimes a cell spot beneath the hindwing. The larvae feed on grasses (Poaceae), and are pests of rice and sorghum.

111. *Pelopidas agna dingo* Evans, 1949

Plate 6, fig. 10 (male), common swift.

DISTRIBUTION.   Northern Territory, and from Cape York and the Jardine River (G. B. Monteith), Queensland, to the Richmond River (G. Mariott), New South Wales; common north of Mackay.

The male can be distinguished from *P. lyelli* by the relative positions of the sex-brand and the two cell spots when present. A line drawn through these two spots in *P. agna* does not cross the sex-brand, whereas in *P. lyelli* it does. In the female of *P. agna* a line similarly drawn through the cell spots does not touch the spot above vein 1A + 2A, whereas in *P. lyelli* it passes through it.

The larva forms a cylindrical shelter by bending round and joining with silk the edges of part of a leaf of the food plant, emerging to eat irregular areas from the leaf edges nearby. Larvae feed on grasses (Poaceae) with fairly broad leaves, including *Paspalum paniculatum* (J. F. R. Kerr).

112. *Pelopidas lyelli lyelli* (Rothschild), 1915

Plates 6, fig. 10A (female), V, Lyell's swift.

DISTRIBUTION.   North-western Australia, Northern Territory, the islands of Torres Strait, and Cape York to Brisbane; common north of Mackay, rare at Toowoomba (J. Macqueen).

This species is very similar to *P. agna*, but may be distinguished by the relative position of the cell spots and the sex-brand in the male, and the spot above 1A + 2A in the female.

The larvae feed mainly at night, hiding during the day in a cylindrical shelter formed by rolling over the edge of a leaf or by joining two or more narrow leaves with silk. The pupa is attached by the cremaster and a silken girdle to a pad of silk spun within the larval shelter. On the Ord River, north-western Australia, the larvae recently defoliated *Sorghum* crops. Some sent to Canberra were successfully reared on *Paspalum dilatatum*. The larvae were recorded on rice at Mareeba, Queensland (R. I. Storey).

Family Papilionidae

SWALLOWTAILS
(Plates 7–9, 14, V–VII; Fig. 13)

MOST of the butterflies of this family are large, and many have brilliant colours. The popular name swallowtail originated more than a century ago in Europe, where the species have tailed hindwings. Many Papilionidae, however, are without tails. The family is most plentiful in the tropics, but extends to the temperate areas of Australia, one species being found even in Tasmania. The species occur more commonly within 200 km of the sea coast, but one is widely distributed in the inland.

The adults fly strongly and rapidly, and some species are thought to be migratory. They feed freely at flowers, especially during the morning, and there they may be captured relatively easily. When feeding the wings are characteristically held nearly erect above the body and are constantly vibrated. Sometimes the adults congregate at damp muddy spots on roads or in gullies to drink water. The males of some species commonly find their way to the tops of hills where they fly back and forth over the summit.

The eggs are usually smooth and almost spherical, but occasionally the surface bears a raised pattern (Plate VII, fig. 2A). They are

Fig. 13. Pupae of Papilionidae; A, *Graphium agamemnon* (New Guinea) (p. 117); B, *Papilio anactus* (p. 118); C, *P. ulysses* (p. 124); D, E, *P. ambrax* (New Guinea) (p. 122); F, *P. canopus* (p. 122); G, *P. fuscus* (p. 121); H, *Pachliopta polydorus* (New Guinea) (p. 126); I, *Ornithoptera priamus euphorion* (p. 128).

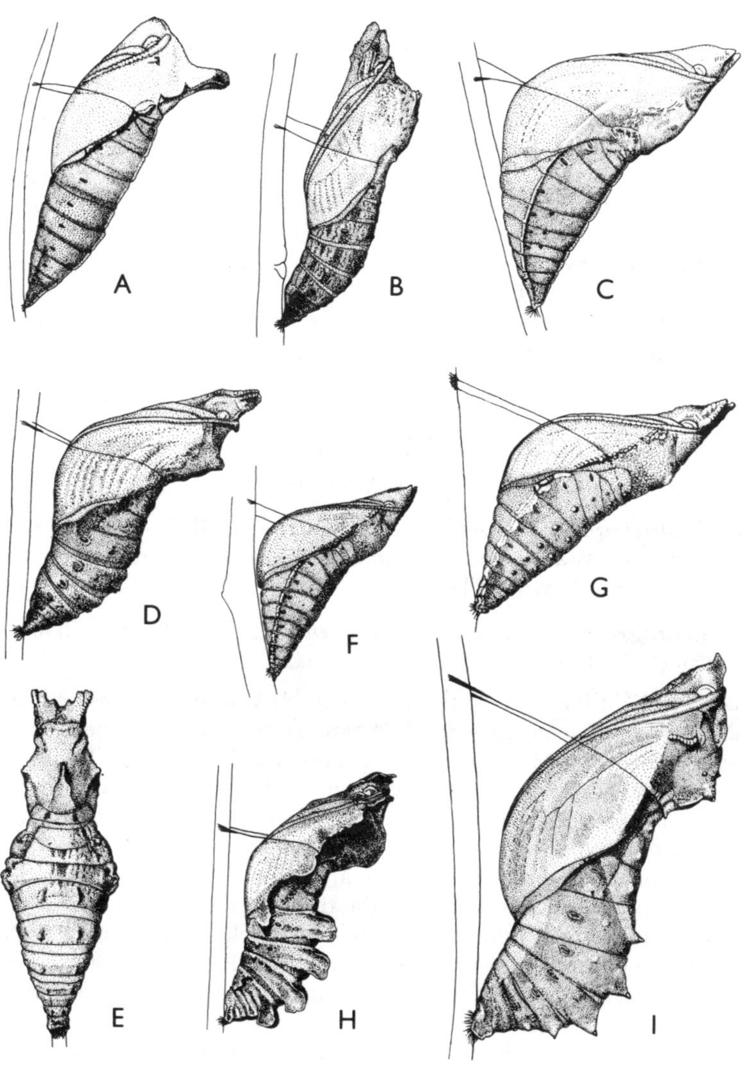

deposited singly on the food plant or sometimes on an object near by. The first instar larva has series of bristle-bearing tubercles, which in later instars may be long and fleshy, or reduced without bristles, or absent. The prothorax has an anterior dorsal slit from which can be protruded a forked fleshy process called the osmeterium (Fig. 1). This organ is protruded if the larva is disturbed, and emits a pungent odour. This often assists a collector to find larvae which are otherwise difficult to detect, as their colours harmonize well with the leaves of the food plant. The pupa is fastened by the cremaster to a pad of silk and is held head upwards by a silken girdle encircling the thorax and wings (Fig. 13; Plate VII, figs. 1D, 2E, 3C).

The larvae of several species now feed commonly on introduced ornamental and orchard trees, and are not often seen on their native food plants.

Genus *PROTOGRAPHIUM* Munroe, 1961

This genus contains only a single Australian species.

1. *Protographium leosthenes* (Doubleday), 1846

Two subspecies are recognized, one from eastern Australia and the other from the Northern Territory.

a. *Protographium leosthenes leosthenes* (Doubleday), 1846

Plates 7, fig. 4 (male), VII, fourbar swordtail.

DISTRIBUTION.   Mainly from near Mt Garnet, Queensland, to the Dorrigo Plateau, New South Wales; farther north specimens have been recorded at Moa Island, Cape York and Coen. W. R. Hindson found it very common in January 65 km south-west of Mt Garnet, and more than 160 km inland. A single very worn specimen was once collected at Sydney. In southern Queensland adults are on the wing from September to April, and there are two generations annually.

This species appears to favour the drier rain forest and soft-wood scrubs. The adults float along with outstretched wings often high above the ground. If disturbed they are capable of strong and rapid flight. D. P. Sands observed several males hill-topping on the summit of Mt White, near Coen.

The larva feeds on *Rauwenhoffia leichhardtii* (Annonaceae), a scrambling rain-forest vine. It moves with a jerky motion and rests upon a pad of silk it spins on the upper surface of a leaf. In shape the pupa bears some resemblance to *Graphium*, but is without the dorsal thoracic horn. The pupal duration is very variable, ranging from as little as eighteen days to about ten months. Neither larvae nor pupae appear to be attacked by birds, as they are found fully exposed on the foliage.

b. *Protographium leosthenes geimbia* (Tindale), 1927

DISTRIBUTION.   Alligator Rivers district, Northern Territory; very few specimens have been taken.

ADULT.   Similar to typical *leosthenes* but much larger, brown-black terminal areas much broader and darker, hindwings with larger yellow spots and smaller white crescentic spots.

Genus *GRAPHIUM* Scopoli, 1777

Included in this genus are many Old-World species which have in common the merging of veins Sc and R_1 in the forewing, which run to the costa as a single vein. The species vary superficially, but are remarkably constant in structure. The structure of the male genitalia provides the basis for the present arrangement of the species.

2. *Graphium macleayanum* (Leach), 1814

Three subspecies of this handsome species are known from the Australian mainland and Tasmania, and a fourth occurs on Lord Howe and Norfolk Islands. The species has recently been recorded from the New Guinea highlands, where the related species *G. weiskei* (Ribbe) occurs.

a. *Graphium macleayanum macleayanum* (Leach), 1814

Plate 7, fig. 6 (male), Macleay's swallowtail.

DISTRIBUTION. From near Rockhampton, Queensland, to Victoria and northern and eastern Tasmania. It is much more common at altitudes above about 300 m. At high altitudes and in the southern parts of its range it is on the wing from late October to early March, but at Sydney it flies from August to April.

Specimens from northern New South Wales and southern Queensland are somewhat intermediate between this and *wilsoni*. Adults of both sexes feed at flowers, and males are commonly observed hill-topping on the summits of mountains in the highlands of the south-east.

A common larval food plant near Sydney is the introduced *Cinnamomum camphora* (camphor laurel, Lauraceae); native food plants include *Doryphora sassafras* (sassafras), *Atherosperma moschatum* (both Monimiaceae), and *Tasmannia lanceolata* and *T. xerophila* (both Winteraceae). At Mt Tamborine, Q., the larvae have been recorded on *Daphnandra micrantha* (Monimiaceae) and on *Cryptocarya hypospodia* and *Endiandra pubens* (both Lauraceae). They are also said to feed on *Geijera salicifolia* (Rutaceae) (McCubbin, 1971).

From eggs laid at Stanwell Park larvae hatched on 20th March, pupae were produced at the end of April, and adults in October. North of Sydney larvae that pupate at the end of March emerge as adults in May. Eggs found at Mt Kosciusko in December on *T. lanceolata* hatched in eleven days and adults emerged the following October and November.

b. *Graphium macleayanum wilsoni* Couchman, 1965

DISTRIBUTION. Mountains and tablelands of north-eastern Queensland, from the McIlwraith Range, near Coen, at about 500 m, and from Mt Misery, north of Mossman to Eungella, near Mackay. Adults are probably on the wing throughout the year at Kuranda. They have been taken more than 160 km inland, 64 km south-south-west of Mt Garnet by W. R. Hindson.

ADULT. Similar to typical *macleayanum* but above with terminal dark areas a little narrower, cell of forewing with only anterior half green, with costal spots usually smaller, subterminal spots on hindwing more consistently obscure in male and very obscure or absent in female. Beneath hindwing with broader and more uniformly dark brown terminal area.

The larva feeds on *Doryphora aromatica* and *Daphnandra repandula* (both Monimiaceae).

c. *Graphium macleayanum moggana* Couchman, 1965
Plate VI.

DISTRIBUTION. Tasmania, in areas of rain forest in the west and south from sea level to 1,500 m.

ADULT. Similar to typical *macleayanum* but smaller and darker, above with smaller more uniformly green basal area, forewing with slightly smaller costal spots, spot near lower end of cell obscure or absent, hindwing with subterminal spots more consistently obscure or absent in male, usually small and whitish in female. Beneath forewing with terminal area sometimes darker, hindwing with rich brown postmedian band often more strongly contrasted with terminal lilac-brown area, and usually with a series of white or greenish subterminal crescentic spots.

The larva (Plate VI, fig. 3) feeds on *Atherosperma moschatum* (Monimiaceae) and the pupa (Plate VI, fig. 4) is attached beneath a leaf of the food plant.

d. *Graphium macleayanum insulanum* (Waterhouse), 1920

DISTRIBUTION. Lord Howe Island and Norfolk Island.

ADULT. Male: similar to typical *macleayanum* but above with broad dark brown margins, subterminal spots in both wings much larger and rather more diffuse, and beneath suffused with ochreous yellow.

Female: similar to male but larger, with subterminal spots in both wings even larger than in male and suffused with yellow above and beneath.

Only a few specimens have been collected, including one on Norfolk Island.

3. *Graphium sarpedon choredon* (C. and R. Felder), 1864

Plates 7, fig. 7 (male), VII, blue triangle.

DISTRIBUTION. Eastern Australia, from the islands of Torres Strait and Cape York to about 160 km south of Sydney. At Sydney adults are on the wing from September to May but in the north they probably fly throughout the year. A single specimen has been sighted, but not captured, as far south as Canberra, A.C.T. This subspecies also occurs in New Guinea and the Aru Islands.

In southern Queensland and New South Wales, the larvae feed commonly on the introduced *Cinnamomum camphora* (camphor laurel). Native food plants include *Cinnamomum oliveri*, *Cryptocarya triplinervis*, *C. hypospodia*, *Litsea reticulata* and *Neolitsea dealbata* (all Lauraceae), *Doryphora aromatica* (Monimiaceae) and *Planchonella laurifolia* (Sapotaceae). On one occasion T. H. Guthrie discovered adults ovipositing on custard apple (*Annona reticulata*, Annonaceae) and succeeded in rearing the larvae on this food plant. Eggs are sometimes laid freely on young avocados, but the larvae which hatch usually fail to develop on this plant. However, at Mackay, Q., Mrs Bird found three large larvae on avocado and these continued to feed on the leaves until pupation and three adults emerged. Although *Tristania laurina* (Myrtaceae) and *Macaranga* (Euphorbiaceae) have been recorded as food plants, G. Sankowsky considers that this is probably not correct.

The larvae rest during the daytime on the upperside of the leaves of the food plant. They are not easily seen, but when the foliage is disturbed they protrude the yellowish osmeterium and the odour emitted reveals their presence. Pupae that are attached beneath the leaves are very difficult to see as they are similar in colour. Pupae concealed on young trees showing signs of larval feeding may sometimes be discovered by carefully feeling the leaves with one's hands. They are often attached to adjacent wooden fences or tree guards.

4. *Graphium eurypylus* (Linnaeus), 1758

This species has a range nearly as great as *G. sarpedon*, some thirteen subspecies being distributed from Sri Lanka and India to southern China and Japan, through south-east Asia to the Moluccas, New Guinea, the Bismarck Archipelago, the D'Entrecasteaux and Aru Islands to northern and eastern Australia. There are two Australian subspecies.

a. *Graphium eurypylus lycaon* (C. and R. Felder), 1865

Plates 7, fig. 8 (male), VI, pale green triangle.

DISTRIBUTION.   Cape York to Sydney. At Sydney it is taken only occasionally, between January and April; adults have been recorded at Rockhampton from August to May.

The larvae feed mainly on plants of the family Annonaceae. The most usual is the cultivated *Annona reticulata* (custard apple), but the larva also feeds on *A. muricata* (sour sop) and *A. glabra*. Native food plants include *Rauwenhoffia leichhardtii*, *Mitrephora froggattii*, *Polyalthia nitidissima*, *Saccopetalum bidwillii* and *Uvaria goezeana* (all Annonaceae). Although *Diploglottis australis* (Sapindaceae) has been reported as a food plant, G. Sankowsky states that this is not correct. Oviposition has been observed on *Magnolia grandiflora* (Magnoliaceae), but the young

116

larvae will not feed on the leaves. However, young larvae placed on *Michelia champaca* (Magnoliaceae) developed normally (G. Sankowsky). T. H. Guthrie also reared larvae experimentally on *Cinnamomum camphora* (camphor laurel, Lauraceae).

During the day the larva usually rests on a pad of silk it spins on the upper surface of a leaf. The pupa is rather like *G. sarpedon* in shape (Plate VII, fig. 3C), but the thoracic horn is shorter. If pupation occurs early in March, adults emerge before the end of that month, but if it occurs as late as April, adults do not emerge until after the winter.

b. *Graphium eurypylus nyctimus* (Waterhouse and Lyell), 1914

DISTRIBUTION. North-western Australia and the Northern Territory; adults have been taken at Koolan Island, W.A., and at Melville Island, Daly River Crossing, Darwin, Alligator Rivers, Cobourg Peninsula (E. D. Edwards) and Groote Eylandt.

ADULT. Similar to *lycaon* but with much narrower central pale bluish green band on both wings.

At East Point, Darwin, the larvae have been found on *Polyalthia nitidissima* (Annonaceae) in November and December.

5. *Graphium macfarlanei macfarlanei* (Butler), 1877

Plate 7, fig. 9 (male), green triangle.

DISTRIBUTION. Islands of Torres Strait, and Cape York to Ingham. It is one of our rarer species, and probably flies mainly in the wet season.

Like other members of this genus, the adults fly very rapidly and are difficult to catch even when feeding, for they alight on each flower only momentarily.

The larva feeds on the cultivated *Annona reticulata* (custard apple) and *A. muricata* (sour sop, Annonaceae), but its native food plants are unknown. The thoracic horn of the pupa is shorter than in *G. agamemnon* (Fig. 13A).

6. *Graphium agamemnon ligatum* (Rothschild), 1895

Plates 7, fig. 10 (male), VI, green spotted triangle.

DISTRIBUTION. Islands of Torres Strait, and Cape York to

Yeppoon. It is on the wing during all except the very dry months of the year.

The short tails of the hindwing vary in length, but are generally longer in females. In old or artificially relaxed specimens the green spots sometimes change to dull greenish yellow or even reddish. There is a progressive fragmentation of the central coloured band into separate spots in the last four species, from *G. sarpedon* through *G. eurypylus* and *G. macfarlanei* to *G. agamemnon*, in which the band is broken up completely into spots.

The adults of *G. agamemnon* fly very rapidly, but feed at *Lantana* and other flowers growing along the margins of rain forest.

The larva feeds on the introduced *Annona reticulata* (custard apple), *A. muricata* (sour sop) and *A. glabra*, and on the native *Mitrephora froggattii*, *Xylopia maccreai*, *Melodorum uhrii*, and *Polyalthia nitidissima* (Annonaceae). Larvae have also been found on the ornamental *Michelia champaca* (Magnoliaceae) (G. Sankowsky).

7. *Graphium aristeus parmatum* (Gray), 1853

Plate 7, fig. 5 (male), fivebar swordtail.

DISTRIBUTION.   Cape York to Rockhampton; not common and probably flying only during the summer months.

Superficially the species resembles *Protographium leosthenes* (Plate 7, fig. 4) but can at once be distinguished by the five bars from the costa of the forewing. The venation of the forewing, the male genitalia and the pupa also differ in the two species.

The larval food plant is not recorded, but is said to be also one of the food plants of *G. eurypylus*.

Genus *PAPILIO* Linnaeus, 1758

Despite a great diversity of coloration and pattern, a remarkable uniformity of structure has resulted in all the species of Papilionini being placed in the one large genus *Papilio*.

8. *Papilio anactus* W. S. Macleay, 1826

Plates 7, fig. 2 (female), V, dingy swallowtail.

DISTRIBUTION.   Eastern Australia, from Kuranda and Cairns to Victoria and westwards to the Flinders Ranges, South Australia; in

Queensland it occurs as far west as Mt Isa and Cunnamulla. It also occurs near Alice Springs in central Australia. In southern areas it flies from October to April, but farther north throughout most of the year. This species is confined to Australia and has no close ally in New Guinea.

The adults fly in sunny places, with a slow gliding type of flight. If disturbed they fly away very rapidly, but often return to the same spot a little later to resume their flight back and forth.

The larval food plants all belong to the Rutaceae. Larvae are commonly found on cultivated *Citrus*, where they cause minor damage to young plants. The wide cultivation of *Citrus*, usually under irrigation, has probably enabled the species to become established in many areas which would otherwise be unsuitable. Native food plants include *Eremocitrus glauca*, *Microcitrus australis*, and *M. australasica*. At Mt Tamborine, Q., larvae have been found on the introduced *Limonia acidissima*.

During summer the life cycle is completed in about two months, but pupae formed in late March and April may overwinter. The pupa is straighter and more slender than *P. aegeus* (Plate VII, fig. 1D), and its colour is probably likewise influenced by the colour of the object to which the pupa is attached.

9. *Papilio aegeus* Donovan, 1805

This species is restricted to the Australian region, occurring in New Guinea and its adjacent islands, the Tanimbar, Kai and Aru Islands, the Bismarck Archipelago, and eastern Australia. Ten subspecies are recognized, two of which are found within Australian limits.

In New Guinea, the islands of Torres Strait, and in far northern Queensland, *P. aegeus* is of special interest as the female is polymorphic, appearing in more than one distinct colour form. The common female form (Plate 8, fig. 1A) occurs almost universally on the Australian mainland, but north of Ingham a much paler but variable form (Plate 8, fig. 1C) has been taken very occasionally; it is more common on Darnley and Murray Islands, and on the New Guinea mainland. A third very rare form (Plate 8, fig. 1B), much darker than the normal form, has been taken on Moa Island. A similar form occurs on Darnley and Murray Islands and in New Guinea.

a. *Papilio aegeus aegeus* Donovan, 1805

Plates 8, figs. 1 (male), 1A, 1B, 1C (females), VII, orchard butterfly.

DISTRIBUTION. Moa, Thursday, Horn and Prince of Wales Islands, and Cape York to Victoria and west to South Australia, where stragglers have been recorded as far as Adelaide and Whyalla (S.A. Dept. Agric.). It has also been reported from Groote Eylandt (J. Waddy) and Alice Springs, N.T., and Lord Howe Island. In southern Queensland it occurs in irrigated homestead gardens as far west as Quilpie (R. W. Guard). It is common north of Sydney, but uncommon and sporadic in the southern parts of its range; it is an irregular visitor to Melbourne. In central and northern New South Wales adults are on the wing from September to April, with two generations annually. In Queensland adults may be taken throughout the year.

The pale form, known from a few specimens taken at Ingham, Cape York, and from Moa, Prince of Wales and Thursday Islands, has been called form *beatrix*. The dark form, which resembles the male much more than either of the other forms, is known from only a single specimen taken at Moa. It has been called form *tullia*.

Normally the adults have a strong though somewhat irregular flight, but if disturbed they fly very rapidly. They are commonly seen in gardens and feed at flowers.

The larvae are most commonly found on cultivated *Citrus* (Rutaceae), and cause minor damage in orchards and gardens. The usual native food plants belong almost entirely to the Rutaceae and include *Halfordia*, *Fagara*, *Phebalium*, *Microcitrus australis* (native orange), *M. australasica* (finger lime), and *Zanthoxylum brachyacanthum*, *Melicopa erythrococca*, and *Micromelum minutum*. The larvae are also known to feed on *Zieria laevigata*, *Z. smithii* and *Eriostemon myoporoides* (D. P. Sands) near Sydney, and *Geijera parviflora* (wilga) and *G. salicifolia* in Queensland. At Burleigh Heads and at Canberra they feed on the introduced ornamental shrub *Choisya ternata*. In Queensland they have been reported on *Flindersia bennettiana*, *F. australis*, *F. schottiana* and *F. collina*, and on the introduced *Murraya koenigii* (all Rutaceae). Larvae have also been reported on *Cryptocarya glaucescens* (Lauraceae). Although there is a record of the larvae feeding on *Morinda citrifolia* (Rubiaceae), G. Sankowsky found that larvae placed on this plant

120

would not eat it. He also found that young larvae died, whereas final instar larvae survived, when placed on various species of *Acronychia* (Rutaceae), which has also been listed amongst the food plants of *P. aegeus*.

The early instars strongly resemble fresh bird droppings. The colour of the pupa matches to some extent the colour of the object to which it is attached, green pupae usually being found on leaves or green stems, whereas grey or brown pupae occur on older branches. The pupa is much more concave dorsally than *P. fuscus* or *P. anactus*; the cremaster is weaker than in *P. fuscus* and the central girdle stronger.

b. *Papilio aegeus ormenus* Guérin-Méneville, 1830

DISTRIBUTION. Within Australian limits known only from Darnley, Murray and Yam Islands, Torres Strait; probably throughout the year.

As in *aegeus* three forms of the female also occur in *ormenus*, but in the latter the pale form is much more common and more variable.

A fresh male specimen taken by D. Franzen at Bamaga, Cape York, in 1976 is said to resemble *ormenus* rather than typical *aegeus*.

10. *Papilio fuscus* Goeze, 1779

This species has a much wider distribution than *P. aegeus*, eighteen subspecies being distributed from the Andaman Islands, Malaya, and Borneo, through the Celebes and Moluccas to New Guinea, the Kai and Aru Islands, north-eastern Australia and the Solomons. It is not known from Sumatra, Java, the Lesser Sunda Islands or the Philippines. Two subspecies occur within Australian limits.

a. *Papilio fuscus capaneus* Westwood, 1843

Plate 8, fig. 3 (male), capaneus butterfly.

DISTRIBUTION. Cape York to Grafton, New South Wales; usually much less common than *P. aegeus*, and rare in some years towards the southern limits of its range. It is probably restricted to areas within about 160 km of the coast.

The flight behaviour of the adults is very similar to that of *P. aegeus*.

The larvae feed on cultivated *Citrus* and several native plants belonging to the Rutaceae, including *Microcitrus australasica, Halfordia scleroxyla, Fagara, Zanthoxylum brachyacanthum, Glycosmis pentaphylla* and

Micromelum minutum, as well as the introduced *Murraya koenigii*. G. Sankowsky found that larvae would not accept *Morinda citrifolia* (Rubiaceae), previously listed as a food plant. Unlike *P. aegeus*, larvae of *P. fuscus* have not been induced to accept parsley or camphor laurel.

The pupa is stouter than *P. aegeus*, with a much less prominent dorsal thoracic projection, less concave dorsum, stronger cremaster and weaker central silken girdle. The last is so weak that it sometimes breaks, but the pupa is still held erect by the strong cremaster. The pupal duration is very variable and is said to be sometimes more than one year.

b. *Papilio fuscus indicatus* Butler, 1876

DISTRIBUTION.   Darnley, Murray, Moa and Thursday Islands, Torres Strait; it also occurs in southern New Guinea.

ADULT.   Similar to *capaneus* but usually darker, with narrower and more obscure creamish bands, and much smaller tornal and sub-terminal orange spots.

11. *Papilio canopus canopus* Westwood, 1842

Plates 8, fig. 4 (female), V, canopus butterfly.

DISTRIBUTION.   North-western Australia north from Broome, and Northern Territory. Specimens come from Broome (M. S. Moulds), Cape Leveque (G. Thomas), Mitchell Plateau (K. F. Kenneally), Ord River, Darwin, Melville Island, Cobourg Peninsula and Cape Wessel (E. D. Edwards), Daly River and Groote Eylandt.

The larva is very similar to *P. fuscus* and, like that species, is without tubercles except for a very short pair on the prothorax and on abdominal segments 7 and 8. At Kununurra, on the Ord River, and at Darwin they feed on cultivated *Citrus* (Rutaceae), and at times cause slight damage. The native food plants at Darwin are *Glycosmis penta-phylla* and *Micromelum minutum* (both Rutaceae). The pupa also re-sembles *P. fuscus*, but the anterior projections may not be as prominent and the abdomen may not be quite as broad; the dorsal abdominal spots are pinkish. The pupal duration at Darwin ranged from fourteen days to a little more than two years.

12. *Papilio ambrax* Boisduval, 1832

Rather similar to *P. aegeus*, but smaller, this species ranges throughout

New Guinea, from Waigeo to the Louisiade Archipelago, and the Aru Islands to north-eastern Queensland. Five subspecies are recognized, two of which have been recorded within Australian limits.

a. *Papilio ambrax egipius* **Miskin, 1876**

Plate 8, figs. 2 (male), 2A (female), ambrax butterfly.

DISTRIBUTION.   North-eastern Queensland, from Cape Tribulation to Mackay; adults are on the wing from January to October.

This is a rain-forest species, which flies mainly in clearings and near forest margins.

The larvae feed on cultivated *Citrus* (Rutaceae), but also on the native *Clausena brevistyla*, *Zanthoxylum ovalifolium* and *Z. brachyacanthum*, and on the introduced plants *Murraya koenigii* and *Limonia acidissima* (all Rutaceae). *Morinda citrifolia* (Rubiaceae) has been listed as a food plant, but G. Sankowsky found larvae would not accept it. The pupa is rather more slender than *P. aegeus*, and its anterior projections are longer and more ornate.

b. *Papilio ambrax ambrax* **Boisduval, 1832**

DISTRIBUTION.   Throughout most of mainland New Guinea; a few specimens have been taken on Darnley Island, Torres Strait.

ADULT.   Similar to *egipius* but much more variable, forewings in both sexes sometimes with white markings absent, hindwing usually with red spots reduced.

13. *Papilio demoleus sthenelus* **W. S. Macleay, 1826**

Plates 7, fig. 1 (male), VI, chequered swallowtail.

DISTRIBUTION.   Groote Eylandt, Prince of Wales Island, throughout most of mainland Australia, and at Lord Howe Island; often very common inland, spasmodic in southern coastal areas, but usually uncommon.

The adults have a direct and rapid flight, usually about one metre above the ground. They feed commonly at flowers, especially those of low-growing herbaceous plants such as lucerne or other legumes. Migrations have been observed in the Northern Territory, south-western Australia and Victoria, and no doubt occur elsewhere.

The native food plants of the larvae belong to *Psoralea* (Fabaceae), including *P. tenax* (emu foot) and *P. patens* in inland Queensland, *P. patens* and *P. cinerea* in South Australia, *P. leucantha* at Carnarvon (M. Hutchison) and Millstream, Western Australia, and *P. pustulata* south-east of Broome, W.A. These are low-growing shrubby perennial legumes with digitate or pinnate leaves and purple or pink flowers. Adults have also been reared from larvae feeding on the introduced *P. pinnata*. Eggs are sometimes deposited on *Citrus* (Rutaceae); the larvae can complete their development on this plant, but sometimes attempts to rear them on *Citrus* have failed and they died in their last instar. Near Texas, Q., the native *Microcitrus australis* has also been recorded as a food plant. Abroad *Citrus* and other Rutaceae are the normal food plants of *P. demoleus*, but near Port Moresby in New Guinea the larvae normally feed on *Psoralea badocana*.

14. *Papilio ulysses joesa* Butler, 1869

Plates 7, fig. 3 (male), VI, Ulysses butterfly.

DISTRIBUTION. Cape York to Mackay and Sarina, north-eastern Queensland; at Kuranda adults are on the wing during most months. A specimen was sighted but not captured at Byfield, central Queensland, by D. P. Sands.

This is primarily a rain-forest species. The adults provide an impressive sight as they fly in sunshine high over the Barron River near Cairns. As only the brilliant blue upperside is visible at any great distance against the dark rain forest, the butterfly is seen as a series of successive bright blue flashes.

The adults feed freely at flowers, such as *Lantana*, growing at the margins of rain forest. They are attracted by bright colours, the males being especially responsive to blue, even at a distance of up to thirty metres. Collectors sometimes exploit this behaviour by pinning a damaged specimen of *P. ulysses*, or a piece of bright blue cloth or paper, to a shrub and catching the males that fly down to investigate. Females are said to be more attracted by bright red objects.

The larvae feed on *Euodia elleryana* and *E. bonwickii* (Rutaceae) growing in rain-forest areas, and especially on the young regrowth foliage produced after the trees have been lopped or felled. One of two larvae found on *Citrus* at Cairns was reared successfully on this food plant by G. Sankowsky.

124

Genus *CRESSIDA* Swainson, 1832

As in the related South American genus *Euryades*, the fertilized female carries a sphragis (Fig. 6A) beneath the abdomen, a stiff bifid structure covering the opening of the female genitalia, and deposited by the male during mating. The genus contains a single species restricted to the Australian region.

15. *Cressida cressida* (Fabricius), 1775

This species is represented by three subspecies, two in Australia and one in New Guinea.

a. *Cressida cressida cressida* (Fabricius), 1775

Plates 7, fig. 12 (male), VII, big greasy.

DISTRIBUTION. Islands of Torres Strait, and Cape York to Sydney; rare south of the Richmond River, although a few specimens have been taken near Sydney. In Queensland it has been taken as far inland as 65 km south-west of Mt Garnet, and at Springsure.

Newly emerged females have a fairly distinct pattern, but the sparse scales which form it are easily denuded and the wings gradually become almost entirely transparent. They then have a greasy appearance, which gives rise to the popular name. Both sexes have a uniformly slow flight, but if disturbed can escape very rapidly. The males show some resemblance to *Pachliopta polydorus*.

In summer eggs hatch in about six days. The mature larva is very variable in colour. In specimens with a heavily mottled ground colour, the pale transverse dorsal markings are very inconspicuous, but in those with a more or less uniform dark reddish brown ground colour, the white transverse markings stand out clearly. The larvae feed on *Aristolochia* spp. (Aristolochiaceae), including *A. pubera*, one that trails on or near the ground in open forest areas in coastal southern Queensland, and *A. thozetii*, a small vine with narrow leaves growing near the beach north of Cairns (G. Sankowsky). Eggs are sometimes laid on the introduced *A. elegans* (Dutchman's pipe), but larvae do not normally develop on this plant. The larvae often remain inactive for some hours and then suddenly move about rapidly.

b. *Cressida cressida cassandra* (Waterhouse and Lyell), 1914

DISTRIBUTION. North-western Australia and Northern

Territory; adults have been taken at Yampi Sound, the Prince Regent River, the Daly River, Adelaide River, Batchelor, Darwin and the Alligator River, from August to April, and at Groote Eylandt.

ADULT.   Male: similar to typical *cressida* but forewing with central black spots smaller and more nearly equal in size.

Female: similar to typical *cressida* but wings rather more heavily scaled, with more distinct pattern.

The males differ only slightly from typical *cressida*, but the females are even darker than newly emerged *cressida*, and much darker than worn specimens of the typical subspecies.

Genus *PACHLIOPTA* Reakirt, 1865

The genus contains thirteen species from the Indo-Australian area, including one from north-eastern Australia and a Timor species, *P. liris* (Godart), reported from north-western Australia early last century, probably in error.

16. *Pachliopta polydorus queenslandicus* (Rothschild), 1895

Plate 7, fig. 11 (male), red-bodied swallowtail.

DISTRIBUTION.   Islands of Torres Strait, and Cape York to Ingham, Paluma and Townsville; common in the islands and at Cape York, sometimes common farther south.

The adults fly slowly and feed commonly at flowers growing at the margins of rain forest. At Magnetic Island D. P. Sands found the adults flying commonly in *Eucalyptus* forest during September and October.

The mature larva is rather like that of *Cressida cressida* (Plate VII, fig. 2C). Recorded food plants are *Aristolochia tagala* and *A. thozetii* (Aristolochiaceae). Larvae reach maturity in about fourteen days. The pupa (Fig. 13H) is generally attached to the food plant and is said to resemble a cluster of flower buds. The pupal duration is about three weeks.

Genus *ORNITHOPTERA* Boisduval, 1832

The species are large and strongly dimorphic, the males being bright green, blue, orange or gold on a black background, and the much larger females black and white with yellow markings on the hindwing.

17. *Ornithoptera priamus* (Linnaeus), 1758

This fine species is represented by fourteen subspecies which range from the Moluccas through New Guinea and the Kai and Aru Islands to the Bismarck Archipelago, the Solomons and eastern Australia. Four subspecies are found within Australian limits.

The males of most of the subspecies are brilliant green, black and gold, but in the subspecies *urvilliana* (Guérin-Méneville) from New Ireland, New Hanover and the Solomons, and *caelestis* (Rothschild) from the Louisiade Archipelago, green is replaced by blue.

The adults, especially the males, have a characteristic sailing flight, usually at tree-top level, but they feed at flowers such as *Lantana* growing at rain-forest margins, and the females often fly close to the ground in search of the larval food plant.

a. *Ornithoptera priamus pronomus* (Gray), 1853

Plate 9, figs. 1 (male), 1A (female), Cape York birdwing.

DISTRIBUTION. Thursday Island, and Cape York north of the Jardine River. Adults have been taken from October to May.

The larvae feed on native *Aristolochia* spp. (Aristolochiaceae) and eggs are also laid on the introduced *A. elegans* (Dutchman's pipe). At Cape York a young larva reached maturity fourteen days after hatching from the egg, and an adult female emerged from the pupa twenty-six days later. After hatching the young larva first devours its egg-shell and later larvae eat their cast skins. If crowded together larger larvae sometimes attack and eat the smaller ones.

b. *Ornithoptera priamus poseidon* Doubleday, 1847

DISTRIBUTION. Darnley, Murray, Yam and Moa Islands, Torres Strait. It also occurs throughout coastal New Guinea, including the D'Entrecasteaux and Trobriand Islands, and in the Kai and Aru Islands.

ADULT. Male: similar to *pronomus* but green bands usually broader, and often a green streak in centre of forewing; hindwing beneath usually with a golden tint.

Female: similar to *pronomus* but markings very variable in size and sometimes almost absent.

c. *Ornithoptera priamus macalpinei* Moulds, 1974

DISTRIBUTION. Claudie River to the McIlwraith Range, Silver Plains and Coen (D. P. Sands and J. F. R. Kerr), Cape York Peninsula.

ADULT. Male: similar to *pronomus* but green stripe above inner margin narrower or broken towards base and not extending as far along termen towards apex; hindwing with broader black areas on termen and inner margin, base more heavily dusted black. Beneath forewing with cell spot and spots beyond and below cell smaller; hindwing with black margin of cell more distinct.

Female: similar to *pronomus* but darker, with smaller white and yellow markings.

The area in which this subspecies occurs is separated from that of *pronomus* by a distance of 240 km in which no *Ornithoptera* has yet been observed (M. S. Moulds).

Larvae have been found at the Claudie River on *Aristolochia*.

d. *Ornithoptera priamus euphorion* (Gray), 1853
Plate V.

DISTRIBUTION. Mt Webb, 50 km north of Cooktown to Mackay; fairly common. At Kuranda adults have been taken during most months.

ADULT. Male: similar to *pronomus* but usually larger, green areas rather more restricted, especially beneath, cell of forewing never outlined with green, abdomen more heavily suffused with brown-black.

Female: similar to *pronomus* but usually larger and darker, forewing with white markings much more restricted above and beneath, hindwing with elongate spots suffused with grey and containing larger black spots, abdomen black above, yellowish beneath.

There is considerable variation in the extent of the markings in the female.

This is the largest and one of the most striking Australian butterflies. In flight, or when feeding at blossom, the adults attract the attention of even the most casual observer. It appears to be mainly a rain-forest species, but at Magnetic Island D. P. Sands found it was common in *Eucalyptus* forest.

The larvae feed on native *Aristolochia* spp. (Aristolochiaceae), especially *A. tagala* in coastal areas, but also on *A. deltantha*. They fail to survive on the introduced *A. elegans* (Dutchman's pipe). Sometimes they completely defoliate their food plants. The pupae are occasionally attached to the foliage, but more often the larvae wander away and pupate elsewhere. In the summer at Kuranda the pupal duration is

128

nearly four weeks, but this may be prolonged to as much as eight weeks if the pupae are removed to more temperate latitudes.

18. *Ornithoptera richmondia* (Gray), 1853

Plate 14, fig. 9 (male), Richmond birdwing.

DISTRIBUTION. Maryborough, Queensland, to the Clarence River, New South Wales; usually rare towards the northern limits of its range, but still fairly common in the Lamington, Tamborine and Burleigh Heads National Parks, and in remnants of rain forest containing its food plant in northern New South Wales. In coastal areas adults have been observed in most months, but especially in September and March; in the mountains they are on the wing mainly in November.

Both sexes vary to some extent. Near Lismore, N.S.W., John Williams has noticed that adults of the spring generation tend to be smaller and duller than those found in the late summer and autumn.

Males may sometimes be attracted by tethering a damaged female with a piece of cotton to a low branch. As in the other subspecies, both sexes feed at flowers. Normally adults are found only in or near rain forest, but on rare occasions they have been observed in drier situations.

The larvae feed on *Aristolochia praevenosa* and *A. deltantha* (Aristolochiaceae), which grow in rain forest, often scrambling high in the tree canopy. Eggs are also laid on the introduced *A. elegans* (Dutchman's pipe), but the larvae will not develop on this plant. They usually leave the food plant to pupate. In northern New South Wales John Williams found that larvae which he collected before they reached the penultimate instar and reared indoors were dark brown or black in the final instar, whereas about half of those that reach maturity under natural outdoor conditions are pale in colour.

Family Pieridae

WHITES AND YELLOWS
(Plates 10–14, VI, VIII, IX, X; Figs. 14–16)

MOST of the butterflies included in this family are of medium size, usually with white or yellow black-margined wings, and sometimes with red and yellow patterns beneath. The hindwings are never tailed. The family has a world-wide distribution but reaches its greatest development in the tropics, particularly in Asia and in South America. In Australia species and individuals are much more numerous in the north, where they constitute a conspicuous element of the butterfly fauna. Few are found in southern Australia, and only the introduced *Pieris rapae* is resident in Tasmania. In coastal New South Wales and in Queensland a few fly commonly during the winter, but most are to be seen during the warmer months.

Adult activity is stimulated by strong sunshine and individuals quickly settle when a cloud obscures the sun. In the tropics they often congregate in moist muddy or sandy spots where each probes for moisture with its haustellum.

Flight is often rapid, but the smaller species usually fly erratically close to the ground. Migration is a well-developed family trait and, at times, enormous numbers take part in directional flights in eastern Australia. The function of such mass movements in butterflies is still obscure and data on their occurrence are currently being collected to throw more light on this fascinating subject. Migratory flights may continue for several days at a time, predominantly in one direction, the butterflies resting at night-time on trees and bushes, often in great

130

numbers, and resuming their flight the following day.

The eggs (Plate VIII, figs. 1D, 2D, 4) are usually narrowly, but sometimes broadly spindle-shaped, with flattened base and vertical and horizontal ribs, and are white or yellow when laid; if white they may change to yellow or orange later. In some species they are deposited singly, but in others are laid in batches of fifty or more. The larvae (Plates VI, VIII, X) are slender and cylindrical, with short fine hairs but without fleshy spines or filaments. They are green to reddish brown in colour, often with longitudinal lateral stripes. Some species are gregarious. The pupae (Figs. 2A–2C, 16; Plates VI, VIII, X) are fastened either horizontally or vertically by the cremaster to a pad of silk and are supported by a central girdle. The head is produced into a median pointed, sometimes forked, projection, and the body may be angular or spined, or the wings may be produced ventrally into a "keel". In colour they are usually green, white, yellow or black.

Subfamily COLIADINAE

(Plates 10, 11, VI, VIII, IX; Figs. 14, 15)

The subfamily includes twelve Australian species in only two genera, *Catopsilia* and *Eurema*.

Genus *CATOPSILIA* Hübner, 1819

This genus contains robust, strong-flying, often dimorphic species, that occur commonly in the tropics of the Oriental and Australian regions.

Three variable Oriental species and one endemic species occur in Australia. Two of them are polymorphic and two distinct colour forms are known, one with pink and the other with grey or black antennae. These may represent seasonal forms.

1. *Catopsilia pyranthe crokera* (W. S. Macleay), 1826

Plates 10, figs. 1 (male, pale form), 1A (female, dark form), VIII, common migrant.

DISTRIBUTION.   North-western Australia, Northern Territory, and throughout most of eastern and central Australia, including the Alice Springs area; very sporadic in Victoria and South Australia. In central New South Wales it usually flies from February to April, but in the north adults are on the wing during most of the year. A specimen has also been recorded from Lord Howe Island.

Dark form. Male: above as in *pale form* (Plate 10, fig. 1), but white tinged with green, forewing with apex and termen black. Beneath as in *pale form*, but without pinkish tinge. Antennae black.

Pale form. Female: similar to male (Plate 10, fig. 1), but forewing with a larger black spot at end of cell.

The pale form is sometimes known as *lacteola*. Both forms of the species may be distinguished from the other two Australian species by the fine pale brown striations on the wings beneath.

At irregular intervals immense flights of this species occur in Queensland, at times reaching as far south as Sydney, but rarely to Victoria. The general direction of observed flights has been from north to south; the individuals keep a steady course, occasionally visiting a flower, or settling if the sun is obscured by clouds. They rest overnight on shrubs.

The larvae feed on several species of *Cassia* (Caesalpiniaceae), including *C. barclayana*, *C. aciphylla* and *C. planiticola*.

2. *Catopsilia pomona pomona* (Fabricius), 1775

Plates 10, figs. 2 (male, pale form), 2A (female, dark form), 2B (female, pale form), VI, lemon migrant.

DISTRIBUTION.   North-western Australia, Northern Territory, the islands of Torres Strait, and from Cape York to Port Macquarie, including most of the northern inland south to the Alice Springs area; usually common. In western Australia specimens have been taken as far south as Exmouth Gulf (S. Wallace) and the Hamersley Range (M. S. Moulds), and have been observed but not caught at Carnarvon (M. Hutchison). In some years specimens are taken much farther south than Sydney, and in 1974 a specimen was sighted by J. C. Le Souef at Blairgowrie, Victoria. A few worn stragglers occasionally reach South Australia.

The form of the female with pink antennae and large reddish blotch beneath the hindwing is sometimes called *catilla*, and the form with black antennae has been called *crocale*. The two forms are variable in markings, especially in the female. At Buderim, Queensland, V. Hesse has reared both forms from eggs laid by a single female of the dark

132

form, thus confirming that both forms belong to the one species.

The larvae feed on the foliage of *Cassia fistula* (Caesalpiniaceae), a large shrub or small tree with pendant clusters of yellow flowers commonly grown in northern gardens; they also feed on several other species of *Cassia*, including *C. odorata*, *C. brewsteri*, *C. tomentella*, *C. marksiana*, *C. queenslandica*, *C. magnifolia*, *C. nodosa*, *C. venusta* (G. Monteith), *C. coronilloides* and *C. pleurocarpa*. At rest they lie along the midrib of a leaflet and are difficult to detect. In summer the larvae mature in from two to three weeks.

3. *Catopsilia gorgophone gorgophone* (Boisduval), 1836

Plate 10, figs. 3 (male), 3A, 3B (females), yellow migrant.

DISTRIBUTION.   Eastern Australia, from Cairns and Mt Garnet, Queensland, to the Illawarra district, New South Wales, sometimes common north of the Richmond River. Specimens are rarely taken at Cairns or as far south as Sydney, but during April and May 1933 it was common in Sydney.

This species has often been regarded as a subspecies of *C. scylla*, but the distribution of the two overlaps in north-eastern Queensland. Both species can be distinguished from *C. pyranthe* and *C. pomona* by the contrasting yellow or orange hindwings and white or cream forewings.

A rare form of *C. gorgophone* (Plate 10, fig. 3B), with pink antennae, has been taken at Brisbane and the Rockhampton district.

The larvae feed on *Cassia surattensis* (Caesalpiniaceae), and females are said to deposit their eggs on this species even when other species of *Cassia* are available. However, at Brisbane they have been recorded on *C. auriculata* (D. Bell). At Rockhampton eggs hatch in about four days in summer. In autumn the pupal duration at Sydney is about two weeks.

4. *Catopsilia scylla etesia* (Hewitson), 1867

Plate 10, fig. 4 (male), orange migrant.

DISTRIBUTION.   North-western Australia, Northern Territory and Cape York to Townsville; at times very common at Darwin. Specimens from central Australia and western Queensland belong to this species rather than to *C. gorgophone*. Occasionally specimens are taken far beyond their normal range, such as one taken near Sydney in April 1975 by E. D. Edwards.

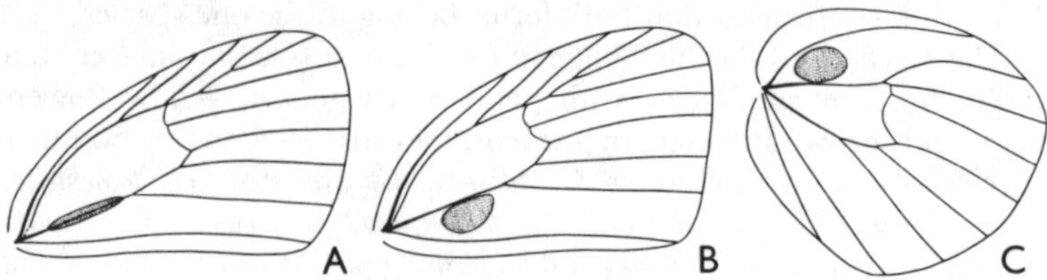

Fig. 14. Sex-brands in male *Eurema* (Pieridae): A, *E. hecabe* (p. 135);
B, C, *E. sana* (p. 138).

Genus *EUREMA* Hübner, 1819

This genus is widely distributed in the tropics of both eastern and western hemispheres, and some of the species are often very common. It contains small sulphur-yellow species with black wing-margins and usually some small reddish brown spots beneath.

The males of most species have patches of specialized scales beneath the forewing near the base, and on the hindwing above. If mounted males are viewed against a strong light, the position and shape of these sex-brands, when present, greatly aid identification. In most Australian species the sex-brands are of two types: *(a)* an elliptical patch of grey scales on vein CuA beneath the forewing (Fig. 14A), and *(b)* an oval patch of pinkish scales lying between CuA and 1A + 2A beneath the forewing, which overlies a similar patch between veins $Sc + R_1$ and Rs on the hindwing above (Figs. 14B, 14C). *Eurema candida, E. hecabe, E. blanda* and *E. smilax* have sex-brands of type *(a)*, whereas *E. sana, E. laeta* and *E. herla* have them of type *(b)*. *E. brigitta* is without a sex-brand. In addition to the sex-brand, the valvae of the male genitalia (Figs. 15A–15H) provide useful characters to separate the species.

The two sexes in *Eurema* also show marked differences in the reflectiveness of their wings to ultraviolet light (Eisner and Common). As the pattern of reflective scales is usually characteristic of each species (Plate IX), it provides an additional means of distinguishing them.

5. *Eurema brigitta australis* (Wallace), 1867

Plates 11, fig. 5 (male), IX, no-brand grass yellow.

DISTRIBUTION. Northern Territory, Moa Island, eastern Australia from Claudie River to Sydney, and at Lord Howe Island. It is common north of the Richmond River, but sporadic farther south.

The species may be recognized by the absence of the sex-brand in the male, by the presence of dark dots at the end of the cell of the forewing beneath, and by marginal flecks of yellow often present near the apex above. The yellow marginal scales are sometimes tinged with pink.

At Byron Bay, New South Wales, E. O. and E. D. Edwards observed adults laying eggs on *Cassia mimosoides* (Caesalpiniaceae), a dwarf species growing on exposed headlands. The young larvae which hatched from the eggs fed on this plant, but would not accept other species of *Cassia* at Sydney.

6. *Eurema candida virgo* (**Wallace**), 1867

Plates 11, fig. 1 (male), IX, broad-margined grass yellow.

DISTRIBUTION.   Cape York to the Claudie River and Coen, Cape York Peninsula; flying in most months of the year.

The species is easily recognized by the uniformly broad black margins, by the sex-brand of type *(a)* in the male, and by the white upperside of the female.

This species flies in rain forest.

7. *Eurema hecabe phoebus* (**Butler**), 1886

Plates 11, figs. 2 (male), 2A (female), VIII, IX, common grass yellow.

DISTRIBUTION.   Western, northern and eastern Australia as far south as Sydney and Stanwell Park in the east, and Geraldton in the west. Adults are common throughout the year in the north but only in the autumn near Sydney and in the summer and autumn near Geraldton (N. McFarland).

The very variable markings of the underside are apparently not of seasonal significance because extreme forms are often collected together. The species is easily recognized by the double indentation of the black margin in the forewing. Some authors refer Australian specimens to subspecies *sulphurata* (Butler), 1875, from the New Hebrides, Loyalty Islands, Fiji and Tonga.

Larval food plants include *Breynia oblongifolia*, *B. stipitata*, *B. cernua*, *B. nivosa*, *Phyllanthus tenellus* (all Euphorbiaceae), *Acacia* spp., including *A. baileyana*, *A. rubida*, *A. spectabilis*, *A. maidenii*, *Albizia lebbeck*, *Leucaena*

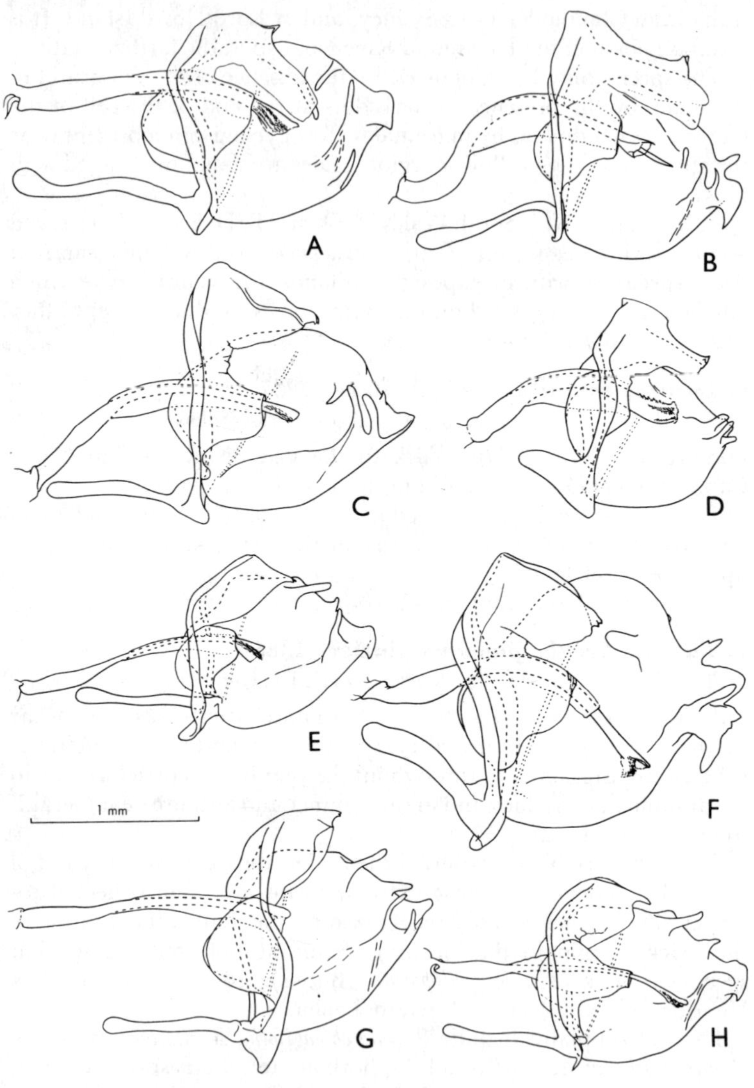

1 mm

leucocephala (all Mimosaceae), *Sesbania cannabina*, *Indigofera* (both Fabaceae), *Cassia surattensis* and *C. coronilloides* (both Caesalpiniaceae). In the autumn at Sydney the pupal duration is from two to three weeks, but in Queensland it is shorter.

8. *Eurema blanda saraha* (Fruhstorfer), 1912

DISTRIBUTION.   Darnley Island, Torres Strait; also from New Guinea, Waigeo, and the Schouten Islands.

Only a single male of this subspecies has been recorded from within Australian limits. *E. blanda indecisa* (Butler), 1898, to which it has been referred, is probably confined to the Moluccas. The species can be separated from others with a sex-brand of type *(a)* by the shape of the black margin of the forewing above, which lacks a double indentation on the inner edge, and by the three brown dots in the cell beneath.

9. *Eurema smilax* (Donovan), 1805

Plates 11, figs. 3 (male), 3A (female), IX, small grass yellow.

DISTRIBUTION.   Friday Island, Torres Strait and throughout mainland Australia; very common in the north, but much rarer in Victoria, South Australia and south-western Australia. It is occasionally taken in Tasmania, but is not established there. The species is restricted to Australia and Lord Howe Island.

In addition to a sex-brand of type *(a)*, the male has a patch of pinkish specialized scales on the hindwing similar to that of type *(b)*. This is the smallest species of the genus in Australia, and may be distinguished by the sex-brand in the male, and by the form of the marginal black areas in both wings. The size of the brown markings on the underside is extremely variable, the figures (Plate 11, figs. 3, 3A) representing the extremes.

Extensive migratory flights of this species have been recorded from time to time in Queensland, New South Wales and Victoria, but they

Fig. 15. Male genitalia of *Eurema*, lateral view with left valva removed:
A, *E. brigitta* (p. 134); B, *E. candida* (p. 135); C, *E. hecabe* (p. 135); D, *E. smilax* (p. 137); E, *E. laeta* (p. 138), F, *E. blanda* (New Guinea) (p. 137); G, *E. herla* (p. 138); H, *E. sana* (p. 138).

have not been studied. Sometimes they coincide with similar flights of *Anaphaeis java teutonia*.

The larvae and pupae are similar in shape to *E. hecabe* (Plate VIII, figs. 3A, 3B). The larvae have been recorded on *Neptunia gracilis* (Mimosaceae), *Cassia fistula*, *C. coronilloides*, *C. tomentosa* and *C. nemophila* (all Caesalpiniaceae).

10. *Eurema sana* (Butler), 1877

Plates 11, fig. 4 (male), IX, spotless grass yellow.

DISTRIBUTION.   North-western Australia, Northern Territory, the islands of Torres Strait, and Cape York to Cairns, Kuranda and Paluma; common in the extreme north, but less so near Cairns and Kuranda in January, Cardwell in March (W.N.B. Quick) and Paluma in April.

This species is easily recognized by the complete absence of spots or other markings beneath the wings.

11. *Eurema laeta lineata* (Miskin), 1889

Plates 11, fig. 7 (male), IX, line grass yellow.

DISTRIBUTION.   North-western Australia, Northern Territory, the islands of Torres Strait, and Cape York to the Paluma Range, south-west of Ingham; not as common as some of the other species.

In this species the apex of the forewing is more acute than in any of the others. The straight brown streaks beneath the hindwing also aid recognition, although these vary considerably in their intensity.

12. *Eurema herla* (W. S. Macleay), 1826

Plates 11, fig. 6 (male), IX, Macleay's grass yellow.

DISTRIBUTION.   North-western Australia, Northern Territory, including Devils Marbles (K. L. Dunn) and Groote Eylandt, Prince of Wales Island, and Cape York to Sydney; rare south of the Richmond River. This species is restricted to Australia.

Well marked specimens are easily identified by the curved brown streaks beneath the hindwing. When these markings are obscure or absent, the shape of the black area of the forewing above is the best guide; it stops at the end of vein CuA_2, the remainder of the termen and the tornus being yellow, except for a small spot at the end of the anal vein.

Subfamily PIERINAE

(Plates 10–14, VIII, X; Fig. 16)

The subfamily includes five genera which have a natural distribution in Australia, and a sixth, *Pieris*, represented by the introduced cabbage white butterfly from Europe. A total of twenty species is known from Australia. An additional species, *Leptosia nina* (Fab.), has just been recorded for the first time by D. Binns from the Mitchell Plateau, Kimberley, north-western Australia.

Genus *ELODINA* C. and R. Felder, 1865

This genus contains rather small and frail white species, with black apex to the forewing.

13. *Elodina parthia* (Hewitson), 1853

Plate 11, fig. 8 (male), chalk white.

DISTRIBUTION.   Cape York to Sydney, rare on the coast south of the Manning River, but fairly common where its food plant grows in the Dungog area and up to 400 km inland. At Millmerran in southern Queensland adults have been taken throughout the year.

The streaked markings on the underside of the hindwing vary in intensity. As in *E. perdita* the underside of the forewing is without a dark patch near the apex. It differs from the other Australian species by its duller chalky white colour.

The larvae feed on several species of *Capparis* (Capparidaceae), including *C. canescens* in southern Queensland. In summer they reach maturity in about three weeks, and the pupal duration is about ten days.

14. *Elodina angulipennis* (P. H. Lucas), 1852

Plate 11, fig. 10 (male), common pearl white.

DISTRIBUTION.   The islands of Torres Strait, and Cape York to Sydney, common; spasmodic at the southern limit of its range, where the food plant occurs only in cultivation.

As in *E. padusa* there is a distinct dark subapical patch beneath the forewing and, compared with that species, the forewing is broader and the termen more convex.

The larvae feed on *Capparis arborea* and *C. canescens* (Capparidaceae).

15. *Elodina padusa* (Hewitson), 1853

Plate 11, figs. 11 (male), 11A (female), narrow-winged pearl white.

DISTRIBUTION. A wide area of northern Australia, including Exmouth (M. S. Moulds), the Hamersley Range (M. Hutchison), Roebourne, Derby, the Drysdale River, Fitzroy Crossing, W.A., the McArthur River, N.T. (E. D. Edwards), and eastern Australia south from Musgrave and Cooktown to central New South Wales and the northern Flinders Ranges, South Australia. It is common in areas where *Capparis* grows. Specimens are occasionally taken in southern New South Wales, the Australian Capital Territory and Victoria, where *Capparis* does not grow naturally. It is a rare visitor to Melbourne (W. N. B. Quick).

This species is superficially similar to *E. angulipennis*, but has narrower wings.

The larvae feed on several species of *Capparis* (Capparidaceae) including *C. mitchellii*.

16. *Elodina perdita* Miskin, 1889

This species is confined to Australia where three subspecies have been recognized. It is the only species without a strong projection on the black apical area of the forewing above. As in *E. parthia*, it is usually without the blackish subapical patch beneath the forewing. The early stages have not been recorded.

a. *Elodina perdita perdita* Miskin, 1889

Plate 11, fig. 9 (male), northern pearl white.

DISTRIBUTION. Moa Island, Weipa and the Claudie River to Mackay, north-eastern Queensland; fairly common.

b. *Elodina perdita walkeri* Butler, 1898

DISTRIBUTION. North-western Australia and Northern Territory; records include Baudin Island, Mitchell Plateau (K. F. Kenneally), Wyndham, Darwin, Cobourg Peninsula (E. D. Edwards), and the Daly and Roper Rivers.

ADULT.   Similar to typical *perdita* but smaller, forewing beneath orange at base, costa brown-black.

c. *Elodina perdita tongura* **Tindale, 1923**

DISTRIBUTION.   Groote Eylandt, Northern Territory. Specimens from Cape Wessel, N.T. (E. D. Edwards) possibly belong here.

ADULT.   Similar to typical *perdita*, but forewing above with base, costa and apex darker, beneath lemon-yellow at base, brown-black costa, and a light brown subapical patch.

This subspecies is larger than *walkeri*.

Genus *DELIAS* Hübner, 1819

This genus contains more than 150 species, distributed from Tibet and India through south-east Asia to Australia and the Solomons.

The eggs (Plate VIII, fig. 1D) are yellow or orange in colour, and are usually laid in clusters on leaves of the food plant. The larvae (Plate VIII, fig. 1A; Plate X, figs. 1, 2) are greenish brown or dark reddish brown in colour, with fairly conspicuous contrasting white hairs. In the pupae (Plate VIII, figs. 1B, 1C; Plate X, figs. 3, 4) the abdomen has a series of spines or short projections, and the anterior end is often produced into a sharp simple or forked spine. Several larvae are usually found on the one plant, and in one species they behave gregariously, spinning a communal silk web upon which they pupate. This web is spun amongst the branches of the food plant, often after most of the leaves have been devoured by the larvae. The food plants are mainly mistletoes (Loranthaceae), but one species also feeds on the related family Santalaceae.

Four of the *Delias* are common winter butterflies in northern Queensland, and *D. nigrina* is often seen on the wing during the winter even in coastal New South Wales. Their flight is slow and often high around the tree-tops, but they visit flowers to feed at nectar. The females have broader black margins on the wings above, but beneath there is little difference between the sexes. The conspicuous patterns, especially on the underside, make identification easy.

17. *Delias argenthona* **(Fabricius), 1793**

Three subspecies are recognized, one in southern New Guinea and two in Australia.

a. *Delias argenthona argenthona* (Fabricius), 1793

Plate 12, figs. 9 (male), 9A (female), northern jezabel.

DISTRIBUTION.   Prince of Wales and Thursday Islands, and Cape York to Wollongong; common north of the Richmond River, but occasionally taken at Port Macquarie, Camden Haven, Dungog, and Sydney. Inland the species ranges west to Longreach and Mitchell, Q., and the Warrumbungle Range and south of Deniliquin, N.S.W. Specimens have been sighted but not captured in the Australian Capital Territory.

The pattern beneath the wings is somewhat variable (Plate 12, figs. 9, 9A) and, in addition, there is some seasonal variation in the extent of the dark markings above. The winter form is smaller than the summer form, and also very much darker, especially in the female. Pupae collected in north Queensland in September and which emerged there soon afterwards produced adults of the summer form. However, larvae collected at the same time, which pupated in Sydney in October and November, produced adults of the darker winter form.

The larvae feed on several species of mistletoe (Loranthaceae), including *Amyema bifurcatum* and *Muellerina celastroides*. Near Coonabarabran, N.S.W., they have been found on *Dendrophthoe vitellina* growing on kurrajong, *Brachychiton populneum* (W. N. B. Quick). The pupa is rather similar in shape to *D. mysis*, but has longer, more conical dorsal thoracic ridge; the ground colour is deep orange instead of the bright yellow of *D. mysis*.

b. *Delias argenthona fragalactea* (Butler), 1869

DISTRIBUTION.   North-western Australia, including Yampi Sound, Prince Regent River, Derby, Fitzroy Crossing (M. S. Moulds), the Drysdale River and Lake Argyle, and Northern Territory, including Melville Island and Groote Eylandt.

ADULT.   Similar to typical *argenthona*, but hindwing beneath with slightly smaller subterminal red spots, and red spot at end of cell not connected to subterminal spots, both wings above slightly darker in both sexes. Some specimens differ little from typical *argenthona*.

18. *Delias mysis* (Fabricius), 1775

This species occurs in New Guinea and its associated islands, the Aru Islands, and northern Australia. Nine subspecies have been distinguished, as well as four from northern Australia. Variation occurs chiefly in the red band beneath the hindwing.

a. *Delias mysis mysis* (Fabricius), 1775

Plate 12, figs. 1 (male), 1A (female), union jack.

DISTRIBUTION. Eastern Queensland, from Cooktown to Mackay and Yeppoon (G. Daniels); a common species north of Mackay, especially during the winter and spring. At Kuranda it has been recorded throughout the year, but few specimens have been taken at Yeppoon.

The adults are more easily caught in the early morning while they feed at flowers, especially *Lantana*, along rain-forest margins.

The larvae are gregarious and feed on a mistletoe with large broad leaves and red flowers which is sometimes parasitic on *Citrus* (McCubbin, 1971). Pupation usually occurs on a leaf of the food plant and several pupae occur together on the one leaf. The pupal duration ranges from two to six weeks.

b. *Delias mysis aestiva* Butler, 1897

DISTRIBUTION. North-western Australia and Northern Territory.

ADULT. Male: similar to typical *mysis* but hindwing above with broader black terminal band, beneath with inner yellow area more restricted, black terminal band narrower and less strongly curved, scarlet band narrower.

Female: similar to male but termen of both wings more convex, above with black margins broader, beneath with black margins and scarlet band broader than in male.

The sexes are more closely similar than in typical *mysis*.

c. *Delias mysis waterhousei* Talbot, 1937

Plates 12, figs. 2 (male), 2A (female), X.

DISTRIBUTION. Cape York to the Claudie River, Cape York Peninsula.

Larvae were reared from eggs collected on a mistletoe by T. H. Guthrie at the Claudie River.

d. *Delias mysis onca* Fruhstorfer, 1910

DISTRIBUTION. Within Australian limits known only from a

pair of specimens in the Australian Museum from Moa Island. This subspecies also occurs in the Milne Bay area of New Guinea.

ADULT.   Male: similar to typical *mysis* but above with subapical spots much smaller, beneath hindwing with black and scarlet bands ending at vein M_1.

Female: above similar to typical *mysis* but black margins narrower with smooth inner edge, subapical spots of forewing almost absent, beneath hindwing with black and scarlet bands ending between veins Rs and M_1, inner edge of black band nearly straight.

19. *Delias ennia* (Wallace), 1867

This species is distributed through New Guinea and its neighbouring islands, and in north-eastern Queensland. Eleven subspecies have been distinguished, two of which occur in Australia.

a. *Delias ennia nigidius* Miskin, 1884

Plate 12, fig. 8 (female), nigidius jezabel.

DISTRIBUTION.   Cooktown to Paluma; at Kuranda it is a common winter species, but also flies there during the summer and autumn.

b. *Delias ennia tindalii* Joicey and Talbot, 1926

Plate 12, fig. 7 (male).

DISTRIBUTION.   Claudie River and Coen district, Cape York Peninsula; very few specimens have been taken.

Larvae have been found on a rambling mistletoe growing on *Melaleuca* at the Claudie River (G. Daniels). Mature larvae had three colour phases.

20. *Delias aganippe* (Donovan), 1805

Plates 12, figs. 3 (male), 3A (female), X, wood white.

DISTRIBUTION.   Throughout southern Australia, except Tasmania. It is the only species of the genus in south-western Australia, where it occurs south from Carnarvon. Specimens have been taken near Alice Springs, Northern Territory (G. Griffin), in New South Wales as far west as Broken Hill (R. H. Fisher), and in Queensland as far west as Cunnamulla and as far north as Peak Downs and Emerald. A specimen is known from Mackay and others from

144

Ravenshoe (A. P. Dodd). In South Australia it has been recorded as far north as the north Flinders Ranges; a dead specimen has even been found on the salt crust of Lake Eyre (McCubbin, 1971). The species is common at times in southern New South Wales, the Australian Capital Territory and Victoria, but has been taken sparingly in Queensland. Adults have been collected from September to April.

The males are often observed flying on hill-tops.

The larvae feed on mistletoes (Loranthaceae), especially *Amyema cambagei*, *A. quandang*, *A. miquelii*, *A. linophyllum* and *A. preissii*, but also on *Exocarpos cupressiformis* (native cherry), *E. strictus*, and very occasionally on the much coarser *E. aphyllus* (Santalaceae). Near Deniliquin, N.S.W., E. D. Edwards has taken larvae on *Santalum lanceolatum* and *S. acuminatum* (Santalaceae), and the latter is the food plant in the Big Desert, north-western Victoria. Many larvae may be found on the one plant, but they do not spin a communal web for pupation as in *D. harpalyce*. Instead they pupate singly, usually on the twigs, branches and trunk of the food plant, or sometimes elsewhere. On more than one occasion C. W. Frazier and Mrs. J. Harslett observed several hundred pupae attached to dead twigs and branches many yards away from the food plant. In midsummer the larvae mature in about six weeks and the pupal duration is two to three weeks.

21. *Delias aruna inferna* **Butler, 1871**

Plate 12, figs. 6 (male), 6A (female), orange jezabel.

DISTRIBUTION. Moa Island, and Cape York to the Claudie River, the McIlwraith Range and Silver Plains, Cape York Peninsula.

The pupa is rather like *D. nigrina* (Plate VIII, figs. 1B, 1C), but is much larger, with very much shorter and blunter anterior and dorso-lateral projections. A. F. Atkins found egg-batches on the leaves of a grey mistletoe growing on *Acacia* and *Eucalyptus* at Mt Lamond, near the Claudie River. On one occasion many pupae were seen together on a broad-leafed mistletoe at Cape York.

22. *Delias harpalyce* **(Donovan), 1805**

Plate 12, figs. 11 (male), 11A (female), imperial white.

DISTRIBUTION. South-eastern Australia; usually rare in southern Queensland and northern New South Wales. At times it is common on the southern tablelands and as far west as Mt Kaputar

(G. Daniels) and Narrandera in New South Wales, and in eastern Victoria.

Two generations occur annually.

The larvae feed on various mistletoes (Loranthaceae), including *Muellerina eucalyptoides* and *Amyema miquelii*, growing at times near the tops of tall trees. They are gregarious and in the second instar spin a silken web over the mistletoe twigs. At maturity they enlarge this web upon which they finally pupate, as many as sixty or seventy pupae being attached to the one web.

23. *Delias nigrina* (Fabricius), 1775

Plates 12, figs. 10 (male), 10A (female), VIII, common jezabel.

DISTRIBUTION. Cape York, Queensland, to far eastern Victoria; usually very common. It is a common winter butterfly even near the southern limits of its range, and in the Cairns district flies in the winter with *D. argenthona*, *D. mysis* and *D. ennia*.

Little variation has been noticed in this species throughout its range. Very occasionally the scarlet markings are replaced by yellow.

The larvae feed on several species of mistletoe (Loranthaceae), including *Amyema congener*, *Muellerina eucalyptoides* and *M. celastroides*. They pupate singly, often on the mistletoe, but sometimes elsewhere on the tree upon which it is growing; at times they lower themselves to the ground on a thread of silk and pupate on a grass stem or other convenient object.

24. *Delias nysa* (Fabricius), 1775

This species is found in eastern Australia, where two subspecies are recognized, and in New Caledonia and the New Hebrides. It is the smallest of our species.

a. *Delias nysa nysa* (Fabricius), 1775

Plate 12, fig. 5 (female), nysa jezabel.

DISTRIBUTION. Cooktown, Cairns and Kuranda, Queensland, to south coastal New South Wales and, occasionally, far eastern Victoria; locally common at times.

Adults are somewhat variable, especially females, in the extent of the black terminal band of the hindwing above, the development of the enclosed greyish spots, and the extent of the yellow basal area of the forewing beneath.

146

The larvae have not been described, but adults were observed by T. H. Guthrie laying eggs on a mistletoe, almost certainly *Korthalsella japonica* (Loranthaceae) growing on *Geijera parviflora* (wilga). More recently M. De Baar also observed a female ovipositing on this mistletoe growing on wilga at Gunnedah, New South Wales, and collected a larva and two pupae from it. Near Pittsworth, Queensland, *Amyema gaudichaudii* (Loranthaceae) has been recorded as the food plant.

b. *Delias nysa nivira* Waterhouse and Lyell, 1914

Plate 12, fig. 4 (male).

DISTRIBUTION.   Claudie River and Coen districts, Cape York Peninsula; only a few specimens have been taken, from September to November.

Genus *ANAPHAEIS* Hübner, 1819

This genus of some nine species is distributed from Africa and the Middle East through the Oriental region to Australia and the south-west Pacific.

25. *Anaphaeis java teutonia* (Fabricius), 1775

Plates 10, figs. 9 (male), 9A, 9B (females), VIII, caper white.

DISTRIBUTION.   Moa Island, throughout mainland Australia, Lord Howe Island, and in Timor; occasionally stragglers have been taken in Tasmania, and a few specimens are known from southern Papua and the Trobriand Islands. It is one of our commonest butterflies.

The females are very variable, Plate 10, figures 9A and 9B representing the extremes of variation in mainland specimens. This variation cannot be seasonal, for all forms fly together.

Immense migratory flights of this species frequently occur in southern Queensland and New South Wales, and sometimes extend into Victoria. Entomologists at the Australian Museum are collecting information about such flights, but it is still too early to reach conclusions about their function in the biology of the species. The larval food plants do not normally grow farther south than about the latitude of Newcastle in the east and Griffith in the south-west, but migratory flights often extend well into Victoria and sometimes to southern Victoria.

The stragglers occasionally found in Tasmania probably originate from the latter. At Canberra it is not uncommon to see this species flying steadily north to north-east during November, although flights in the opposite direction have also been noticed.

The larvae feed on various Capparidaceae, including *Capparis mitchellii*, *C. spinosa* var. *nummularia* (P. A. Lascelles), *C. arborea*, *C. lasiantha* and cultivated *C. nobilis*, sometimes completely defoliating the plant. In central and north-western New South Wales the larvae commonly feed on *Apophyllum anomalum*, the choice between this plant and *Capparis* probably depending on the availability of young succulent foliage at the time the females are ovipositing. Pupation occurs on a twig of the food plant, or upon some other nearby object.

Genus *CEPORA* Billberg, 1820

The genus is widespread in the Oriental and Australian regions, and includes some twenty species. Only one species is found in Australia.

26. *Cepora perimale* (Donovan), 1805

This species occurs from the Celebes and Lombok in the west, through New Guinea and its associated islands to New Caledonia, the New Hebrides, Fiji, and Norfolk Island in the east. Some thirty subspecies have been distinguished, two of which occur within Australian limits.

a. *Cepora perimale scyllara* (W. S. Macleay), 1826

Plate 11, figs. 12 (male), 12A (female), Australian gull.

DISTRIBUTION. North-western Australia, Northern Territory, Warraber (Sue) (E. D. Edwards), Moa, Badu, Horn and Thursday Islands, Torres Strait, and from Cape York to Newcastle, and in some years even to Sydney; one specimen has been taken as far south as Nowra (E. D. Edwards), and another 32 km from Melbourne (A. N. Burns). Very common in the north.

The colour of the hindwing beneath is extremely variable, ranging from pale yellow (paler than in Plate 11, fig. 13), through various shades of yellow and orange to dark brown (darker than Plate 11, fig. 12). Specimens with white hindwings come from near Darwin. There may be a seasonal basis for some of the colour differences, yellow specimens being taken in summer and brown in winter.

The larvae feed on various species of *Capparis*, including *C. mitchellii*

(Capparidaceae). They prefer the young shoots, especially the young regrowth produced after a tree is felled. The pupa (Fig. 16A, 16B) may be green or brown, with dark brown markings.

b. *Cepora perimale latilimbata* **(Butler), 1876**

Plate 11, fig. 13 (male).

DISTRIBUTION. Darnley and Yam Islands (G. B. Monteith), Torres Strait, and New Guinea. Only a few specimens have been taken at Darnley Island, but it is common in New Guinea. Some of the specimens from Moa and Murray Islands show intermediate characters.

ADULT. Similar to *scyllara*, but above with narrow dark margins and restricted spots, forewing with only two subapical spots. Beneath as in *scyllara*, but all specimens taken have hindwings yellow beneath.

Genus *APPIAS* Hübner, 1819

This genus occurs throughout the tropical parts of the Australian and Oriental regions. Some thirty species are known, many of which have a wide range and are very variable, especially in the female.

27. *Appias paulina ega* **(Boisduval), 1836**

Plate 10, figs. 5 (male), 5A (female), common albatross.

DISTRIBUTION. North-western Australia, Northern Territory, including Groote Eylandt, Darnley, Thursday and Moa Islands, Cape York to far eastern Victoria, and at Lord Howe Island. Rarely taken in the Australian Capital Territory and Victoria, usually common in coastal Queensland and northern New South Wales. A specimen was once taken at Launceston, but the species is not established in Tasmania.

Variation in this species is apparently not seasonal.

The food plant is stated to be *Drypetes australasica* (Euphorbiaceae), but Oriental species of *Appias* feed on *Capparis* and other Capparidaceae. The pupa (Figs. 16C, 16D) has a long thin anterior spine and a pair of flattened lateral spines on the abdomen.

28. *Appias albina albina* **(Boisduval), 1836**

Plate 10, figs. 6 (male), 6A (female), white albatross.

DISTRIBUTION. Darwin, Cobourg Peninsula and Cape Wessel

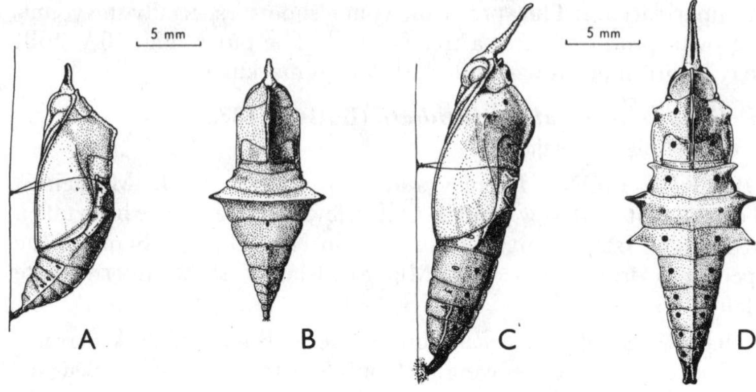

Fig. 16. Pupae of Pieridae, lateral and dorsal views: A, B, *Cepora perimale* (p. 148); C, D, *Appias paulina* (p. 149).

(E. D. Edwards), Northern Territory, and Moa Island, Torres Strait; only a few specimens have been taken.

The life history is not known from Australia. In Malaya larvae have been recorded on *Capparis* (Capparidaceae) (Corbet and Pendlebury, 1978).

29. *Appias melania* (Fabricius), 1775

Plate 10, figs. 7 (male), 7A (female), grey albatross.

DISTRIBUTION.   Cooktown to the Paluma Range, south-west of Ingham; taken chiefly in summer. The species is confined to Australia.

The life history has not been recorded.

30. *Appias celestina* (Boisduval), 1832

Plate 13, figs. 4 (male), 4A (female) (New Guinea).

DISTRIBUTION.   New Guinea and its neighbouring islands, including Waigeo, the Bismarck Archipelago, and the Kai and Aru

150

Islands. Only three specimens in very poor condition have been recorded from Cape York, but to which of the seven subspecies they belong has not been determined. Further specimens are needed to confirm the occurrence of the species in Australia.

31. *Appias ada caria* **Waterhouse and Lyell, 1914**

Plate 10, figs. 8 (male), 8A (female), rare albatross.

DISTRIBUTION.   Cape York Peninsula, from the Dulcie River to the McIlwraith Range, and farther south between Cape Tribulation and Daintree; one specimen has been reported from the Innisfail area.

The life history and early stages have not been described from Australia. A pupal skin from the Aru Islands in the Australian Museum is similar to *A. paulina ega* but broader, with the lateral spines of the abdomen more prominent and black, and the pointed anterior projection not as long.

Genus *PIERIS* Schrank, 1801

This genus has numerous species in the Northern Hemisphere and a few in Central and South America, but is represented in the Australian region by only one introduced European species which some authors refer to a separate genus *Artogeia* Verity.

32. *Pieris rapae rapae* **(Linnaeus), 1758**

Plate 14, fig. 17 (male), cabbage white.

DISTRIBUTION.   Widely distributed in the Palaearctic region, where several subspecies are recognized. The nominotypical subspecies from Europe was introduced accidentally to North America in 1860, to New Zealand in 1930, and to Australia in 1937 (C. G. L. Gooding). The first Australian record was from Victoria, and this infestation probably originated in New Zealand. The species reached Tasmania early in 1940 and quickly spread to South Australia and southern New South Wales later the same year. By early 1941 specimens were captured in Sydney and by 1942 the species was recorded from Western Australia. The following year it was reported in Queensland and in Perth. In the east it now occurs on the Atherton Tableland, in an area 65 km south-west of Mt Garnet (W. R. Hindson), in coastal areas from Port Douglas to Tully, and south from Rockhampton; in the west it has been recorded from Mt Newman (M. S.

Moulds) and is established south from Carnarvon (M. Hutchison). It has also been reported from Ernabella Mission, South Australia (P. McQuillan) and the Alice Springs area, Northern Territory. It is common in gardens or on farms where the food plants grow.

P. rapae is a well known migrant in the Palaearctic, but in North America migration was noticed for the first time in 1916, and in Australia and New Zealand it has still not been recorded definitely.

The larva is a pest of cruciferous crops. Females show a strong preference to oviposit on Brassicaceae, including cultivated *Brassica* spp. (cabbage, cauliflower, broccoli, turnip, mustard, rape and radish), and various weeds such as *Lepidium hyssopifolium* (peppercress). Eggs are laid less frequently on *Tropaeolum* (garden nasturtium, Tropaeolaceae), *Reseda* (mignonette, Resedaceae), and *Cleome* spp. (Capparidaceae), upon each of which the larvae are able to develop. Pupation may occur on the food plant, but more frequently on neighbouring objects such as wooden fences.

Family Nymphalidae

DANAIDS, BROWNS, NYMPHS
(Plates 13–20, 25, X–XV; Fig. 17)

THE butterflies of this large family vary in size from small to large, and show a great diversity in colour and pattern. They have a world-wide distribution. Although most of the subfamilies are best represented in the tropics, the Satyrinae are usually more abundant in temperate areas. Many of the latter are shade-loving insects, whereas most of the remainder of the Nymphalidae fly mainly in hot sunshine and frequently bask in the sunlight with their wings partly outspread. The Satyrinae are rather weak-flying insects, usually remaining near the ground, whereas the Nymphalinae are strong fliers, and often rest on trees high above the ground. The Danainae and the Nymphalinae include some of the best known butterfly migrants.

Seven of the eight known subfamilies occur in Australia, but only three, the Danainae, the Satyrinae and the Nymphalinae, are well represented.

Subfamily DANAINAE
DANAIDS
(Plates 13, 15, XI, XII)

This subfamily contains large butterflies of slow and laboured, or gliding flight, which when disturbed or migrating are capable of rapid sustained flight. They have a distribution primarily through the tropics of the Oriental and Australian regions. All of the Australian species are northern, but two occur commonly in the south of the continent, and a couple more are taken occasionally as far south as eastern Victoria.

The eggs are subconical in shape, greater in height than diameter, flattened at the base and slightly truncate at the top, with broad vertical ribs and much finer horizontal lines or ribs. The first instar larva has a few hairs on each segment and two to four pairs of short tubercles, but the mature larva (Plate XI, fig. 1B; Plate XII, fig. 1) has a nearly smooth cylindrical body, usually conspicuously banded, and with two, three or four pairs of long fleshy filaments bearing short hairs. These organs probably have a sensory function, and the pair on the thorax is movable. The larvae feed during the day fully exposed, and their brightly banded pattern probably acts as a warning to potential predators that they are unpalatable. The food plants invariably have a milky sap and belong to the families Asclepiadaceae, Apocynaceae and Moraceae. The pupae (Plate XI, fig. 1C; Plate XII, fig. 2) are short, stout and smooth, usually ornamented with golden or silver markings, and are sometimes objects of great beauty. They are suspended head downwards by the cremaster from a pad of silk previously spun by the larva.

The subfamily includes sixteen species in Australia, placed in two genera *Danaus* and *Euploea*. Some of the species recorded from the far north are represented in collections by very few specimens. Collectors in the more remote northern areas should therefore keep a careful watch for these species. In addition there is still relatively little available information about the adult behaviour of rarer species, and their early stages and food plants are largely unknown.

Genus *DANAUS* Kluk, 1802

The large and handsome members of this genus may be distinguished at once from *Euploea* by the prominent raised sex-mark on the male hindwing near vein CuA_2, which may form a pouch beneath the wing.

1. *Danaus plexippus plexippus* (Linnaeus), 1758

Plate 15, fig. 1 (male), wanderer.

DISTRIBUTION. Eastern and south-eastern Australia; usually common along the east coast south from Cairns, and near Adelaide, rare on Cape York Peninsula, uncommon in Victoria, a rare visitor to

Tasmania. It is known from the McDonnell Ranges near Hermannsburg, central Australia, and has recently become established in the Perth area, and has been recorded from Kununurra (K. L. Dunn), Western Australia. It also occurs in New Guinea and most of the islands of the south-west Pacific, including Lord Howe Island, New Zealand, and Norfolk Island. This species has gradually extended its range in recent times across the Pacific from North America, where it is known as the Monarch.

Adult variation appears to be mainly in size, brightness of colour, and in the number and size of the pale spots in the dark marginal areas.

In North America the species has a remarkable annual two-way migration, between the southern United States and southern Canada. Well defined migration has not been observed in Australia but winter assemblages have been reported from the Mt Lofty Ranges, from near Burra, about 140 km north of Adelaide, and from near Sydney.

Larval food plants include the introduced cotton bushes *Asclepias fruticosa*, *A. curassavica*, *A. rotundifolia*, *A. semilunata*, *Calotropis gigantea* and the moth-plant *Araujia hortorum* (all Asclepiadaceae).

2. *Danaus genutia alexis* (Waterhouse and Lyell), 1914

Plate 15, fig. 2 (male), orange tiger.

DISTRIBUTION. North-western Australia and Northern Territory. Specimens have been taken at Derby, Daly River, Darwin, and Mataranka (J. C. Le Souef).

During the dry season at Daly River adults usually flew in open shady places, often in company with *D. chrysippus* with which they could be easily confused.

There is no record of the life history from Australia.

3. *Danaus chrysippus petilia* (Stoll), 1790

Plate 15, fig. 5 (male), lesser wanderer.

DISTRIBUTION. The islands of Torres Strait, and throughout mainland Australia, including the interior, and at Lord Howe Island; sporadic in the south. A few specimens have been taken in Tasmania, where it is not established. It is plentiful at times on the mainland.

The size and number of white spots varies considerably, but the black and tawny markings are remarkably constant.

In coastal areas the larvae feed on the introduced cotton-bushes *Asclepias fruticosa*, *A. curassavica*, *A. rotundifolia*, *Calotropis gigantea*, and *C. procera*, and the native *Leichardtia australis* and *Ischnostemma carnosum*, but inland they have been reported on the native *Pentatropis atropurpurea* and *P. quinquepartita*; in the Flinders Ranges, South Australia, they feed on *Cynanchum floribundum*. All of these food plants belong to the Asclepiadaceae. On *Asclepias* they are sometimes found in company with the larvae of *D. plexippus*, from which they may be separated by the presence of three instead of two pairs of fleshy filaments. The pupal duration in summer is about two weeks.

4. *Danaus philene gelanor* (Waterhouse and Lyell), 1914

Plate 15, fig. 3 (male), brown tiger.

DISTRIBUTION.   Darnley Island, Torres Strait; only a few specimens have been taken. A male in the Australian Museum is labelled "North Queensland".

Nothing is known of the life history or food plants.

5. *Danaus affinis affinis* (Fabricius), 1775

Plate 15, fig. 4 (male), black and white tiger.

DISTRIBUTION.   North-western Australia, Northern Territory, the islands of Torres Strait and from Cape York to Port Macquarie, rarely to Sydney and the lower Blue Mts (T. J. Hawkeswood). North of the Richmond River it is a common coastal butterfly, but only a few specimens have been taken as far south as Sydney.

There is some variation in the number and shape of the white spots.

The larvae feed on *Ischnostemma carnosum* (Asclepiadaceae), a trailing plant which grows on the margins of salt creeks and swamps, always among reeds, to which it clings for support. Pupation occurs on the reed stems, rather than on the food plant. The early stages have been recorded only rarely.

6. *Danaus hamatus hamatus* (W. S. Macleay), 1826

Plates 15, fig. 6 (male), XII, blue tiger.

DISTRIBUTION.   North-western Australia, Northern Territory, and from Cape York and the islands of Torres Strait to Sydney, and rarely inland New South Wales, the Australian Capital Territory and far eastern Victoria. It is usually common north of the Richmond River, and at times is extremely abundant.

Some authors (Corbet and Pendlebury, 1978) place this and related species in a separate genus *Tirumala* Moore, 1880. The elongate blue spots and streaks are variable in size and shape. In Queensland the adults are sometimes abundant and in certain areas tend to accumulate in vast numbers.

The larval food plant is usually *Secamone elliptica*, but near the coast the larvae also feed on *Ischnostemma carnosum* (both Asclepiadaceae).

Genus *EUPLOEA* Fabricius, 1807

The adults of this genus are all large, velvety black or brown in colour with white markings, sometimes, though rarely in Australian species, with a blue gloss above. They are often difficult to distinguish, especially the females. The males of most species have characteristic sex-marks above, either on the forewing or near the costa of the hindwing, which greatly assist species recognition. The males of *E. climena*, *E. batesii*, *E. alcathoe* and *E. eichhorni* (Nos. 7–10) are without sex-marks; *E. core* and *E. algea* (Nos. 11, 12) have a single and *E. sylvester* (No. 13) a pair of elliptical sex-marks near the inner margin of the forewing; *E. tulliolus* and *E. darchia* (Nos. 14, 15) are without sex-marks on the forewing, but have a large sex-patch towards the costa of the hindwing, normally covered by the bowed inner margin of the forewing; and *E. usipetes* (No. 16) has a short broad sex-mark on the forewing and a large sex-mark, rather similar to that of *E. tulliolus*, on the hindwing.

The hair-pencils near the tip of the male abdomen are yellow in colour and are often expanded when the living insects are handled. Both sexes have a characteristic odour when captured.

The genus is essentially tropical and includes more than forty species distributed through the Oriental and Australian regions.

Five of our species are common, whereas the others are represented in collections by few specimens from remote northern localities. The adults tend to shelter in shady places during the heat of the day and are sometimes extremely abundant.

7. *Euploea climena macleari* Butler, 1887

DISTRIBUTION.   North-western Australia and Christmas Island in the Indian Ocean. A few specimens are known from Derby and Roebourne.

ADULT.   Male: above forewing uniform velvety brown, without sex-mark; hindwing brown, termen broadly whitish with brown veins. Beneath as above but with bluish spots near centre of wings.

8. *Euploea batesii belia* **Waterhouse and Lyell, 1914**
Plate 13, fig. 11 (female).

DISTRIBUTION.   Cape York, and Darnley and Murray Islands, Torres Strait.

9. *Euploea alcathoe monilifera* **(Moore), 1883**
Plate 13, fig. 3 (male).

DISTRIBUTION.   Thursday Island and Cape York; very few specimens have been taken.

10. *Euploea eichhorni* **Staudinger, 1884**
Plate 15, fig. 8 (male), Eichhorn's crow.

DISTRIBUTION.   Cape York to Proserpine, northern Queensland; common in swampy places near Cairns. A specimen has been recorded from Lindeman Island, near Proserpine. The species is confined to Australia.

This is the only common Australian species without a sex-mark in the male. The female has only one white spot above the end of the cell in the forewing, whereas the female of *E. sylvester* has three.

The life history and early stages have not been described, but both larvae and pupae are said to resemble *E. core corinna*. They feed on *Ficus eugenioides* (Moraceae), *Hoya australis*, *Gymnanthera nitida* and *Asclepias* spp. (Asclepiadaceae), and the introduced ornamental *Nerium* (oleander, Apocynaceae). Detailed information about the early stages is required. The superficial resemblance of the larvae of this and other species of *Euploea* to the very common *E. core corinna* has no doubt caused their neglect.

11. *Euploea core corinna* **(W. S. Macleay), 1826**
Plates 15, fig. 12 (male), XI, common Australian crow or oleander butterfly.

DISTRIBUTION.   North-western Australia north from Onslow (Mrs Milne), Northern Territory, the islands of Torres Strait, and

Cape York to southern New South Wales, the Australian Capital Territory, and rarely central and eastern Victoria. It also occurs near Alice Springs and Hermannsburg, central Australia, and at Lord Howe Island. Very occasionally specimens have been collected or reared from near Adelaide, South Australia. It is spasmodic in its appearance in the south, but is common in most months in the north.

Specimens from the southern parts of its range show little variation, but those from Cape York are much more variable.

The larvae are not often seen on their native food plants, but feed commonly on the cultivated ornamentals *Nerium* (oleander), *Mandevilla*, *Trachelospermum* (all Apocynaceae), *Stephanotis*, and even *Asclepias* spp. (all Asclepiadaceae). Native food plants are *Ficus* spp., including *F. platypoda* and *F. obliqua* (Moraceae), *Hoya australis*, *Ischnostemma carnosum*, *Secamone elliptica*, *Sarcostemma australe*, *Leichardtia australis*, *Gymnanthera nitida*, *Cryptostegia madagascariensis grandifolia* (M. S. Moulds) (all Asclepiadaceae), and *Parsonsia straminea* and *Carissa ovata* (both Apocynaceae).

The larvae are considered to be a minor pest of the ornamental food plants. At Sydney as many as three generations have been observed on *Mandevilla* in the one season, the pupal duration being about two weeks. The larvae usually pupate on the food plant, and when numerous the pupae are a striking sight.

12. *Euploea algea amycus* Miskin, 1890
Plate 13, fig. 2 (male).

DISTRIBUTION. Darnley, Warraber (Sue) (E. D. Edwards), Booby, and Thursday Islands, and Cape York. Few specimens have been collected.

13. *Euploea sylvester* (Fabricius), 1793
This species is distributed from Sri Lanka to Australia, New Caledonia and the New Hebrides. There are two variable subspecies in Australia, where it is the only species with a pair of elliptical sex-marks on the forewing in the male.

a. *Euploea sylvester sylvester* (Fabricius), 1793
Plate 15, figs. 13, 13A, 13B (males), two-brand crow.

DISTRIBUTION. Islands of Torres Strait, and Cape York to

Rockhampton; common north of Mackay, rare at Rockhampton. Inland a specimen has been taken at Forsayth, Q., about 250 km west of Ingham (G. R. Brown).

South of Cooktown the markings are fairly constant, but in the Torres Strait islands and at Cape York the size and number of the white spots may be considerably reduced (Plate 15, fig. 13A). Occasionally the spots are almost entirely lost (Plate 15, fig. 13B).

The life history and early stages have not been described, but the larva is reported to feed on *Ficus racemosa* (Moraceae). The larva of the subspecies from Sri Lanka, which feeds on *Ichnocarpus* (Apocynaceae), has three pairs of fleshy filaments, one each on the meso- and metathorax, and on the eighth abdominal segment. A native species of *Ichnocarpus* occurs in north Queensland.

b. *Euploea sylvester pelor* Doubleday, 1847
Plate 15, fig. 13C (male).

DISTRIBUTION. North-western Australia and the Northern Territory; common at Darwin, but also taken as far west as Yampi Sound, W.A.

This subspecies is less variable than typical *sylvester*. However, the white spots are sometimes reduced, although no example has been found without markings. Occasionally the spots on the forewing are greatly enlarged.

14. *Euploea tulliolus tulliolus* (Fabricius), 1793
Plate 15, fig. 9 (male), eastern brown crow.

DISTRIBUTION. Darnley and Yam Islands, Torres Strait, and Cape York to the Clarence River (J. Williams) and Urunga (M. S. Moulds), New South Wales; rare on Cape York Peninsula, but common from Cairns to Gladstone, rare south of Brisbane.

This species is similar in size to *E. darchia*, these two being the smallest Australian species. The males of both species have a strongly bowed inner margin of the forewing which covers the sex-mark, a patch of specialized pale scales near the costa of the hindwing. The markings are fairly constant, but occasionally the hindwing has some more or less obscure white spots above. Specimens from Darnley Island have been named *turneri* Butler, but are not distinguishable from typical *tulliolus*. As in other species of the genus, *E. tulliolus* often congregates in large numbers in shady patches of rain forest or in forest-fringed gullies.

The larvae feed on *Malaisia scandens* (Moraceae).

15. *Euploea darchia* (W. S. Macleay), 1826

This species is believed to have entered Australia from the north-west, whereas *E. tulliolus* is of Papuan origin. *E. darchia* occurs in Timor and in Australia; in the latter, two subspecies are recognized, one in the north-west and the other in the north-east.

a. *Euploea darchia darchia* (W. S. Macleay), 1826

Plate 15, figs. 10 (male), 10A (female), Darwin brown crow.

DISTRIBUTION. North-western Australia and the Northern Territory. It has been taken at the Prince Regent River, Kimberley, W.A., and at the Daly River, Darwin, Cobourg Peninsula and Cape Wessel (E. D. Edwards); it is common at Darwin.

b. *Euploea darchia niveata* (Butler), 1875

Plate 15, fig. 11 (male).

DISTRIBUTION. Islands of Torres Strait, and Cape York to Burleigh Heads, Queensland; usually not common, rare at Mackay and farther south. Specimens have been taken at Yeppoon (S. Brown), Caloundra (J. Olive) and Burleigh Heads (T. Lambkin).

The life history and early stages have not been described, but the larva is reported to feed on *Malaisia scandens* (Moraceae). It is said to be very similar to *E. tulliolus*, but rather paler and relatively shorter.

16. *Euploea usipetes usipetes* Hewitson, 1858

Plate 13, fig. 10 (male from New Guinea).

DISTRIBUTION. Thursday Island and Cape York; only a couple of specimens have been taken in Australia.

Subfamily ITHOMIINAE

(Plates 15, X)

The subfamily includes medium-sized butterflies of slow, gliding flight, found mainly in Central and South America, but with one genus *Tellervo* in the tropics of the Papuan and Australian areas.

The mature larva of *Tellervo* is smooth, with a single pair of long fleshy filaments on the thorax, and feeds on Apocynaceae. The pupa is short and stout, and is suspended head downwards by the cremaster.

Genus *TELLERVO* Kirby, 1894

This genus contains only one variable species ranging from the Celebes to New Guinea, the Bismarck Archipelago, north-eastern Australia and the Solomons. The fore tarsus in the female has five segments. The eyes of living adults are yellow.

17. *Tellervo zoilus* (Fabricius), 1775

This is a rain-forest species, which in Australia is restricted to north-eastern Queensland, where two of the several known subspecies occur. Their flight is relatively slow but, like other black and white species that fly in rain forest, the adults are very difficult to see in the inter-mittent deep shade and sunshine.

a. *Tellervo zoilus zoilus* (Fabricius), 1775

Plate 15, fig. 7A (female), Cairns hamadryad.

DISTRIBUTION. Mt Webb, 50 km north of Cooktown to Ingham, and the Atherton Tableland to the Paluma Range; common in the Cairns area.

At Cairns the larva feeds on *Parsonsia velutina* (Apocynaceae).

b. *Tellervo zoilus gelo* Waterhouse and Lyell, 1914

Plates 15, fig. 7 (male), X, Cape York hamadryad.

DISTRIBUTION.   The islands of Torres Strait, and Cape York to Coen; fairly common.

M. S. Moulds reared an adult from a larva he collected on *Parsonsia* sp. (Apocynaceae) at Mt Lamond, near the Claudie River.

Subfamily SATYRINAE
BROWNS
(Plates 13, 16–18, X, XII–XIV)

The adults of this subfamily are mostly of medium size, but many small dainty species are known. They are world-wide in distribution but are especially well represented in temperate latitudes. They are generally shade-loving insects but sometimes visit flowers to feed. Most of them frequent the undergrowth, long grass, dense forest, wooded gullies, and the shade of trees and cliffs. Some occur in open grassy areas, especially at high altitudes, where they may be very abundant. A few are common in suburban gardens.

Their flight is weak and irregular, but when disturbed the adults are often difficult to capture. Towards evening, or during very cloudy weather, they rest amongst tall grasses, in cliff cavities, on the leaf litter on the ground, or less commonly on the foliage of trees. While the insect is at rest, with wings held above the body, the protectively patterned underside makes detection very difficult.

The wings of most species bear prominent eye-spots (ocelli), especially on the underside, the size and number of which may vary considerably within any one species. In several Australian species the underside features silvery white bands and spots.

Some species have only one generation annually and the adults are on the wing for only a few weeks, the date of their appearance varying but slightly from year to year. The males are usually much more abundant early in the season, whereas towards the end of the season females usually predominate.

The eggs (Plate XIV, figs. 6, 6A) are mostly subspherical in shape, with a flattened base and slight vertical ribs. In colour they may be cream, yellowish, or some shade of green, and are usually deposited singly on the food plant. In a few species they are dropped at random as the female flies over or rests on the grass. The first instar larva is hairy and cream or green in colour with a large brown or black head. The mature larva (Plates X, XII–XIV) is covered with dense, usually very short, setae or hairs which give it a roughened appearance. It is broadest near the middle with the anal segment prominently forked. This bifid posterior end provides the most practical means of recognizing Satyrinae larvae in the field. The shape of the prominent head, which often has dorsolateral projections or horns, is a reliable guide to the identity of most species. In colour the mature larvae may be consistently either green or brown, but in some species both green and brown individuals occur. The food plants are mainly grasses, but include sedges and palms. The pupae (Plates X, XII–XIV) may be fairly stout and rounded, or more slender and often somewhat angular; they are either green or brown in colour. They are usually suspended head downwards by the cremaster on the food plant or on the undersides of stones or logs, but some lie unattached on the ground, usually beneath debris, or at the base of the food plant.

During the day the grass-feeding larvae usually conceal themselves deep in the tussocks or under debris, and are difficult to find. At night they climb up to feed on the grass blades, when they may be collected by searching carefully with a light.

Much additional information on the life histories, food plants and behaviour of most Australian species of Satyrinae is still needed, and collectors can make a worthwhile contribution in this field. Botanists interested in the taxonomy of grasses and sedges are anxious to know the natural food preferences of Satyrinae and Hesperiidae in the hope that this will assist them to establish plant relationships. It is therefore desirable, whenever the natural food plants are discovered, to have samples identified by an authority. Whatever their preferences, the larvae of many species will accept common introduced grass species. Using these, complete life histories can be worked out in cages containing pots of growing grass. If fertilized field-collected females are confined in them, they will often deposit their eggs, and the larvae are not difficult to rear. This rearing technique could also be used effectively for experiments on seasonal variation in our northern species.

Genus *MELANITIS* Fabricius, 1807

This genus includes thirteen species distributed through the Ethiopian, Oriental and Australian regions.

18. *Melanitis leda bankia* (Fabricius), 1775

Plates 16, figs. 7, 7A (males), XII, evening brown.

DISTRIBUTION.   North-western Australia, Northern Territory, including Groote Eylandt, the islands of Torres Strait, and Cape York to Sydney and Lord Howe Island; common north of Port Stephens. It is occasionally taken near Sydney, but is rare there. It is mainly a coastal species, but in Queensland has been taken as far inland as the Carnarvon Range.

Two distinct seasonal forms occur in Australia, a summer form (fig. 7) and a winter form (fig. 7A), but variation in weather conditions may be responsible for some variability in their occurrence from year to year. Both forms are extremely variable.

164

As in other species of the genus, the adult flies mainly towards dusk and at dawn. It rests on the ground and, as the underside closely resembles the dead leaf litter, it is very difficult to see. When disturbed it flies erratically for a short distance and once more settles on the ground.

The larva feeds on various coarse grasses (Poaceae), such as *Imperata* (blady grass), *Paspalum*, *Stenotaphrum secundatum* (buffalo grass), millet, and sugar cane, and remains during the day on the foliage.

19. *Melanitis amabilis valentina* **Fruhstorfer, 1908**

DISTRIBUTION.   Within Australian limits, only a single male is known from Darnley Island.

ADULT.   Male: above chocolate-brown with a broad yellow band across forewing. Beneath forewing as above but paler, with a small ocellus; hindwing red-brown with whitish markings and five small ocelli.

Female: similar to male, but forewing with basal area reddish brown and with broad white band.

Genus *ELYMNIAS* Hübner, 1818

The genus includes more than forty species distributed through the Oriental and Australian regions, from Sri Lanka to New Guinea, the Bismarck Archipelago and north-eastern Australia. Only five species occur in New Guinea, one of which is also found in the Cape York Peninsula.

The adults are shade-loving, rain-forest insects, and their larvae feed on palms (Arecaceae), especially *Calamus* (lawyer palm).

20. *Elymnias agondas australiana* **Fruhstorfer, 1900**

Plate 13, figs. 12 (male), 12A (female), palmfly.

DISTRIBUTION.   Cape York, the Claudie River, McIlwraith Range and Rocky River, Silver Plains, Cape York Peninsula.

The life history and early stages have not been described.

Genus *MYCALESIS* Hübner, 1818

Represented by many species in the Ethiopian and Oriental regions, the genus contains more than thirty in New Guinea, three of which also extend to northern and north-eastern Australia.

The three Australian species are usually common where they occur and the adults fly close to the ground. They usually show some seasonal variation in markings, distinctive wet- and dry-season forms being recognized.

21. *Mycalesis sirius sirius* (Fabricius), 1775

Plate 16, fig. 10 (male), cedar bushbrown.

DISTRIBUTION.   Northern Territory, the islands of Torres Strait, and Cape York to Mackay; common.

The ocelli vary in size and number, but seasonal variation is not as marked as in the other species of the genus.

The larvae feed on *Imperata* (blady grass) and other coarse grasses.

22. *Mycalesis terminus terminus* (Fabricius), 1775

Plate 16, figs. 8, 8A (males), orange bushbrown.

DISTRIBUTION.   Islands of Torres Strait, and Cape York to near Bundaberg, Queensland. It has been recorded at Bingera Weir, near Bundaberg, and there are specimens in the Australian National Insect Collection from the Kolan River, about 48 km north of Gin Gin. The species is very common in the northern parts of its range, where adults are on the wing throughout the year.

During the drier months, from September to November, the adult (Plate 16, fig. 8A, male) is paler above and beneath, the pale quadrate patch on the forewing above is ill-defined and the ocelli beneath both wings are minute. Each form is not entirely confined to either season, and intermediate forms may be found.

The larvae feed on *Imperata* (blady grass) and other grasses. Some taken in north Queensland in September continued to feed on common grasses at Sydney. They pupated in October and the adults emerged in three weeks. At Cairns during April and May the pupal duration is only twelve days.

23. *Mycalesis perseus perseus* (Fabricius), 1775

Plate 16, figs. 9, 9A (males), dingy bushbrown.

DISTRIBUTION.   Northern Territory, the islands of Torres Strait, and Cape York to Yeppoon, Maryborough and Kingaroy, Queensland; taken only recently near Yeppoon and farther south.

The number and size of the ocelli beneath the wings are extremely variable. In the dry-season form (Plate 16, fig. 9) the ocelli are much smaller than in the wet-season form, and may be reduced in number. In the much smaller wet-season form (Plate 16, fig. 9A, male) there are four prominent ocelli beneath the forewing and seven beneath the hindwing, a more distinct white transverse line across the wings, and it is without the fine striae found in the other form. The female is similar to the male, but of a paler brown colour. There is a gradual transition from one form to the other as the year progresses.

Only a cast larval head capsule and an empty pupal skin have been described from Australia.

Genus *ORSOTRIAENA* Wallengren, 1858

The genus contains only two species from the Oriental and Australian regions.

24. *Orsotriaena medus moira* Waterhouse and Lyell, 1914

Plate 16, fig. 11 (male), nigger.

DISTRIBUTION.   Moa, Prince of Wales, and Darnley Islands, and Cape York. Adults have been taken throughout the year, except from November to January.

Additional small ocelli sometimes occur beneath both wings, and the transverse white line varies in width.

The life history and early stages have not been described from Australia.

Genus *HYPOCYSTA* Westwood, 1851

The genus contains twelve species confined to New Guinea, the Aru Islands, and northern and eastern Australia. Five of six in Australia are endemic. The adults are small, dainty, weak-flying insects.

25. *Hypocysta euphemia* Westwood, 1851

Plate 16, figs. 12 (male), 12A (female), rock ringlet.

DISTRIBUTION.   Southern Queensland to far eastern Victoria. It has been taken mainly at Springbrook, Lamington National Park, near Stanthorpe, Sydney, the Blue Mountains and the Illawarra, but is also known from Ebor, Clyde Mountain, Pambula, and near Cann

River in Victoria. At Sydney adults are common from August to May, and there are said to be two generations annually.

Additional small ocelli are sometimes present. The adults nearly always fly in rocky areas, usually near rock faces, and are frequently seen in railway cuttings. They often rest on the walls of houses and, towards sunset, may be found commonly resting in the small caves along cliffs of the Hawkesbury Sandstone near Sydney.

The larva and pupa somewhat resemble those of *H. pseudirius* (Plate XIII, figs. 4A–4E), but the larva is darker and its head more rounded when viewed from the front, and the pupa is more angular and less mottled. The larvae feed on common grasses (Poaceae).

26. *Hypocysta irius* (Fabricius), 1775

Plate 18, fig. 16 (male), northern ringlet.

DISTRIBUTION.   Cape York, Queensland to Grafton, New South Wales; larger and more common in northern areas.

This species appears to be mainly confined to rain forest and flies commonly along the margins of roads through the forest. It is distinguished from its allies by the orange patch on the forewing. The larva is reported to feed on *Imperata* and other coarse grasses (Poaceae).

27. *Hypocysta metirius* Butler, 1875

Plate 18, fig. 17 (male), common brown ringlet.

DISTRIBUTION.   Claudie River and Coen, Queensland, to southern New South Wales. In north Queensland it is found mainly on the tablelands, at Kuranda, Atherton, Lake Barrine, Ravenshoe, Herberton, and on the Eungella Range. In southern Queensland it also occurs more commonly on the tablelands, and has even been recorded on the Maranoa River, 600 km west of Brisbane; in New South Wales it is taken mainly along the coast to as far south as Pambula. At Sydney adults occur from September to April.

The species is allied to *H. irius*, but is smaller and is without an orange patch on the forewing.

In general appearance the larva resembles a small specimen of *Geitoneura acantha* (Plate XIII, figs. 3A, 3B), the head being similar to fig. 3B, but with more pointed horns. However, in shape the larva agrees more closely with *H. pseudirius* (Plate XIII, fig. 4A). The pupa is

168

similar to *H. pseudirius* (Plate XIII, figs. 4C, 4D), but is not mottled, and the anterior end has a pair of pointed projections and is not cut off so abruptly. The larva feeds on common grasses, including *Cynodon dactylon* (couch grass) and *Imperata* (Poaceae).

28. *Hypocysta pseudirius* Butler, 1875

Plates 18, fig. 18 (male), XIII, dingy ringlet.

DISTRIBUTION. Atherton Tableland, and Mackay, Queensland, to Lake Conjola, near Ulladulla (G. Daniels), New South Wales. In central Queensland it has been taken as far inland as Springsure, and in New South Wales as far as the Warrumbungle Range. At Sydney adults have been taken from August to May.

The species may be distinguished from *H. metirius* by its smaller size, paler coloration, more rounded tornus and more restricted orange area of the hindwing. The ocelli are also smaller and there are sometimes additional minute ocelli beneath the hindwing. The early stages of the two species also differ. The larva feeds on common grasses.

29. *Hypocysta adiante* (Hübner), 1827–1831

The species is confined to Australia, where two subspecies occur, one in north-western Australia and the Northern Territory, and the other in eastern Australia.

a. *Hypocysta adiante adiante* (Hübner), 1827–1831

Plate 18, fig. 14 (male), orange ringlet.

DISTRIBUTION. Islands of Torres Strait, and Cape York, Queensland to Jervis Bay and Lake Conjola, near Ulladulla (G. Daniels), New South Wales. In the south it occurs mainly on the coast, but also in the lower Blue Mts (T. J. Hawkeswood). At Sydney adults are common in spring and autumn, whereas at Cairns they are common throughout the year.

Specimens from the islands of Torres Strait, Cape York and the islands off the north Queensland coast are paler in colour and may approach *antirius*.

At Sydney in February the eggs hatch in about ten days, the larvae mature in about thirty-six days, and the pupal duration is about fourteen days. The food plant is grass. The pupa is much longer and more slender than in the other species.

b. *Hypocysta adiante antirius* **Butler, 1868**

Plate 18, fig. 15 (male), Darwin ringlet.

DISTRIBUTION. North-western Australia and Northern Territory, including Bathurst and Melville Islands and Groote Eylandt. In Western Australia it occurs at Yampi Sound, Prince Regent River and commonly at Kununurra and Wyndham on the Ord River. It is also common at Darwin, and has been taken at Brocks Creek, Katherine, Daly River, Cobourg Peninsula (E. D. Edwards) and Mataranka (M. S. Moulds).

This subspecies is smaller and paler than *adiante*, and the ocelli beneath are visible from above through the flimsy wing. Sometimes the subtornal ocellus on the hindwing above has pigmented scales.

30. *Hypocysta angustata angustata* **Waterhouse and Lyell, 1914**

Plate 18, fig. 19 (male), black and white ringlet.

DISTRIBUTION. Claudie River and Coen, Cape York Peninsula.

Adults fly in rain forest and, like other black and white rain-forest species, are difficult to see in the mottled light and shade.

Genus *ARGYNNINA* Butler, 1867

Two species are referred to this genus, one of which occurs in Tasmania and the other on the mainland, ranging from south-eastern Queensland to Victoria.

31. *Argynnina cyrila* **Waterhouse and Lyell, 1914**

Plates 18, fig. 10 (male), XIV, Cyril's brown.

DISTRIBUTION. Southern Queensland, New South Wales, Australian Capital Territory, and Victoria. Adults have been collected at Cunninghams Gap and Springbrook in Queensland, the New England National Park (1,580 m), the Blue Mountains, Clyde Mountain, Mt Kosciusko, and on the central and south coast of New South Wales, the Brindabella Range (1,200 m), A.C.T., and in Victoria chiefly in the Dandenong Ranges, Mt Macedon, Mt Cole (W. N. B. Quick), and recently also in the Grampians (R. C. and N. Manskie). In coastal areas adults fly from late August to October, but in the mountains from September to December, depending on the altitude.

This species is larger than *A. hobartia* and brighter beneath, with narrower and longer forewings in the male. A more striking difference is the large black sex-mark in the male forewing of *A. cyrila*, which is absent in *A. hobartia*.

During the morning adults fly close to the ground but, later in the day, are more difficult to catch when they usually fly much higher. If disturbed their flight is surprisingly rapid and erratic. Males tend to collect on hill-tops (E. D. Edwards), a most unusual habit in Australian Satyrinae.

Eggs laid in early September hatched in twenty days. The larvae were fed on common grasses (Poaceae) and pupated thirteen to twenty weeks later. The pupae remained dormant throughout autumn and winter, and emerged in August and September. The black pupa closely resembles a piece of charcoal and its skin appears to be tougher than in other satyrines.

32. *Argynnina hobartia* (Westwood), 1851

This species occurs widely in Tasmania, where three subspecies have been recognized.

a. *Argynnina hobartia hobartia* (Westwood), 1851

Plate 18, fig. 9 (male), Hobart brown.

DISTRIBUTION.   Northern, central and eastern Tasmania, below about 900 m. As the name implies, the original specimen probably came from Hobart, where it is found near sea-level and on the slopes of Mt Wellington. Near Hobart this is one of the early spring butterflies flying from late September to early November; at higher altitudes it may appear a little later.

The yellow markings of the forewing contrast with the much darker markings of the hindwing. There is only slight variation in specimens taken at different altitudes.

The larva feeds on grasses (Poaceae) and, near Hobart, especially *Lolium perenne* (perenial ryegrass).

b. *Argynnina hobartia tasmanica* (Lyell), 1900

Plate 18, fig. 11 (male), Tasmanian brown.

DISTRIBUTION.   The high rainfall areas of western and south-western Tasmania, below about 900 m. It occurs on the coastal plains, the western ranges and some inland localities from Tullah, near Rosebery, in the north to the Florentine Valley and Mt Anne in the south. It is on the wing in late October and November.

This is a much darker insect than *hobartia*, with smaller and paler spots and with subterminal ocelli more prominent.

The life history is similar to that of typical *hobartia*.

c. *Argynnina hobartia montana* L. E. and R. Couchman, 1977

DISTRIBUTION. Tasmania, from Cradle Mt south to Lake Petrarch, at altitudes above 900 m. It is on the wing throughout December and early January.

ADULT. Above similar to *tasmanica*, but forewing with spots larger, cream; hindwing with markings reddish brown, but more restricted than in *hobartia*. Beneath forewing similar to *tasmanica* but cream spots larger; hindwing intermediate in markings and colour between *tasmanica* and *hobartia*.

The life history is similar to that of typical *hobartia*.

Genus *GEITONEURA* Butler, 1867

The genus contains three species confined to southern Australia. One is found throughout the southern mainland and Tasmania, another is restricted to the south-east, except Tasmania, and the third is restricted to the south-west. The species are found both on the lowlands and on the tablelands. As in *Heteronympha*, the cryptic pattern of the hindwings beneath, which is exposed when the adult is at rest amongst dead leaves on the ground, make detection difficult.

33. *Geitoneura acantha* (Donovan), 1805

This is a very common species occurring both on the coast and the tablelands of south-eastern mainland Australia, from central Queensland to South Australia. Two subspecies are recognized.

a. *Geitoneura acantha acantha* (Donovan), 1805

Plates 17, fig. 9 (male), XIII, eastern ringed xenica.

DISTRIBUTION. Central and southern Queensland, south from Expedition Range (A. F. Atkins), Carnarvon Range and the Bunya Mountains, New South Wales, and the Australian Capital Territory. In New South Wales it has been taken as far west as the Warrumbungle Range. Adults are on the wing from late October to March, and are usually common.

Specimens from the tablelands of southern Queensland are larger than those from the central coast of New South Wales. A series from Talbingo, near Tumut, N.S.W., taken by J. F. R. Kerr, is much paler than typical *acantha* and even paler than *ocrea*.

Eggs laid in March produced adults in November, the pupal duration being about four weeks. The larvae have been reared on *Themeda australis* (kangaroo grass) (E. D. Edwards) and on other common grasses (Poaceae). During the day they hide deep inside the tussocks, emerging to feed at night.

b. *Geitoneura acantha ocrea* (Guest), 1882

Plate 17, fig. 10 (female).

DISTRIBUTION.   Victoria and South Australia, including Kangaroo Island. The adults, which are on the wing from November to March, are common in Victoria, but much rarer in South Australia, where it now seems to be confined to the ·Mt Lofty Ranges and Kangaroo Island.

The larvae have been collected on *Poa tenera* (Poaceae), but no doubt also feed on other grasses.

34. *Geitoneura minyas* (Waterhouse and Lyell), 1914

This species is confined to south-western Australia, where it occurs early in the spring, as much as four weeks before *G. klugii*. Superficially it might be taken for a pale dwarf specimen of that species, but may be distinguished by the orange areas of the forewing extending to the inner margin. Two subspecies are recognized, but forms intermediate between the two are known.

a. *Geitoneura minyas minyas* (Waterhouse and Lyell), 1914

Plate 17, fig. 12 (male), western xenica.

DISTRIBUTION.   South-western Australia, from Moora to the Stirling Range and to a point east of Albany. At Perth adults are on the wing from September to November; they are often common locally in October.

This subspecies is variable in size and in the markings beneath the hindwing. Specimens from the Stirling Range are smaller than usual, and those from Perth approach *mjobergi* in pattern. All may be distinguished from *G. klugii*, which flies in the same localities, by their size, the more rounded wings in both sexes, the narrow straight sex-mark and by the orange inner margin of the forewing. The early stages have not been described.

b. *Geitoneura minyas mjobergi* (Aurivillius), 1920
Plate 17, fig. 13 (male).

DISTRIBUTION.   Geraldton to Moora, Western Australia.

35. *Geitoneura klugii* (Guérin-Méneville), 1830

The species has a wide distribution in south-eastern and south-western Australia, and in Tasmania. Though somewhat variable, no constant differences have been detected in specimens taken throughout this range. However, two small island populations, one in South Australia and the other in south-western Australia, have been distinguished as subspecies.

a. *Geitoneura klugii klugii* (Guérin-Méneville), 1830
Plates 17, fig. 11 (male), X, Klug's xenica.

DISTRIBUTION.   The tablelands of south-eastern Queensland, New South Wales and the Australian Capital Territory, extending to the coast in southern New South Wales, throughout much of southern and eastern Victoria, on the coast and in the mountains, in southern South Australia, to as far north as Wilpena Pound in the Flinders Ranges, throughout Tasmania up to more than 900 m and in south-western Australia, south from Northampton and Geraldton. Adults are on the wing from November to March in the south-east, but from late October to January in the south-west.

This subspecies is remarkably constant throughout its range, variation being largely confined to the cryptic underside of the hindwing. The male sex-mark is broader than in the other two species.

The larvae feed during the winter on common grasses (Poaceae), including *Brachypodium distachyon*, *Poa tenera* and *Themeda australis* (kangaroo grass). The pupal duration is about a month. In shape both larva and pupa resemble *G. acantha* (Plate XIII, figs. 3A–3F), but the head of the larva is without pointed dorsolateral horns. Unlike

G. acantha the larvae remain during the daytime fully exposed on the foliage.

b. *Geitoneura klugii mulesi* (**Burns**), **1948**

DISTRIBUTION.   Wardang Island, Spencer Gulf, South Australia. Adults have been taken from late October to December.

ADULT.   Similar to typical *klugii* but much smaller, paler, brownish black with more extensive orange areas, especially in female, sex-mark in male much narrower.

As in the typical subspecies, the inner margin of the forewing is brownish black, not orange as in *G. minyas*.

c. *Geitoneura klugii insula* **Burns, 1951**

DISTRIBUTION.   Rottnest Island, 15 km from the mainland, near Perth, Western Australia. Adults are common in November.

ADULT.   Similar to typical *klugii*, but orange markings richer in colour and more restricted, hindwing beneath more uniform in colour and pattern, the markings darker and more extensive.

Genus *HETERONYMPHA* Wallengren, 1858

The genus contains seven species largely confined to southern Australia, especially the south-east. All are found in New South Wales and Victoria, three in Queensland, three in Tasmania, and only one in South Australia and south-western Australia. Some of the species are much more plentiful in the mountains, but occur at lower altitudes in the south. Most have but a single generation annually, and the adults usually appear very consistently at the same time each year. Most of the species are relatively restricted to characteristic habitats, but the ubiquitous *H. merope* occurs in grassy areas throughout its range and even flies and breeds commonly in suburban gardens.

36. *Heteronympha merope* (**Fabricius**), **1775**

This is the commonest and most widely distributed species, and is found mainly in coastal areas from Rockhampton to South Australia, in Tasmania, and in south-western Australia. It also occurs on the tablelands and in some localities farther inland. Three subspecies are recognized.

a. *Heteronympha merope merope* (Fabricius), 1775

Plate 17, figs. 5 (male), 5A (female), common brown.

DISTRIBUTION.   Eastern Australia, mainly from Rockhampton to South Australia; in Queensland it is common only on the south-east tablelands, but farther south both on the coast and tablelands. It has been taken as far north as Mossman and Kuranda, and as far west as the Carnarvon Range, Q., Warrumbungle Range and Deniliquin, N.S.W., and Mildura, V. In South Australia it occurs in the Flinders Ranges and on Eyre Peninsula and Kangaroo Island as well as in the south-east.

Males appear in numbers at Sydney in October, when few females may be seen. As summer advances the numbers of males diminish and females increase until in February only occasional males can be found. Females continue to fly even as late as early May in some localities.

The hindwing beneath in both sexes is rather variable but, as in other species of the genus, the pattern is always highly protective. At rest all but the greyish tip of the forewing is covered by the hindwing, making the insect difficult to detect amongst fallen leaves.

It has been stated that the eggs are dropped by the female during flight, but E. D. Edwards has observed females depositing them on foliage of the grass food plant. In South Australia these include *Themeda australis, Poa tenera, P. poaeformis* and *Brachypodium distachyon* (all Poaceae). The early larval instars are always green in colour. Both larva and pupa resemble *H. paradelpha* (Plate XIII, figs. 1A−1D), but the head of the larva is straight instead of concave above. The pupal duration at Sydney is from four to six weeks.

Larvae of this and other satyrines should be sought in grassy places along creeks, or in mountain gullies, or amongst grass growing along fences. Pupae are more difficult to find.

b. *Heteronympha merope salazar* Fruhstorfer, 1911

DISTRIBUTION.   Flinders Island and Tasmania, from sea level to about 760 m but especially in the drier eastern half of the island. Males appear in mid-December; females are on the wing from about the end of December until late April.

ADULT.   Similar to typical *merope*, but in male the brownish orange areas of both wings more restricted; in female the large yellow patches in outer half of forewing much duller in colour.

176

c. *Heteronympha merope duboulayi* (Butler), 1867

Plate 17, fig. 6 (female), western brown.

DISTRIBUTION. South-western Australia, from Geraldton to Esperance, common.

The life history is similar to that of typical *merope*.

37. *Heteronympha mirifica* (Butler), 1866

Plates 17, figs. 7 (male), 7A (female), XIII, wonder brown.

DISTRIBUTION. Queensland, New South Wales and far eastern Victoria, from the Bunya Mts (M. De Baar) and the Maroochy River to Mallacoota (A. D. Bishop); coastal in southern New South Wales, but on both the coast and tablelands farther north. The males always appear much earlier in the spring than females, and few males are still on the wing in January.

In shape the female is very similar to *H. merope*. The two sexes of *H. mirifica* were for years regarded as separate species, a view apparently confirmed by the observation that they were seldom seen flying together. However, both sexes have more recently been reared from similar larvae and even from the same batch of eggs. As in other species of the genus the resting adult is very difficult to detect, even when it is seen to settle on nearby leaf litter. Both sexes are usually found in moist gullies, especially in rain forest, but males sometimes fly high in more open areas and settle on trees.

The larva feeds on grass and may be distinguished from those of allied species by the dark dorsal line, which is expanded twice on each segment. The oblique posture of the pupa, instead of the more usual vertical suspension, such as in *Geitoneura acantha* (Plate XIII, figs. 3E, 3F), may be an adaptation to the probable habit of the larvae to pupate in confined spaces close to the ground, under the edges of stones or under logs.

38. *Heteronympha paradelpha* Lower, 1893

This species occurs in the mountains of southern Queensland, New South Wales and Victoria, reaching the coast in the south. Two subspecies have been recognized.

a. *Heteronympha paradelpha paradelpha* Lower, 1893

Plates 17, figs. 3 (male), 3A (female), XIII, spotted brown.

DISTRIBUTION. Tablelands and coast of New South Wales, south from the Blue Mountains and Sydney, the Australian Capital Territory, and eastern and central Victoria as far west as Lorne. Adults are on the wing from January to March, and are very local in their occurrence.

The species may be distinguished from *H. banksii* and *H. solandri* by the rather more convex outer margins of the wings, the presence of an ocellus near the apex of the forewing, the absence of a white subterminal spot on the forewing of the female, and by the underside of the hindwing. The early stages are very distinct in these three species.

As in *H. penelope* and *H. merope*, the larval head lacks short dorsolateral horns, but the dull black underside at once distinguishes the larva of *H. paradelpha*. The larvae eat common soft grasses (Poaceae), including the native *Poa tenera*. They reach maturity in December and the adults emerge about a month later.

b. *Heteronympha paradelpha deervalensis* Burns, 1946

DISTRIBUTION.   Deer Vale, Dorrigo and Ebor, New South Wales, (from 1,200 to 1,500 m) during January and February. A single female taken in March near Stanthorpe, Queensland, may belong to this subspecies.

ADULT.   Similar to typical *paradelpha*, but black areas rather more extensive.

39. *Heteronympha penelope* Waterhouse, 1937

This species is found in the mountains of New South Wales, south from Ben Lomond, in the mountains and lowlands of Victoria and in Tasmania. It is absent from Queensland and Western Australia. Six subspecies are recognized.

a. *Heteronympha penelope penelope* Waterhouse, 1937

Plates 17, figs. 4 (male), 4A (female), XII, shouldered brown.

DISTRIBUTION.   The mountains of New South Wales and the Australian Capital Territory. It has been taken at Stonehenge, Ebor (1,460 m), Yarrowitch, Barrington Tops, Blue Mountains (above 730 m), Mt Canobolas (near Orange), Moss Vale, Clyde Mountain (610 m), Batlow, Mt Kosciusko (1,500 m), the Brindabella Range (900 to 1,370 m) and Mt Majura (890 m). Adults are on the wing from January to April.

This is the largest subspecies, with the brownish orange markings in the male richer in colour, and the black rather more extensive than in the others. In wing shape it resembles *sterope*, the forewing being relatively long in the male, and the hindwing of the female with tornus produced.

The larva is easily recognized by the head which is convex above. It feeds on *Poa* sp. (snow grass) and other common grasses. The pupa is very similar to *H. paradelpha* (Plate XIII, fig. 1D).

b. *Heteronympha penelope sterope* Waterhouse, 1937.

DISTRIBUTION.   The mountains and foothills of eastern and central Victoria. Adults have been taken at Mt Buffalo, Hazelwood, Moe, Fern Tree Gully, Gisborne, Mt Buangor, and the Brisbane Ranges, south-west of Bacchus Marsh, from January to March.

ADULT.   Male: similar to typical *penelope* but slightly smaller, brownish orange markings usually paler and larger, forewing with brownish orange patch in outer half of cell often not separated by a black band from area near inner margin. Beneath paler and more uniform in colour.

Female: similar to typical *penelope*, but slightly smaller, orange markings a little larger, beneath with less pronounced purplish or pinkish suffusion.

c. *Heteronympha penelope alope* Waterhouse, 1937

DISTRIBUTION.   Lorne, Victoria. Adults have been taken in February and March.

ADULT.   Similar to *sterope* but smaller, with larger brownish orange areas, smaller ocelli, and with subapical ocellus of hindwing small or absent; female with pale spot below subapical ocellus of forewing which beneath is nearly white.

A study of larger numbers of this subspecies might indicate that it intergrades with *sterope*.

d. *Heteronympha penelope maraia* Tindale, 1951

DISTRIBUTION.   Grampians, and the lower Glenelg River, south-western Victoria, and the south-eastern corner of South Australia. Adults have been taken at Fyans Creek (240 m), Mt Rosea, and the

Mt Difficult Range (790 m), in the Grampians during January and February. In South Australia it has been taken as far north as Millicent and may be fairly common during March.

ADULT. Similar to *sterope*, but with brownish orange areas slightly larger in both sexes; female similar in colour to male, with spot below subapical ocellus of forewing orange-brown, subapical ocellus of hindwing small, beneath with apex of forewing and hindwing orange-brown.

Pupation occurs on the ground amongst leaves and debris held together by a few strands of loose silk. Larvae have been collected in November at Fyans Creek in the Grampians, feeding at night on soft native grasses, *Poa* sp., *Danthonia pilosa*, and *Themeda australis*. When reared at Adelaide the pupal duration ranged from thirty-seven to sixty days.

e. *Heteronympha penelope diemeni* Waterhouse, 1937

DISTRIBUTION. The lowlands of north-western, northern and eastern Tasmania, up to about 760 m. Adults have been collected on the lower slopes of Mt Wellington, and at Hobart, Dunalley, Maria Island, Launceston, Burnie and King Island, from January to March.

ADULT. Similar to *sterope* but smaller, forewing shorter, apex and termen more rounded, tornus of female hindwing less produced; brownish orange markings more restricted in both sexes, in male the brownish orange area in outer half of forewing cell usually separated by a black bar from area near inner margin, subapical ocellus of hindwing often small or sometimes an additional small ocellus beneath it; female with cream spot below subapical ocellus of forewing.

The sexes in this subspecies and in *panope* resemble each other more closely than in any of the mainland subspecies.

f. *Heteronympha penelope panope* Waterhouse, 1937

DISTRIBUTION. Tasmania, in alpine areas from 600 to 1,200 m and near sea level between the Pieman River and Strahan. It has been taken at Miena (1,030 m), the Steppes (700 m), Lake St Clair (700 m), Derwent Bridge, St Valentines Peak (790 m), the Cradle Mountain area (850 m), Waratah (610 m), and Mt Barrow (760 to 1,200 m). Adults are on the wing from late January to early March.

ADULT. Similar to *diemeni* but smaller and darker above, with brownish orange areas even more restricted, spot below subapical ocellus of forewing almost white, subapical ocellus of hindwing prominent, and below it usually a second prominent ocellus, both white-centred. Beneath with spot below subapical ocellus of forewing white, ocelli in both wings as above, prominent, hindwing in male reddish brown, in female suffused with purplish.

Specimens from the west coast are brighter in colour, with very prominent ocelli above and beneath.

40. *Heteronympha banksii* (Leach), 1814

This species occurs on the tablelands and mountains from southern Queensland to eastern Victoria, and in the Grampians, western Victoria. It also occurs on the coast in New South Wales and eastern Victoria. Three subspecies are recognized. The species appears to favour cool grassy slopes in sheltered humid areas, where the larvae are protected from hot sunlight and their grass food plants remain green. It differs in wing shape from *H. solandri* and others in the genus, and the purplish sheen on the underside of the hindwing also distinguishes it.

a. *Heteronympha banksii banksii* (Leach), 1814

Plate 17, fig. 1 (male), Banks' brown.

DISTRIBUTION. New South Wales, the Australian Capital Territory and Victoria; common in the mountains from 300 to 900 m; rarer on the coast from the Manning River, New South Wales, to Loch, Trafalgar and the Dandenong Ranges, Victoria. Adults have been taken from February to April.

Female: similar to male but brownish orange markings a little larger, a small pure white spot below small subapical ocellus, sex-mark absent. Beneath similar to male.

The larvae usually feed on grasses (Poaceae) but in eastern Victoria have also been found on *Carex longebrachiata* (Cyperaceae).

b. *Heteronympha banksii mariposa* Tindale, 1953

DISTRIBUTION. South-eastern Queensland, at 600 to 900 m from February to April. Specimens have been taken at Mt Glorious (J. F. R. Kerr), Macpherson Range (about 760 m) in March, Bunya Mountains from February to April, and Glen Aplin and Eukey, near Stanthorpe, in February.

ADULT.   Similar to typical *banksii*, but hindwing with larger ocelli.

c. *Heteronympha banksii nevina* **Tindale, 1953**

DISTRIBUTION.   Mt Rosea, Grampians, Victoria, 300 to 600 m. Adults have been taken in late January and February.

ADULT.   Similar to typical *banksii*, but hindwing with brownish orange areas more extensive, especially in female, grey-brown basal area not extending to tornus, which is orange. Beneath forewing in female with basal half orange, outer spots yellow and surrounded with black; hindwing in male orange in basal half, outer half suffused with chocolate-brown, in female basal half fawn with rich red-brown markings, outer half blotched with rich brown.

Larvae have been taken in late November feeding after 9 p.m. on native grasses, *Poa* spp., including *P. tenera*, and pupated in late December and January. The pupal duration ranged from twenty to thirty-two days.

41. *Heteronympha solandri* **Waterhouse, 1904**

This species occurs in the mountains of southern New South Wales, the Australian Capital Territory, and eastern Victoria, and in the Grampians, western Victoria. Two subspecies are recognized.

a. *Heteronympha solandri solandri* **Waterhouse, 1904**

Plate 17, figs. 2 (male), 2A (female), Solander's brown.

DISTRIBUTION.   New South Wales, near Hampton (1,300 m), Brown Mountain (1,060 m), and Mt Kosciusko; Australian Capital Territory (1,200 to 1,400 m); eastern Victoria from Poowong, Mt Erica (1,370 m), Mt Hotham, Mt St Bernard (1,500 m), Mt Donna Buang (1,200 m), Mt Buffalo and the Otway Ranges. Adults are on the wing from the end of December to early March.

The species is readily distinguished from *H. banksii* by the shape of both wings and by the colour of the underside of the hindwing. Adults fly earlier in the season and the species is usually found at higher altitudes.

The early stages are rather like those of *H. banksii banksii* and *H. mirifica* (Plate XIII, figs. 2A–2E). The larvae feed on *Poa* sp. (snow grass) and other grasses (Poaceae).

b. *Heteronympha solandri angela* **Tindale, 1953**

DISTRIBUTION. Mt Rosea, Grampians, western Victoria, at 600 m. Adults have been collected in late December and January.

Similar to typical *solandri* but smaller, with reduced orange markings.

The larva may be distinguished from that of *H. banksii nevina* by a conspicuous triangular black marking on the back of the head. Larvae have been collected feeding late at night on native grasses, *Poa* spp., during November. The pupal duration is about three weeks.

42. *Heteronympha cordace* **(Geyer), 1832**

This species occurs at many localities on the tablelands and mountains of New South Wales, the Australian Capital Territory, and Victoria, at Dartmoor in western Victoria, and in the mountains and on the west and south coast of Tasmania. Five subspecies have been recognized.

a. *Heteronympha cordace cordace* **(Geyer), 1832**

Plate 17, fig. 8 (female), bright-eyed brown.

DISTRIBUTION.   The mountains of New South Wales and the Australian Capital Territory, usually above about 900 m, the mountains and hills of eastern Victoria, and at Mt Macedon and Mt Buangor. Adults are on the wing from December to early March, usually in swampy places.

Specimens from Dorrigo and Barrington Tops tend to be darker, with smaller orange markings, than those from southern New South Wales and Victoria. Even those from the Blue Mountains tend to be darker than southern specimens, but no constant characters have been noted to separate northern from southern populations. Occasionally an additional ocellus is present on the hindwing near the subapical ocellus.

Under natural conditions the larvae may be dependent on *Carex appressa*, as in *wilsoni*.

b. *Heteronympha cordace wilsoni* **Burns, 1948**

DISTRIBUTION.   Nelson, Dartmoor, on the lower Glenelg River, and the Wannon River, Grampians (425 m), western Victoria. Adults have been taken from late November to January, in swampy places where the food plant grows. It has been suggested that Grampians specimens should perhaps be referred to the typical subspecies.

ADULT. Similar to typical *cordace* but smaller, above with orange markings paler and larger, ocelli smaller. Beneath paler with much more obscure markings, hindwing with subapical ocellus absent in male, very small in female, subtornal ocellus small in both sexes and often absent in male.

The larvae feed at night on a tall sedge, *Carex appressa* (Cyperaceae), hiding during the day in the base of the plant or in debris. The life history was studied at Melbourne, where the eggs hatched in fourteen to twenty days, the larvae reaching maturity in from twenty-five to twenty-nine weeks. The pupal duration was twenty-five to thirty-eight days.

c. *Heteronympha cordace legana* Couchman, 1954

DISTRIBUTION. North-eastern Tasmania, at altitudes up to 760 m. The original specimens were collected in January near Lake Leake at about 610 m, but the subspecies has also been taken near Lilydale, between Saddleback and Moorina, at Mt. Barrow in February (530 to 730 m), Storys Creek, near Murderers Tier and, most recently, at Birralee and north-west of Frankford (D. Binns), west of the Tamar River, at altitudes of 50 to 200 m.

ADULT. Similar to typical *cordace* but smaller, with orange subterminal spots on both wings larger, and in hindwing of female more numerous, subapical ocellus of forewing and subtornal ocellus of hindwing small, subapical ocellus of hindwing reduced to a minute black dot; hindwing beneath paler, with subtornal ocellus small and subapical ocellus minute, and the two white spots between them very small.

The adults fly in swampy areas where *Carex appressa* (Cyperaceae), the larval food plant, grows.

d. *Heteronympha cordace kurena* Couchman, 1954

DISTRIBUTION. Central plateau of Tasmania, from the Dee River and east of Great Lake to Derwent Bridge and Cradle Mountain, at 610 to 1,030 m, and north-west to St Valentines Peak and Hampshire at about 455 m. Adults are on the wing in January, flying in the "button grass" (*Gymnoschoenus sphaerocephalus*) swamps.

ADULT.   Similar to typical *cordace* but smaller; male with subapical ocelli of both wings and subtornal ocellus of hindwing small, round, subapical ocellus of hindwing without a surrounding dark ring; female hindwing with postmedian band of orange spots only slightly constricted in middle, and a minute white dot below subapical ocellus; female beneath usually with an extra small ocellus below subapical ocellus.

This is the smallest subspecies. Specimens from the Hampshire district are very dark above, the orange markings being reduced to narrow bands, and beneath they are darker than in other subspecies.

e. *Heteronympha cordace comptena* Couchman, 1954

DISTRIBUTION.   Western and southern Tasmania, from Mt Agnew near Zeehan to Dover and South Bruny Island, at altitudes up to 395 m; its range may well extend north to near Marrawah. Adults are on the wing in January and February.

ADULT   Male: similar to typical *cordace*, but orange markings brighter and rather larger, subapical ocelli in both wings with large blue centres, that of hindwing not surrounded by a dark ring, subtornal ocellus of hindwing with very large blue centre. Hindwing beneath with white spots between ocelli large and distinct.

Female: similar to male, both wings above with a clear white dot below subapical ocellus, subtornal ocellus of hindwing very large, oval, with very large blue centre. Beneath similar to male, but hindwing with a black-ringed white spot below subapical ocellus.

This subspecies is distinguished by its richer coloration, the large blue centres to the ocelli and the large and distinct white spots beneath the hindwing.

Genus *NESOXENICA* Waterhouse and Lyell, 1914

This genus contains a single alpine species confined to Tasmania.

43. *Nesoxenica leprea* (Hewitson), 1864

This is a small narrow-winged species with two Tasmanian subspecies.

a. *Nesoxenica leprea leprea* (Hewitson), 1864

Plate 18, fig. 12 (male), leprea brown.

DISTRIBUTION.   From Hastings north to the Mt Field National

Park and Tarraleah, Tasmania, and in small areas near Walls of Jerusalem (E. D. Edwards) and west of Smithton in the north-west. At Mt Wellington it is on the wing from late December to early March.

Near the Springs on Mt Wellington adults have been observed flying commonly close to the ground about 9 a.m., whereas later in the day they were flying around the tree-tops.

b. *Nesoxenica leprea elia* **Waterhouse and Lyell, 1914**

Plate 18, fig. 13 (male), elia brown.

DISTRIBUTION. Western Tasmania, from the Hellyer River south to Macquarie Harbour, and from Cradle Mt south to Mt Anne and Mt Eliza.

This subspecies is usually common where *Nothofagus cunninghamii* grows. During cold and wet weather the adults rest on the beech twigs, where the exposed pattern beneath their wings harmonizes so effectively with the lichens growing on the trees that they are very difficult to see unless flushed.

The mature larva closely resembles the leaves and stems of the food plant and is difficult to detect. Near Lake St Clair (790 m) a female was noticed by L. E. Couchman laying an egg on a small sedge *Uncinia tenella* (Cyperaceae), which grows in small openings in stands of *Nothofagus*.

Genus *OREIXENICA* **Waterhouse and Lyell, 1914**

The genus contains small dainty butterflies restricted to south-eastern Australia and Tasmania. They occur mainly at high altitudes, and are sometimes very abundant.

The larvae of the various species are difficult to distinguish, but the pupae offer useful points of difference. The pupae of three species, *O. orichora*, *O. latialis* and *O. ptunarra*, lie unattached on or near the ground, whereas those of the remaining species, *O. lathoniella*, *O. correae* and *O. kershawi*, are attached by the cremaster and hang head downwards. With the exception of *O. kershawi*, the adult males are without a sex-mark on the forewing.

Four species occur commonly at altitudes above 1,500 m in the Australian Capital Territory, and at Mt Kosciusko and Mt Hotham. *O. orichora* is the earliest species, and flies mainly in December, January and early February, although a few specimens may be taken in late

186

February and even in March. On the other hand, adults of *O. latialis* are not seen until the latter half of February and in March. In *O. correae*, which flies in January and February, the sexes are very different from one another. Where *O. lathoniella* occurs with these species, it may be recognized by its large size and bright silvery markings on the underside of the hindwing. Both it and *O. latialis* are found at lower altitudes in the A.C.T., New South Wales and Victoria, where *O. kershawi* also occurs. Three species, *O. lathoniella*, *O. orichora*, and *O. ptunarra*, are known from Tasmania, the last being restricted to the alpine swamps of that island.

44. *Oreixenica orichora* (Meyrick), 1885

This species is confined to the higher mountains of southern New South Wales, the Australian Capital Territory, Victoria and Tasmania. Two subspecies are recognized.

a. *Oreixenica orichora orichora* (Meyrick), 1885

Plate 18, figs. 6 (male), 6A (female), orichora brown.

DISTRIBUTION. Brindabella Range, Australian Capital Territory, Tinderry Mountains, Brown Mt, Black Range, Tumbarumba, Kiandra, and Mt Kosciusko, New South Wales, at altitudes above about 1,200 m, and Mt Hotham, Mt Tamboritha, Mt Reynard and Mt Buller (A. F. Atkins), and at Shaws Creek, north of Heyfield, Victoria. Adults are on the wing in December, January and early February, very few being found in late February; a few females have been taken even in March.

In most males and some females the orange spots in the basal two-thirds of the forewing tend to be connected to one another. Occasionally there is a small black spot above the subapical ocellus of the forewing, and rarely this has a white centre forming a double ocellus. Specimens from the A.C.T. and nearby localities are usually much larger than those from Mt Kosciusko, with richer brownish orange markings. The subspecies can be readily recognized by the dull whitish postmedian and subterminal spots on the underside of the hindwing.

The larval head is narrower than in *O. latialis*, and the pupa differs by the lateral spikes at its anterior end. The larvae feed on *Poa* sp. (snow grass), but will accept other soft grasses (Poaceae).

b. *Oreixenica orichora paludosa* (T. P. Lucas), 1892

DISTRIBUTION.   Central plateau of Tasmania, from Cradle Mt to the northern end of Lake St Clair and east to the Great Lake and north of Arthurs Lake, at altitudes of 760 to 1,525 m.

ADULT.   Similar to typical *orichora* but darker, with smaller spots, and usually a double ocellus both above and beneath near apex of forewing, white postmedian and subterminal spots on hindwing beneath slightly silvery in appearance.

This subspecies has until recently been known as *flynni* Hardy, 1916.

45. *Oreixenica ptunarra* Couchman, 1953

This is a small alpine species, restricted to areas bordering mountain lakes and swamps at altitudes of 450 to 1,060 m in Tasmania. Adults may be common locally, and are on the wing from early March to early April, when few other species remain active. Three subspecies are recognized, each of which is very variable. The wings are more thinly scaled than in other species.

a. *Oreixenica ptunarra ptunarra* Couchman, 1953

DISTRIBUTION.   Great Lake district, Tasmania, at about 900 m. Adults fly close to the ground over rough grassy slopes bordering swampy areas.

ADULT.   Male: above brownish black, markings pale orange-yellow; forewing with markings paler towards costa, a white-centred black subapical ocellus, ringed with orange; hindwing with spots small, sometimes almost absent, a prominent subtornal white-centred black ocellus, ringed with orange-yellow. Beneath forewing grey-brown, becoming dull reddish brown towards termen, markings as above, a narrow subterminal line of silvery white spots; hindwing dull grey suffused with reddish brown, spots silvery white, prominent subapical and subtornal ocelli, each bordered on inner side by a large crescentic silvery white spot.

Female: differs greatly from male; apex of forewing less acute, termen of hindwing more rounded, markings much more extensive; above forewing brownish black, largely obscured by light orange markings, a short cream bar from costa preceding subapical ocellus, and a narrow cream subterminal line; hindwing with basal one-third brownish black enclosing an orange spot, outer two-thirds light orange, a row of cream subterminal spots and a terminal brownish black line. Beneath similar to male.

Sexual dimorphism is more pronounced, and both sexes are smaller than in either of the other two subspecies. The males, which are the darkest of the three, contrast strongly with the females, which are the palest, with obscure markings in the forewing and both wings apparently thinly scaled.

b. *Oreixenica ptunarra roonina* Couchman, 1953

Plate 13, figs. 8 (male), 8A (female).

DISTRIBUTION.   Oatlands district, Tasmania, at 450 to 640 m.

This is the largest of the three subspecies, with the pale markings in the male as extensive as in *angeli*, and intermediate orange markings in the female. Sexual dimorphism is less pronounced than in the typical subspecies, but a little more than in *angeli*.

c. *Oreixenica ptunarra angeli* Couchman, 1953

DISTRIBUTION.   Headwaters of the Macquarie River, and the Lake Leake district, Tasmania, at about 610 m. Adults are usually found in the grassy areas bordering swamps, but sometimes amongst *Lepidosperma* growing in the swamps.

ADULT.   Male: similar to *roonina* but rather smaller; above with spots dull orange, a little darker and smaller than in *roonina*, paler towards costa, forewing with white centre of subapical ocellus sometimes minute or absent. Beneath hindwing with postmedian silvery white band narrower and twice interrupted in middle.

Female: similar to *roonina* but smaller; above with spots reddish orange, prominent, in forewing extending as a band across wing and often continuing along veins to termen.

This is intermediate in size between the other two subspecies. The markings in the male also differ in colour, being a little darker. This and typical *ptunarra* are more variable than *roonina*. Often there is an extension of the subapical ocellus of the forewing towards the apex, especially in females, and occasionally this ocellus is doubled, as is normal in *O. orichora paludosa*. Sometimes there is a small ocellar spot below the subapical ocellus of the forewing, and more rarely there is a white centre to this spot. Females very occasionally have the subapical ocellus developed on the upperside of the hindwing as well as on the underside. Both sexes fly very close to the ground, and the male has a more direct and rapid flight than *O. lathoniella lathoniella*, which occurs in the same localities.

The early stages were reared at Hobart from eggs laid in March by Lake Leake females. The eggs hatched in forty-one days; the larvae developed slowly, reaching maturity early the following February. The pupal duration was twenty-eight to thirty-four days. The larvae feed at night at the tips of the grass blades, hiding during the day in the base of the tuft. The pupa is less humped than in *O. lathoniella*, and is without prominent projections.

46. *Oreixenica latialis* Waterhouse and Lyell, 1914

This species is restricted to the tablelands and mountains of southern New South Wales and the Australian Capital Territory at altitudes above about 1,000 m, and the higher mountains of eastern Victoria. Three subspecies are recognized.

a. *Oreixenica latialis latialis* Waterhouse and Lyell, 1914

Plates 18, figs. 1 (male), 1A (female), XII, alpine silver xenica.

DISTRIBUTION. Near Kanangra Walls (1,225 m) (E. D. Edwards), the Tinderry Mountains, and at Mt Kosciusko, New South Wales, Brindabella Range, Australian Capital Territory, and Mt Tamboritha, Mt Reynard (A. F. Atkins) and Mt Hotham, Victoria, at altitudes above 1,500 m. Adults fly in the latter half of February and in March, when most individuals of *O. orichora orichora* have disappeared. This subspecies is very common at Mt Kosciusko in March, and at dusk large numbers at times may be seen resting on grasses and shrubs.

This subspecies is smaller than any form of *O. lathoniella*, with narrower wings and much duller silvery markings on the underside. Specimens from near Kanangra Walls and from the Brindabella Range are darker than those from Mt Kosciusko.

Larvae feed at night on *Poa* sp. (snow grass). Some collected at Mt Kosciusko in December, and reared at Sydney, produced adults at the end of February. The pupal duration was sixteen days. Eggs laid at Mt Kosciusko in March hatched in Sydney in twelve days. The larvae pupated in December and adults emerged in about three weeks. No doubt the life cycle would be appropriately prolonged at the lower temperatures of Mt Kosciusko.

b. *Oreixenica latialis nama* **Couchman, 1953**

DISTRIBUTION.   Near Nimmitabel, south-eastern New South Wales, from about 1,000 to 1,200 m. Adults are on the wing in late March.

ADULT.   Both sexes were stated to be larger and darker, above and beneath, than typical *latialis* from Mt Kosciusko. Series of specimens collected recently from various localities near Nimmitabel show considerable variation, the darkest being similar to specimens from the Brindabella Range and the lightest similar to those from Mt Kosciusko. Closer study is needed of the various populations of *latialis*.

c. *Oreixenica latialis theddora* **Couchman, 1953**

DISTRIBUTION.   The Mt Buffalo plateau, Victoria, at 1,236 to 1,370 m. Adults have been taken in February and March.

ADULT.   Male: much larger than typical *latialis*, with orange markings deeper in colour and rather more restricted in size, but with orange markings larger than in *nama*.

Female: larger than male, with more rounded wings and paler markings.

In size and brightness this subspecies superficially resembles *O. lathoniella herceus*, but the markings are those of *O. latialis*.

47. *Oreixenica lathoniella* (Westwood), 1851

This is the most widespread species in the genus, being found in the mountains of most of New South Wales and the Australian Capital Territory, and in the mountains and coastal areas of Victoria and Tasmania. Hobart is accepted as the type locality. The species is very variable, but adults can usually be recognized by the colour and shape of the brownish orange or orange markings of the upperside and the bright silvery white pattern of the underside. Sometimes the orange markings are reduced in size, more commonly in specimens taken at the higher altitudes both in Tasmania and on the mainland. Four subspecies are recognized.

a. *Oreixenica lathoniella lathoniella* (Westwood), 1851

Plate 18, fig. 2 (male), common silver xenica.

DISTRIBUTION.   Northern, eastern and south-eastern Tasmania, from sea level to about 760 m. Adults are on the wing from January to March.

This subspecies is usually paler and smaller than *herceus* but larger and paler than *laranda* and *barnardi*. The orange-yellow subbasal and postmedian spots of the forewing are often reduced in size, especially in specimens taken in the mountains. The small subterminal spots of both wings may also be somewhat reduced in size, but not to the same extent as in some specimens of *herceus*.

b. *Oreixenica lathoniella laranda* Waterhouse and Lyell, 1914

Plate 18, fig. 3 (male).

DISTRIBUTION. Western and south-western Tasmania, from north of Mt Farrell to south of Mt Darwin, and probably around the south-western and southern areas to south of Hastings. Adults have been taken from mid-February to early April.

This is usually much darker and brighter than typical *lathoniella* but the two tend to grade into one another. The reddish brown ground colour beneath the hindwings is distinctive. Some specimens show a tendency to develop a double subapical ocellus above and beneath the forewing, and occasionally the white centre is doubled. The species is found in the "button-grass" swamps.

c. *Oreixenica lathoniella barnardi* Turner, 1926

DISTRIBUTION. Central northern Tasmania, from the Middlesex Plains across the central plateau to the Great Lake and the Shannon River, at altitudes between 760 and 1,065 m.

ADULT. Smaller than *laranda*, with orange markings above reduced in size, usually with black ground colour extending as a bar across middle of forewing; beneath light greenish brown.

Towards the western limits of its range this subspecies merges into *laranda* and towards the eastern limits into typical *lathoniella* (Couchman and Couchman, 1977).

d. *Oreixenica lathoniella herceus* Waterhouse and Lyell, 1914

Plates 18, fig. 4 (female), XII.

DISTRIBUTION. The tablelands and mountains of New South

192

Wales, south from Dorrigo, the mountains of the Australian Capital Territory, and throughout much of eastern, south-central and south-western Victoria, from sea level to about 1,500 m. Adults fly from February to April.

This is the largest and brightest subspecies, but is also very variable. Specimens from New South Wales and the A.C.T. are usually larger than those from Victoria. The series of subterminal orange spots on both wings are often reduced in number and size. Occasionally the silvery spots beneath are missing.

The larvae feed on *Poa* sp. and *Microlaena stipoides* (Poaceae).

48. *Oreixenica correae* (Olliff), 1890

Plate 18, fig. 5 (male), 5A (female), correa brown.

DISTRIBUTION. The mountains of south-eastern New South Wales, the Australian Capital Territory and eastern Victoria, above about 1,200 m. Specimens have been taken at Mt Kosciusko, often commonly above 1,500 m, the Brindabella Range, Mt Hotham, Mt St Bernard, Mt Erica, Mt Buffalo, Mt Baw Baw, and Mt Donna Buang. Adults fly from late December to mid-March, and a few females until late March.

In size this species is similar to *O. lathoniella*, but the markings are different and the hindwing beneath is sexually dimorphic, the females bearing silvery white spots. Specimens from the Australian Capital Territory and Victoria are larger than those from Mt Kosciusko and have more conspicuous markings on the underside.

The larvae feed on *Poa* sp. (Poaceae). Young larvae reared at Melbourne from eggs laid by Mt Baw Baw females fed only during the day, ascending the foliage about noon (W. N. B. Quick), but in the Brindabella Range larvae have also been observed feeding at night (E. D. Edwards).

49. *Oreixenica kershawi* (Miskin), 1876

This species may be distinguished from others in the genus by the more elongate hindwings. Its early stages indicate that it is related to the other species. Four subspecies are recognized. It occurs very locally at Barrington Tops and on the Southern Tablelands in New South Wales, in the mountains of the Australian Capital Territory, in Victoria, and in south-eastern South Australia.

a. *Oreixenica kershawi kershawi* (Miskin), 1876

Plate 18, fig. 7 (male), Kershaw's brown.

DISTRIBUTION.   The mountains of Victoria, including Mt St Bernard, Mt Erica, the Dandenong and Otway Ranges, the Grampians, and at Loch and Lorne. Adults may be taken from January to early April. Odd specimens from near Kiandra (J. F. R. Kerr), Yaouk (J. W. C. d'Apice) and Nimmitabel, N.S.W., may prove to belong to this subspecies.

Eggs from the Dandenong Ranges laid in late March hatched in nineteen days. The larvae moulted to the second instar five weeks later. They remained in this instar throughout the winter, the moult to the third instar not occurring until the last week in September. The remaining instars lasted about one month each, pupation occurring during the third week in December. The early stages resemble those of *O. lathoniella*, but the head of the larva is slightly concave above and the pupa has fewer projections. The larvae also feed on grasses (Poaceae). In captivity they have been reared successfully on *Tetrarrhena juncea*.

b. *Oreixenica kershawi ella* (Olliff), 1888

Plate 18, fig. 8 (male).

DISTRIBUTION.   Barrington Tops, New South Wales, above about 900 m. The original locality was stated to be Liverpool Plains, but this is almost certainly an error. Adults are on the wing from late December to February.

ADULT.   Similar to typical *kershawi* but larger; above forewing with larger brownish orange spots; hindwing with a broad uninterrupted postmedian band. Beneath hindwing dark reddish brown, postmedian silvery white band broad, only slightly constricted near middle.

This is larger than the typical subspecies, with more prominent spots on the forewing, a broad orange band above and a broad silvery white band beneath the hindwing.

c. *Oreixenica kershawi phryne* Tindale, 1949

DISTRIBUTION.   Brindabella Range (about 1,200 m) an Gibraltar Falls (870 m) (N. B. Tindale), Australian Capital Territory, and Black Range, New South Wales (east of Canberra) (D. Ferguson). Adults have been taken in January and February.

ADULT.   Similar to typical *kershawi* but larger, above with larger brownish orange spots, hindwing with postmedian band broader. Beneath hindwing with ground colour pale brown, postmedian silvery white band broader than in *kershawi*, but narrower than in *ella*, strongly constricted near middle.

This is the largest subspecies, with the spots of the forewing more closely resembling those of *ella*, but the postmedian band of the hindwing above and beneath more similar to that of *kanunda*.

A nearly mature larva, collected in late December by E. D. Edwards on *Poa* sp. (snow grass), resembled a green *O. lathoniella* larva, but was much larger. The pupa was similar to typical *kershawi*, but was pale brown with slightly darker markings. The pupal duration was fourteen days.

d. *Oreixenica kershawi kanunda* Tindale, 1949

DISTRIBUTION.   Near Millicent, south-eastern South Australia, and in the Dartmoor and Nelson (A. F. Atkins) districts, south-western Victoria. Adults have been taken in January and February.

ADULT.   Similar to typical *kershawi* but smaller, above with much larger brownish orange markings, hindwing with postmedian band broader. Beneath hindwing with ground colour rich brown, postmedian silvery white band broader than in *kershawi*, but narrower than in *ella*, strongly constricted near middle.

This is the smallest subspecies, with the largest spots, but with the silvery white band beneath the hindwing intermediate in width between those of *kershawi* and *ella*.

Genus *TISIPHONE* Hübner, 1819

The genus is confined to Australia but, unlike the other endemic genera, is found in tropical north Queensland as well as in southern Australia. The two species occur in relatively humid localities on the coast and tablelands, although the southern species *T. abeona* also survives in one or two restricted residual moist localities in western Victoria and south-eastern South Australia. The larval food plant is *Gahnia* (Cyperaceae). The larvae and pupae of the two species are said to resemble one another closely. In *T. abeona* the larva is green and has a smooth appearance although the head and body are densely clothed with minute setae. The pupa is smooth and green in colour. The adults are fairly large handsome slow-flying insects, with smooth eyes and the bases of the main veins of the forewing swollen. The genus does not occur in Tasmania, despite the presence of *Gahnia*.

50. *Tisiphone helena* (Olliff), 1888

Plate 16, fig. 6 (male), helena brown.

DISTRIBUTION. Daintree River to Paluma, north-eastern Queensland. Near the coast adults have been recorded at the Daintree River, near Gordonvale (150 m), and between Innisfail and Tully (W. R. Hindson); at higher elevations they have been taken at Kuranda, Mt Misery and Mt Spurgeon (above about 1,200 m), near Herberton, Mt Baldy, near Atherton, near Millaa Millaa (G. R. Brown), Tully Falls, Gray Range south of Cairns (about 400 m), and at Paluma. The original specimen came from Mt Bellenden Ker. At Kuranda adults are on the wing from September to April, but are usually not common. There are probably two generations annually.

The life history and early stages have not been described, but are said to resemble those of *T. abeona*. Larvae have been observed on *Gahnia* spp. at Kuranda (A. D. Bishop), at the Daintree River, and at Mt Baldy, near Atherton.

51. *Tisiphone abeona* (Donovan), 1805

This species occurs in various forms in mainland south-eastern Australia from Gympie, Queensland, to south-eastern South Australia. Seven subspecies are recognized which fall naturally into two groups, a southern one including variants of *T. abeona abeona*, and a northern one including variants of *T. abeona morrisi*. The ranges of the *abeona* and *morrisi* groups meet in the Port Macquarie district, where there is an extremely variable population (Plate 16, figs. 3, 3A), known as "*joanna*" and thought to be maintained by hybridization of the *abeona* and *morrisi* forms.

The larvae of *T. abeona* (Plate XII, fig. 10) feed only on *Gahnia* or sword-grass (Cyperaceae), usually tall-growing sedges with sharp edges to the leaves, which grow in moist places. They include *G. sieberana*, *G. melanocarpa*, *G. clarkei* and *G. erythrocarpa* (B. A. Conroy).

a. *Tisiphone abeona abeona* (Donovan), 1805

Plate 16, fig. 4 (male), sword-grass brown.

DISTRIBUTION.   Coast and tablelands of New South Wales, from the Hunter River to Moruya. It occurs in the Blue Mountains, the Illawarra Range, Black Range, east of Canberra (D. Ferguson), and Clyde Mountain, but adults from Brown Mountain show some of the characteristics of *albifascia*. In coastal areas a gradual transition to *albifascia* occurs between Batemans Bay and Victoria. There are two generations annually, at least in coastal areas.

At Sydney the eggs hatch in nine or ten days in December, but this may be increased to sixteen days in October or March. Larvae from eggs laid in spring reach maturity in three to five months, but those from autumn eggs develop in five to eight months. The pupal duration at Sydney is twenty-four to forty-five days in August and September, nineteen to twenty-six days in October and November, eighteen days from December to February, and thirty-four days in April.

The larvae of typical *abeona* appear to prefer *G. sieberana* which is found growing in dense clumps in coastal areas, or along creeks on higher ground and on the tablelands. The larvae remain hidden head downwards deep in the centre of the *Gahnia* clumps during the day. The first instar larva feeds in the early morning, but later instars usually feed in the early evening, eating angular pieces from the edges of the leaves.

b. *Tisiphone abeona albifascia* Waterhouse, 1904

Plate 16, fig. 5 (male).

DISTRIBUTION. South-eastern New South Wales, south from Merimbula, the escarpment south from Brown Mountain and at Geehi; Victoria, from sea level to about 760 m, west to Wilsons Promontory, the Mornington Peninsula, the Dandenong Ranges, Healesville, Mt Macedon, Anglesea, Lorne, Apollo Bay, Cape Otway, and the slopes of Mt Buffalo. There are two generations annually at lower altitudes.

Specimens of this subspecies are usually larger and brighter than typical *abeona*. Gradual clinal changes in several characters have been found in populations distributed from south coastal New South Wales to central Victoria (B. A. Conroy). Populations at Geehi and localities in the Victorian mountains probably have only a single generation annually.

The life history and early stages are similar to those of typical *abeona*.

c. *Tisiphone abeona antoni* **Tindale, 1947**

DISTRIBUTION.   The Grampians and the Dartmoor and Portland districts, western Victoria, and Lake Edward, south-eastern South Australia. It is rare in South Australia, but more common in the Grampians. Adults have been collected from November to February, and there may be two generations annually.

ADULT.   Similar to *albifascia*, but forewing with subocellar costal band cream in both sexes, broad orange band with costal half distinctly cream in female. Beneath with postmedian band similar in width to that of *albifascia*, but inner subterminal line more conspicuous.

This is apparently an extreme form of *albifascia*, from which it is now geographically isolated, preventing any further interbreeding with that subspecies.

The early stages are similar to those of typical *abeona*. The larvae feed on *Gahnia sieberana* (Fisher, 1978).

d. *Tisiphone abeona aurelia* **Waterhouse, 1915**

DISTRIBUTION.   Coastal New South Wales, from Camden Haven to the mouth of the Hunter River, near Newcastle. Adults are common in spring and autumn.

ADULT.   Very similar to typical *abeona*, but subtornal ocellus of hindwing usually with a broader orange, rather than orange-red, ring. Beneath with more prominent markings.

This subspecies is geographically isolated from typical *abeona* by the mouth of the ·Hunter River and the city of Newcastle, where *Gahnia* is now absent. Its wing pattern does not appear to be consistently different from that of *abeona*.

e. *Tisiphone abeona morrisi* **Waterhouse, 1914**
Plate 16, fig. 2 (male).

DISTRIBUTION.   Coastal New South Wales from Tweed Heads to the Macleay River. It formerly occurred in coastal southern Queensland, from Southport to Coolangatta, but has disappeared from this area. It is common farther south, where adults are on the wing from September to April.

In coastal areas north of the Macleay River this subspecies has very constant markings; the specimens from an altitude of about 450 m near

Lismore do not differ significantly from those in coastal areas.

f. *Tisiphone abeona regalis* Waterhouse, 1928

Plate XII.

DISTRIBUTION. Mountains of northern New South Wales, above about 450 m, from Dorrigo to Barrington Tops, and at Glen Aplin and Mt Norman, near Stanthorpe, Queensland, at about 900 m. Adults have been taken in January and February. They are fairly common in the New England National Park at about 1,370 m. There is only one generation annually, at least at the higher altitudes.

ADULT. Male: similar to *morrisi*, but often larger with more conspicuous cream markings; forewing with cream patch near tornus nearly surrounding ocellus. Beneath with cream bar in cell broader and connected to cream patch near tornus.

Female: similar to *morrisi*, but with much broader cream markings, and forewing above with cream bar in cell.

In its extreme form, especially in the female, this is the brightest subspecies, but the markings may be reduced, especially in the more northern populations.

g. *Tisiphone abeona rawnsleyi* (Miskin), 1876

Plate 16, fig. 1 (male).

DISTRIBUTION. Coastal south-eastern Queensland, from Gympie to Caloundra; common on the Mooloolah River. Adults have been taken from October to December, and in March and April.

This is a small form of *morrisi* that has lost most of the cream markings above. It is now geographically isolated from *morrisi* by a broad coastal area which includes the lower valley of the Brisbane River.

Genus *XOIS* Hewitson, 1865

This small genus is known from Fiji, Samoa, New Guinea and Australia.

52. *Xois arctoa arctoa* (Fabricius), 1775

Plate 16, fig. 13 (male), dingy ring.

DISTRIBUTION. North-western Australia, Northern Territory, the islands of Torres Strait, and Cape York to eastern Victoria; very common north of Sydney, less common in southern coastal New South Wales to the Nowa Nowa district of eastern Gippsland.

The species is easily recognized by the large twin-centred ocellus of the forewing. Some specimens have additional ocelli on the hindwings. The adults fly close to the ground, resting on grasses but never on the ground.

The larvae feed on common grasses, including *Imperata* (blady grass) (Poaceae).

Subfamily MORPHINAE
OWLS
(Plate 13)

The subfamily contains large and fairly robust butterflies, usually confined to tropical rain forests. It includes the brilliant blue Neotropical *Morpho*, and the much duller but nevertheless handsome *Amathusia* and its allies from the Oriental region and *Taenaris* and its allies mainly from the Australian region.

As in the Satyrinae, the head of the larva sometimes carries a pair of horns, but the body is usually covered with a fine pubescence and tufts of long hair. The anterior end of the pupa in some genera is produced to a pair of pointed processes enclosing the labial palpi; the pupa is suspended head downwards by the cremaster.

The adults are said to be most active in the early morning and late afternoon and are difficult to capture in dense forest as they weave through the undergrowth and between the tree trunks. In the Indo-Malayan area they are attracted to baits of over-ripe fruit.

Genus *TAENARIS* Hübner, 1819

The genus occurs in southern Malaya, Palawan, Borneo, Sumatra, Java, and the Moluccas, and from New Guinea to the Solomons and the north-eastern tip of Australia.

The early stages of the Australian species have not been described. However, larvae of *T. myops* (C. and R. Felder), 1860, from New Guinea are yellowish green in colour and have fairly dense pale secondary hairs; the light brown head has a pair of black sclerotized slightly clubbed and spiny horns. The larvae are said to be gregarious and lie side by side on a leaf of the food plant feeding systematically on a front. The pupa is similar to that of many Satyrinae, smooth, light green in colour, with a pair of short pointed anterior projections and black cremaster.

Adults of *Taenaris* fly mainly in and near rain forest, but are sometimes seen in savannah woodland in southern Papua. They have a fairly direct flight, usually well above the ground. In flight the forewings appear to provide most of the power, the hindwings apparently scarcely being used at all.

53. *Taenaris catops turdula* **Fruhstorfer, 1914**

DISTRIBUTION. Within Australian limits, only at Darnley Island, Torres Strait.

ADULT. Female: above forewing white, costa dark grey; hindwing white, base of cell and along inner margin with long yellow hairs, ocelli beneath visible from above. Beneath forewing as above; hindwing white, dark greyish on costa, yellowish towards inner margin, a subapical and a large white-centred black subtornal ocellus, with yellow and dark grey outer rings.

54. *Taenaris artemis* **(Snellen van Vollenhoven), 1860**

This is a variable species distributed throughout most of mainland New Guinea and its associated islands, from Waigeo to the Louisiade Archipelago, as well as the Aru Islands.

a. *Taenaris artemis jamesi* **Butler, 1876**

Plate 13, fig. 5 (male).

DISTRIBUTION. Within Australian limits, known with certainty only from eleven specimens taken at Darnley Island.

b. *Taenaris artemis zetes* **Brooks, 1944**

DISTRIBUTION. Known from only one pair in the British Museum taken at Murray Island. Both specimens are in poor condition and additional examples are needed for study.

ADULT. Until further specimens are collected it is difficult to distinguish this from Darnley Island specimens. However, T. G. Howarth states that the ocelli beneath the hindwing are larger and there is an additional small ocellus at the costa, joined to the subapical ocellus.

Subfamily CHARAXINAE

(Plates 20, 29, X, XII, XIV)

The large and robust butterflies of this subfamily are closely related to the Nymphalinae, in which many authors include them.

The larval head usually bears one or two pairs of prominent tapering horns, and at its posterior end the larva has a pair of short pointed projections. Its body lacks prominent hairs, except on the ventral and anal prolegs, and is without branched or other spines. The pupa is rounded and smooth.

Genus *POLYURA* Billberg, 1820

The genus ranges through the Oriental and Australian regions from Sri Lanka, India and Taiwan to New Guinea, Australia and Fiji. The adults have a rapid and direct flight and are attracted by fermenting juices upon which they feed. All the larval instars have two pairs of long pointed horns on the head, and the body is devoid of noticeable hair except on the prolegs. The stout smooth pupa is suspended head downwards by the cremaster.

55. *Polyura pyrrhus* (Linnaeus), 1758

This species ranges from the Lesser Sunda Islands, the Moluccas and the Kai and Tanimbar Islands through New Guinea, the Bismarck Archipelago, the Solomons, northern and eastern Australia and Lord Howe Island. Many subspecies have been describd including one from Australia and another from Lord Howe Island.

a. *Polyura pyrrhus sempronius* (Fabricius), 1793

Plates 20, fig. 1 (male), X, XII, XIV, tailed emperor.

DISTRIBUTION. North-western Australia, Northern Territory, Thursday Island, and Cape York to southern New South Wales and the Australian Capital Territory, and spasmodically in Victoria and at Adelaide, South Australia. It occurs on the coast, tablelands and inland.

This fine species can be easily identified by its size and the pointed tails to the hindwing. The adults usually settle high on the foliage of a tree, with wings closed and the head directed downwards. Like the exotic species of the subfamily, they are attracted to the fermenting juices of over-ripe fruit, and have been reported feeding at sap exuding from shrubs, and even at the leaking bunghole of a wine cask. In this situation the adults are easily captured.

The larva covers one or more leaves of the food plant with silk, upon which it rests during the daytime, during moults, or during the coldest part of the winter. The winter is passed either as a larva or a pupa, neither of which appears to be adversely affected by frost in inland areas. The usual native food plants are wattles (Mimosaceae), including *Acacia decurrens*, *A. maidenii*, *A. baileyana* (Cootamundra wattle), *A. dealbata*, *A. longifolia*, *A. podalyriifolia* (Queensland wattle), *A. neriifolia* and *A. spectabilis*, but they also feed on *Abarema sapindoides* (Mimosaceae), *Brachychiton populneum* (kurrajong), *B. discolor*, *B. acerifolium* (flame tree, Sterculiaceae), *Cinnamomum camphora* (camphor laurel, Lauraceae), *Robinia pseudoacacia* (false acacia, Fabaceae), *Delonix regia* (poinciana), *Cassia fistula*, *C. alata*, *Caesalpinia gilliesii*, *C. ferrea* (all Caesalpiniaceae), *Albizia lophantha* (Mimosaceae), *Celtis paniculata*, *C. philippensis*, *C. sinensis* (Ulmaceae), *Lagerstroemia indica* (crepe myrtle, Lythraceae), and *Guilfoylia monostylis* (Simaroubaceae).

b. *Polyura pyrrhus tiberius* (**Waterhouse**), 1920

DISTRIBUTION.   Lord Howe Island.

ADULT.   Similar to *sempronius* but forewing with more concave termen, dark margin much broader and browner, central area much more yellow; hindwing with dark margin much browner, greenish terminal line narrower and paler, orange tornal patch reduced in size. Beneath much paler in both wings, hindwing with reddish brown markings near tornus and orange terminal line reduced.

Only a few specimens have been taken.

Genus *CHARAXES* Ochsenheimer, 1816

Apart from a rich development in Africa, several species of this genus are known from the Oriental and Australian regions. It had not been recorded from Australia until recently.

56. *Charaxes latona* **Butler**, 1865

Plate 29, fig. 8 (male).

DISTRIBUTION.   Claudie River, Cape York Peninsula. This fine species is known in Australia from only a few specimens taken since June–July 1978 by M. De Baar and S. J. Johnson.

ADULT. Male: above orange-brown, forewing with apex and termen broadly black, and a double black bar at end of cell; hindwing with termen slightly produced at ends of veins, a narrow pointed tail at end of vein M_3 and a shorter tail at end of CuA_2, a thick black terminal line and a series of black subterminal spots, those towards tornus each containing a small bluish white spot.

Female: larger than male, paler orange-brown in colour, basally darker orange-brown.

The adults are similar in size and shape to *Polyura pyrrhus*, but the orange-brown and black coloration is very distinctive. The males were collected on a hill-top, but may well be attracted to carrion or fermenting fruit baits.

Subfamily NYMPHALINAE
NYMPHS
(Plates 14, 19, 20, X, XI, XV; Fig. 17)

This large subfamily includes mainly medium-sized to large butterflies, but there are a few smaller species. They have a world-wide distribution, but are much more numerous in the tropics. Many of them are brightly coloured and, with few exceptions, have a rapid direct flight. Most fly in bright sunshine, feed freely at flowers, and frequently settle in sunshine on a prominent leaf or twig, or on the ground, with the wings expanded. A few species are shade-loving forest insects with procryptic leaf-like patterns beneath the wings, which effectively simulate dead leaves. The adults also feed at fermenting sap-flows or fruit juices, or at fresh animal dung. Migration is a regular feature of the adult behaviour in some species, although little is yet known about its function in Australia. Here migratory flights have been observed chiefly in *Vanessa kershawi*, but also in *V. itea* and *Junonia villida*.

Some species exhibit sexual dimorphism and several exotic species mimic distasteful Danainae or other butterflies. The female of *Hypolimnas misippus*, which resembles *Danaus chrysippus*, provides the only striking example of this type of mimicry in the Australian nymphalines.

The height of the eggs usually exceeds their diameter; they have vertical or horizontal ribs, or are pitted or rarely have small projections. The larvae (Plates X, XI, XV) are diverse in structure, but usually bear series of branched spines. The pupae (Fig. 17; Plates XI, XV) are always suspended head downwards by the cremaster, usually have angular projections, and are sometimes grotesque in shape. They often have metallic markings.

Most of the twenty-seven species recorded from Australia are tropical or subtropical insects, and only four occur south of Sydney. A specimen of *Apaturina erminea*, one of the species occurring in the far north, has only just been captured, although they have been sighted on several occasions. Three other northern species are known in Australia from very few specimens. All of the Australian species belong to genera which also occur in New Guinea or the Oriental region, and in fact there is a considerable overlap in species. Australian populations of these widespread species usually represent separate subspecies.

Genus *APATURINA* Herrich-Schäffer, 1864

This genus contains large robust butterflies belonging to two species in the Australian region. One of them is known to occur on Cape York Peninsula.

57. *Apaturina erminea* (Cramer), 1779

Plate 14, fig. 10 (male, New Guinea).

DISTRIBUTION. Claudie River, Cape York Peninsula. Specimens, very probably of this species, have been observed on separate occasions by D. P. Sands, G. Daniels and S. J. Johnson and one has been captured by G. Wood.

This large species has a stout body and forewing similar in shape to *Polyura*. Its flight is rapid and the adults are probably attracted to fermenting fluids. A. P. Dodd has noted that in New Guinea the adults settle on tree trunks, head downwards, with wings closed above the body. In New Guinea females may be similar in colour and pattern to the males or have the greenish blue areas replaced by brown.

Genus *LEXIAS* Boisduval, 1832

This genus contains two species, one of which occurs in the Philippines and the other ranges from the Moluccas to the Bismarck Archipelago.

58. *Lexias aeropa* (**Linnaeus**), 1758

Plate 14, fig. 11 (male).

DISTRIBUTION.   Near Shelburne Bay, Cape York Peninsula. The Queensland specimens may belong to subspecies *eutychias* Fruhstorfer, 1913, which also occurs in the Kai and Aru Islands and on the New Guinea mainland.

Females of *eutychias* may have the bands above cream-white or yellow. The species is known in Australia from only two males and a female collected in or near rain forest. Another female was sighted but not captured. The female was collected feeding at the skins of over-ripe bananas, and in New Guinea the species is often taken at fermenting fruit baits. The males were noticed flying only a short distance before settling with wings outspread in the sunshine. The food plant of the larvae is stated to be *Calophyllum* (Guttiferae), a genus which occurs on Cape York Peninsula.

Genus *PHAEDYMA* Felder, 1861

This genus is widely distributed from India to China and through south-east Asia, including the Philippines and Indonesia, to New Guinea, north-eastern Australia and the Solomons. The adults of this and the next two genera, *Neptis* and *Pantoporia*, have a characteristic gliding flight, usually in sunny openings in rain forest, and frequently settle with the wings outspread.

The larvae are relatively smooth, with a pair of short pointed projections on the head above, and with one or two pairs of long and slender, or rounded, often spiny processes on the thorax, and one or two pairs of similar but shorter processes on the abdomen. The pupae are curved ventrally and the wings are expanded laterally (Figs. 17A, B).

59. *Phaedyma shepherdi* (**Moore**), 1858

This species ranges through New Guinea, the Kai and Aru Islands, the D'Entrecasteaux and Trobriand Islands and eastern Australia. Nine subspecies are recognized. The two Australian subspecies differ chiefly in the width of the central band of the hindwing.

a. *Phaedyma shepherdi shepherdi* (**Moore**), 1858

Plate 19, fig. 4 (male), common aeroplane.

DISTRIBUTION.   Shute Harbour and Mackay to the Manning

River; more common in Queensland, where it flies from September to May. In central Queensland it has been taken as far inland as the Expedition Range (R. C. Manskie). A specimen was observed, but not collected, by K. D. Fairey as far south as Newcastle.

The larvae have been reported feeding on several trees, including *Pongamia pinnata* and *Mucuna gigantea* (both Fabaceae), *Aphananthe philippinensis* (Ulmaceae), *Brachychiton acerifolium* (flame tree, Sterculiaceae), *Celtis philippensis* and the introduced ornamental *Celtis sinensis* (European nettle tree) (Ulmaceae).

b. *Phaedyma shepherdi latifasciata* (Butler), 1875

DISTRIBUTION.   Prince of Wales, Moa and Thursday Islands, and Cape York to Ingham; fairly common. Inland it has been taken 65 km south-west of Mt Garnet (W. R. Hindson).

ADULT.   Similar to typical *shepherdi* but white markings larger, especially central band of hindwing; in male white markings tinged with green.

At Cairns the larval food plants include *Pongamia pinnata* (Fabaceae), *Ehretia acuminata* (Boraginaceae), *Aphananthe philippinensis*, *Celtis paniculata* and *C. philippensis* (Ulmaceae). At Weipa the larvae have been found on *Bombax ceiba* (Bombacaceae) (G. B. Monteith). Soon after hatching the young larva cuts out small pieces of the leaf and attaches them by one edge so that they hang down on each side of the midrib. The larva rests along the midrib between these brown leaf fragments and is difficult to detect. However, the presence of small irregular pieces of leaf suspended beneath the leaves of the food plant indicates that larvae may be present. They develop very slowly and at maturity leave their hiding places and pupate beneath another leaf near by.

Genus *NEPTIS* Fabricius, 1807

This genus ranges very widely from southern Europe, Africa and Asia to New Guinea and Australia.

60. *Neptis praslini staudingereana* de Nicéville, 1898

Plate 19, fig. 3, (male), black and white aeroplane.

DISTRIBUTION.   Cape York to Cardwell, Queensland; sometimes

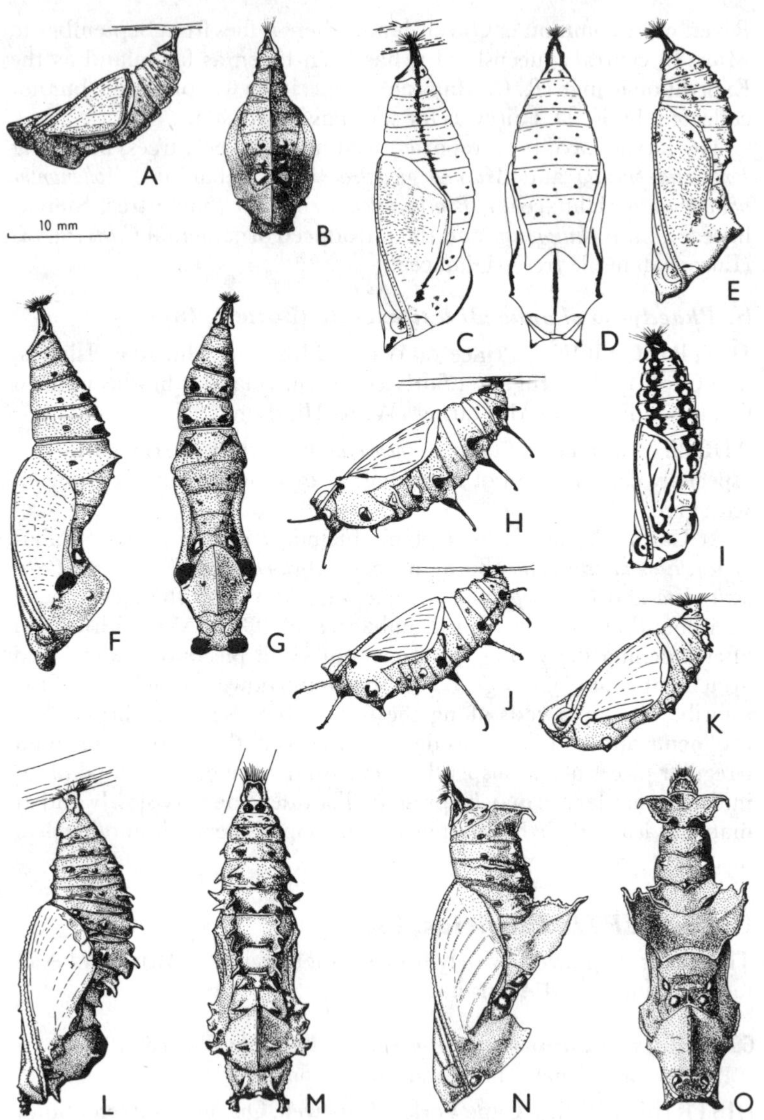

common. At Cairns adults have been taken from June to December.

The adult bears a superficial resemblance to *Tellervo zoilus* (Plate 15, figs. 7, 7A), but differs in structure and in flight behaviour. In life the eyes of both species are yellow. The early stages have not been described.

Genus *PANTOPORIA* Hübner, 1819

Distributed from India to China and through south-east Asia to New Guinea and north-eastern Australia, this genus contains fourteen species.

The early stages are very similar to those of the last two genera. Two species occur in north-eastern Queensland.

61. *Pantoporia venilia moorei* (W. J. Macleay), 1866

Plate 19, fig. 2 (male), Cape York aeroplane.

DISTRIBUTION. Cape York to Port Stewart, Cape York Peninsula; fairly common in most months at Cape York.

The early stages have not been described.

62. *Pantoporia consimilis consimilis* (Boisduval), 1832

Plate 19, fig. 1 (male), orange aeroplane.

DISTRIBUTION. Moa, Prince of Wales, Horn and Thursday Islands, and Cape York to Mackay; common. It has also been recorded at Kowanyama, near the Gulf of Carpentaria, west of Cooktown, and may have a patchy distribution along the west coast of Cape York Peninsula. Near Cairns adults have been taken throughout the year.

The larval habits are very similar to those of *Phaedyma shepherdi*; the larva rests beneath the midrib of a leaflet of the food plant between several irregular leaf fragments that it has attached and hang freely on either side of it. Unlike *P. shepherdi*, however, it pupates at the extreme end of the dead portion of the damaged leaf upon which it fed and

Fig. 17. Pupae of Nymphalidae, lateral and dorsal views: A, B, *Phaedyma shepherdi* (p. 206); C, D, *Dolleschallia bisaltide* (p. 211); E, *Vanessa itea* (p. 217); F, G, *Mynes geoffroyi* (p. 210); H, *Vagrans egista* (p. 221); I, *Acraea andromacha* (p. 224); J, *Cupha prosope* (p. 223); K, *Phalanta phalantha* (p. 222); L, M, *Cethosia cydippe* (p. 220), N, O, *Vindula arsinoe* (p. 221).

sheltered, the brown pupa being very difficult to detect. The larva feeds on *Kuntsleria blackii* and *Derris* sp. (both Fabaceae).

Genus *ARGYREUS* Scopoli, 1777

One of several genera formerly included in *Argynnis* Fabricius, 1807, this genus contains only one Australian species, *A. hyperbius* Johannsen, which is widely distributed in the Oriental region, as well as New Guinea and a small area of eastern Australia.

63. *Argyreus hyperbius inconstans* (Butler), 1873

Plates 20, fig. 10 (male), XV, Australian fritillary.

DISTRIBUTION.   Eastern coastal Australia from Gympie to 5 km north of Port Macquarie; very local, usually rare. Adults are probably present in northern localities throughout the year.

Adults fly in moist or swampy places where the food plant grows, feeding at flowers and settling on low vegetation or on the ground. Their appearance in flight somewhat resembles that of *Danaus chrysippus* and they may thereby gain some protection from predators.

In midsummer the eggs hatch in about five days and the pupal duration is seven to nine days. The larvae feed on the wild violet, *Viola betonicifolia* (Violaceae), growing in the shelter of clumps of *Lomandra longifolia* and grasses. They apparently feed at night, sheltering away from the small violet plants during the day.

Genus *MYNES* Boisduval, 1832

This genus includes six species distributed from Flores and the Moluccas through New Guinea to north-eastern Australia and the Solomons. Only one is represented in Australia.

64. *Mynes geoffroyi guerini* Wallace, 1869

Plate 19, figs. 5 (male), 5A (female), white nymph.

DISTRIBUTION.   Claudie River, Queensland to Ballina, New South Wales; fairly common at times north of Mackay, usually less common farther south.

Various gradations between forms with white or black undersides are found, the examples figured representing nearly the extremes. In New Guinea white forms are much less common than melanic ones. The species has a superficial resemblance to a *Delias* (Pieridae), but the flight behaviour and structure of both adults and immature stages are very distinct.

The larvae feed gregariously on the foliage of *Dendrocnide moroides*, *D. photinophylla* (stinging trees) and *Pipturus argenteus* (all Urticaceae), and several pupae (Figs. 17F, G) may occur together beneath the one leaf.

Genus *DOLESCHALLIA* C. and R. Felder, 1860

This small genus of about nine species occurs from India and Sri Lanka through south-east Asia, the Philippines and Indonesia to New Guinea, Australia, the Bismarck Archipelago, New Caledonia, the Solomons, the New Hebrides and Fiji. It is best developed in New Guinea, where six species are found; only one occurs in eastern Australia. The adults are strong fliers, being found near shady rain-forest areas, settling on foliage or flowers in the sunshine with their wings outspread. They rest with wings closed, exposing the protective undersides, which in shape and pattern closely resemble dead leaves.

65. *Doleschallia bisaltide australis* C. and R. Felder, 1867
Plate 19, figs. 10 (male), 10A (female), Australian leafwing.

DISTRIBUTION.   Islands of Torres Strait and Cape York to the Clarence River; common at times in the north, rarer farther south.

The central line across the underside of the wings, together with the variable brownish markings (Plate 19, figs. 10, 10A) enhance the resting insect's resemblance to a dead leaf.

The larva feeds on *Pseuderanthemum variabile* (Acanthaceae), especially small plants, or the regrowth of older plants, on the rain-forest floor. It has also been found on several ornamental plants including *P. bicolor*, *Graptophilum pictum*, *Strobilanthes isophyllus* and *Asystasia gangetica* (all Acanthaceae). The pupa (Figs. 17C, D) is smooth with a pair of small anterior projections.

Genus *HYPOLIMNAS* Hübner, 1819

The genus and several of the species have a very wide distribution in the Old-World tropics and one species is believed to have been introduced into the tropics of the Western Hemisphere. Within the Oriental and Australian regions it is best developed in New Guinea, where six species are found. Three of these also occur in northern and eastern Australia. Some of the species are noted for their great variability and sexual dimorphism; the females of one species, *H. misippus*, are a mimic of *Danaus*. The larvae have numerous finely branched spines, and a pair of similar but longer spines on the head. The anterior half of the pupa is angular and the abdomen carries dorsal rows of short pointed processes. The larvae feed on Acanthaceae, Amaranthaceae, Portulacaceae, Malvaceae, Rubiaceae, Urticaceae and Asteraceae.

66. *Hypolimnas bolina nerina* (Fabricius), 1775

Plates 19, figs. 7 (male), 7A (female), XI, common eggfly.

DISTRIBUTION. North-western Australia north from Onslow (Mrs Milne), Northern Territory, the islands of Torres Strait, Cape York and much of the eastern half of Australia, to far eastern Victoria, the Murray Valley, Adelaide, and the Flinders Ranges, South Australia, and Lord Howe Island.

It is common in the tropics and subtropics of Queensland, and thrives in irrigated homestead gardens as far west as Quilpie (R. W. Guard). It has also been taken near Alice Springs, and two females were observed even in the Simpson Desert (E. Slater). The species is usually rare near Sydney, but in some years adults are more common and larvae have been collected. It is very spasmodic near the southern limits of its range and specimens are rarely taken. At Brisbane and farther north adults are on the wing almost throughout the year.

In the male the white spots are occasionally completely obscured by purplish blue. The female is extremely variable, although usually of the form described above. Rarely the orange patch of the forewing is absent and the large white markings of both wings are almost entirely absent.

The adults are common in gardens, and feed freely at *Lantana* and other flowers.

The larvae usually feed on *Alternanthera denticulata* (joyweed, Amaranthaceae) and *Sida rhombifolia* (Paddy's lucerne or sida-retusa, Malvaceae), but other food plants include *Asystasia scandens*, *A. gangetica*, *Pseuderanthemum variabile*, *Dipteracanthus* spp., *Ruellia* spp. (all Acanthaceae), *Portulaca* (Portulacaceae), *Polygonum prostratum* (Polygonaceae), *Richardia brasiliensis* (Rubiaceae), and *Synedrella nodiflora* (Asteraceae). The young larvae are gregarious until they moult to the third instar, after which they are usually found singly. During the day they often conceal themselves at some distance from the food plant.

67. *Hypolimnas misippus* (Linnaeus), 1764

Plate 19, figs. 9 (male), 9A (female), danaid eggfly.

DISTRIBUTION.   North-western Australia, Northern Territory, Darnley and Thursday Islands, and Cape York to Brisbane, and occasionally to northern and central New South Wales. In some years it is very common in north Queensland, but only a few specimens have been taken at Brisbane. Occasional specimens have been collected in New South Wales, even as far south as Penrith (E. O. Edwards). In the north the most western record is Yampi Sound.

There is a striking difference in the two sexes, the males being rather similar to *H. bolina* but smaller, with narrower wings and larger white spots. The female, on the other hand, bears a strong superficial resemblance to *Danaus chrysippus* (Plate 15, fig. 5), a species which is probably distasteful to birds and other predators. The mimic is thought thereby to derive some measure of protection, but the flight behaviour of the two species may be distinguished by a keen observer.

The larvae feed on *Portulaca oleracea* (pigweed), (Portulacaceae), *Pseuderanthemum variabile* and *Asystasia gangetica* (both Acanthaceae).

68. *Hypolimnas alimena* (Linnaeus), 1758

The species occurs in the Moluccas, the Kai and Aru Islands, New Guinea, the Bismarck Archipelago, the Solomons, and northern Australia. Numerous subspecies are recognized, including two from Australia.

a. *Hypolimnas alimena lamina* Fruhstorfer, 1903

Plates 19, figs. 8 (male), 8A (female), XV, blue-banded eggfly.

DISTRIBUTION.   Islands of Torres Strait, and Cape York to Gympie and Nambour, Queensland, and rarely near Murwillumbah and Ballina, New South Wales (G. Sankowsky). Common at times north from Rockhampton and Yeppoon.

The usual female form is very similar to the male, but two others are known. Intermediates between the three forms also occur. One of these forms (Plate 19, fig. 8A) is brown-black often with a broad terminal area of the hindwing orange-brown; it is without the pale post-median band, but has the band of four large white spots on the forewing. The other form is orange-brown, with basal areas dark brown and with the two rows of subterminal spots, but lacks the white band towards the apex of the forewing. Both of these occur at Cape York and on the islands of Torres Strait, but the former is the more abundant.

In early spring the eggs hatch in about a week. The larva feeds on *Pseuderanthemum variabile*, growing in rain forest, but has also been recorded by G. Sankowsky on the introduced ornamentals *Asystasia gangetica* and *Graptophilum pictum* (all Acanthaceae). The pupal duration is about ten days.

b. *Hypolimnas alimena darwinensis* Waterhouse and Lyell, 1914

DISTRIBUTION. Northern Territory, including Groote Eylandt.

ADULT. Male: similar to *lamina*, but above with blue band enclosing four white spots towards costa of forewing, hindwing with outer series of white or bluish white subterminal spots much larger.

Female: above forewing brown, with a band of four large white spots towards apex, and a series of faint white subterminal spots; hindwing brown, outer half much paler, two series of subterminal spots, the outer ones large, a terminal white line. Beneath rather similar to above but paler, hindwing with traces of a postmedian band.

The male of this subspecies is very like the blue female form of *lamina*, whereas the female resembles the brown female form of *lamina*, but has larger subterminal spots of the hindwing.

69. *Hypolimnas anomala albula* (Wallace), 1869

Plate 14, fig. 8 (male).

DISTRIBUTION. Within Australian limits only two males have been collected, one at Darwin in March and the other at Cape York in April (W. F. Gibb).

This species has until recently been recorded as *H. antilope* (Cramer), but is now correctly referred to *H. anomala*. It may not breed in Australia. Collectors in the far north of the continent should keep a careful look out for further specimens, especially to determine if there are resident populations. The female bears a superficial resemblance to certain Oriental species of *Euploea*, and it has been suggested that this may represent a case of mimicry. In Malaya the larvae of *H. anomala* feed gregariously on *Pipturus argenteus* and in the Philippines on *Pouzolzia* (both Urticaceae); these plants also grow in northern Australia. The pupa is rather similar to that of *Vanessa*.

Genus *YOMA* Doherty, 1886

The genus contains a single species distributed from Burma and Taiwan, through the Philippines and Indonesia to New Guinea, northern Australia, the Bismarck Archipelago, and the Solomon Islands.

70. *Yoma sabina parva* (Butler), 1876

Plate 19, fig. 6 (male), Australian lurcher.

DISTRIBUTION.   Northern Territory, including Groote Eylandt (J. Waddy), the islands of Torres Strait, and Cape York to Cairns and the Atherton Tableland; not common.

This species is easily recognized by the shape and pattern of the wings. The very variable underside may have a greenish tint or may lack the pale median band. A smaller dry-season form has been taken at Cape York.

The life history and early stages have not been recorded from Australia.

Genus *VANESSA* Fabricius, 1807

World-wide in distribution, the genus includes several wide-ranging migratory species. Most occur in the Palaearctic and Nearctic regions, but a few are found in temperate and tropical latitudes of other regions. Three species occur in Australia, two of which are also known from New Zealand. The adults, which are often very abundant, have hairy eyes. The larvae have numerous short branched spines, and the head is hairy without long spines. The pupae are rather angular, with pairs of short dorsal projections and often with metallic coloration or metallic dorsal spots.

71. *Vanessa kershawi* (McCoy), 1868

Plates 20, fig. 5 (male), X, Australian painted lady.

DISTRIBUTION. Throughout much of mainland Australia, Tasmania, and Lord Howe Island; common.

Until recently this species was treated as a subspecies of the very widespread *V. cardui*, from which it may be distinguished by the male genitalia and by the blue centres to the black subterminal spots of the hindwing. As would be expected in a migratory species, little variation has been noted in specimens taken throughout Australia, although melanic or aberrant specimens without the white markings on the forewing, or with the orange areas of the hindwing partly obscured, are known.

The adults fly rapidly, but frequently settle with partly outspread wings on the ground or on flowers to feed. During migration they maintain a rapid and consistently direct flight, within about two metres of the ground, and often at a wide angle to the direction of the wind. Such flights often include specimens of *Junonia villida* or *V. itea*.

The larvae feed at night and remain concealed during the day. The young larvae hide in a shelter formed in a curled leaf of the food plant, or formed by loosely joining a couple of adjacent leaves with silk. The older larvae usually shelter beneath the food plant, on or near the ground. The larvae usually feed on Asteraceae, including the native *Helichrysum bracteatum*, *Helipterum roseum*, and *Ammobium alatum*. The introduced *Arctotheca calendula* (Cape weed), *Onopordum acanthium* (Scotch thistle), and ornamental *Artemisia* are also common food plants. E. D. Edwards found larvae on *Gnaphalium* near Sydney, and at Canberra T. E. Bellas found them on garden lavender (*Lavandula officinalis*; Lamiaceae).

72. *Vanessa cardui* (Linnaeus), 1758

Plate 14, fig. 13 (male), painted lady.

DISTRIBUTION. Within Australian limits known from Perth, Bunbury and Rottnest Island, south-western Australia. The first Australian specimens were taken many years ago by F. L. Whitlock and A. N. Burns, and W. E. Wright took a specimen in Perth in January, 1958. In November, 1974 C. G. Miller found adults flying commonly on the sand dunes near Bunbury, and others were collected

216

there by A. F. Atkins in late October, 1977. Adults were common near Fremantle in early November, 1978 (M. Hutchison).

ADULT.   Similar to *V. kershawi* but larger, with more elongate forewing, hindwing above with subterminal spots black, rarely one or two near tornus with minute blue centres.

Evidence is accumulating that this species is established in the south-west. Most specimens have been taken flying in the same situations as *V. kershawi* and some of those collected have been in very fresh condition, suggesting that they may have bred locally, and several pairs have been observed in courtship flight. *V. cardui* is a well-known migrant wherever it occurs, and the Australian specimens could have originated in Africa.

Abroad the larvae have been recorded feeding on more than a hundred different food plants, but chiefly Asteraceae, Malvaceae and legumes.

73. *Vanessa itea* (Fabricius), 1775

Plates 20, fig. 6 (male), XV, Australian admiral.

DISTRIBUTION.   Atherton Tableland, north Queensland, and throughout south-eastern, southern and south-western Australia, Tasmania and Lord Howe Island; common. It also occurs on Norfolk Island, in New Zealand, and east to Rapa Island. The species occurs wherever nettles grow, especially on the coast and tablelands, but has been taken as far inland as Cunnamulla, south-western Queensland, and Kulgera and the MacDonnell Ranges, Northern Territory. It is apparently migratory, a few specimens having been taken in migratory flights of *V. kershawi*.

Adults are remarkably constant throughout the range of the species. They frequently settle in the sunshine with wings outspread, and rest with closed wings and head directed downwards on tree-trunks and fences, the underside of the wings effectively matching the grey bark or wood. Sometimes they congregate to feed at sap flows on the trunks of eucalypts and at times may hill-top.

The larvae feed at night on stinging nettles, both the native *Urtica incisa* and the introduced *U. urens* (Urticaceae). They are found during the day in a curled leaf or in a loosely formed shelter on the plant formed by joining a couple of leaves with a few strands of silk. They seldom pupate on food plants; more usually pupae (Fig. 17E) are attached to near by walls or fences. In Tasmania larvae have been recorded feeding on the introduced ornamental *Soleirolia soleirolii* (Urticaceae).

Genus *JUNONIA* Hübner, 1819

Like *Vanessa* this genus has a very wide distribution, but is especially numerous in the Old-World tropics. It may be distinguished from *Vanessa* by the smooth eyes of the adults. Four species occur in Australia, all of which also range widely beyond this continent. The larvae have numerous short branched spines, and a hairy head with a pair of short branched spines. The larval food plants are distributed through many families including Plantaginaceae, Scrophulariaceae, Verbenaceae, Gentianaceae, Asteraceae, Portulacaceae, Convolvulaceae, Goodeniaceae and Acanthaceae. The pupae are more rounded than in *Vanessa*, but have a series of very short blunt dorsal processes.

74. *Junonia erigone walkeri* (Butler), 1901
Plate 14, fig. 1 (female).

DISTRIBUTION.   Within Australian limits known from a single female collected at Rimbija Island, Wessel Islands, Northern Territory, in January 1977. This specimen has been tentatively assigned to the Timor subspecies.

The tornus of the hindwing is less produced than in *J. hedonia*, and the termen has a stronger projection at the end of vein M_3.

75. *Junonia hedonia zelima* (Fabricius), 1775
Plate 20, fig. 4 (male), brown soldier.

DISTRIBUTION.   Northern Territory, the islands of Torres Strait, and Cape York to Southport; common north of Mackay, only occasionally collected south of Rockhampton.

This species exhibits some seasonal variation, but the extent of this in Australian populations requires study.

The larvae feed on a small herbaceous plant, *Hygrophila salicifolia* (Acanthaceae), growing in swampy areas and in gullies near the coast, usually amongst *Melaleuca* (paper bark) trees (G. Sankowsky). At Cairns they have also been recorded on *Hemigraphis alternata* (Acanthaceae), a prostrate ornamental plant with purplish leaves. At Darwin they have been found on a broad-leafed plant of this family (D. Bell).

218

76. *Junonia villida calybe* (**Godart**), **1819**

Plates 20, fig. 2 (male), XI, meadow argus.

DISTRIBUTION. Throughout Australia and Tasmania, and at Lord Howe Island; very common both in coastal areas and far inland. It occurs commonly even at Hughenden, Charleville and Cunnamulla in western Queensland, and has also been taken on Norfolk Island and in New Zealand. Migratory flights have been observed from time to time, and occasional specimens take part in flights of *Vanessa kershawi*.

The adults are somewhat variable, especially on the underside but this apparently has neither a seasonal nor a geographic basis. They are a familiar sight as they fly about in gardens or on roadways, and settle frequently on the ground with their wings outspread.

The larvae feed on various native and introduced herbaceous plants belonging to several families. They include *Plantago* spp. (plantains, Plantaginaceae), including the introduced *P. lanceolata*, *Antirrhinum* (garden snapdragon), *Russelia equisetiformis* (both Scrophulariaceae), *Verbena* spp., including *V. bonariensis* (purple-top, Verbenaceae), *Centaurium spicatum* (Australian centaury, Gentianaceae), *Epaltes australis* (Asteraceae), *Portulaca oleracea* (pigweed, Portulacaceae), *Evolvulus alsinoides* (Convolvulaceae), *Scaevola aemula* and *Goodenia* sp. (both Goodeniaceae). Near Adelaide eggs are sometimes laid on *Kickxia elatine* (Scrophulariaceae), but larvae have not been reared on this plant.

77. *Junonia orithya albicincta* **Butler, 1875**

Plate 20, figs. 3 (male), 3A (female), blue argus.

DISTRIBUTION. North-western Australia, Northern Territory, islands of Torres Strait, and Cape York to Brisbane, Millmerran and Tenterfield (W. R. Hindson); common north of Gladstone, but rare farther south.

The wings above show some variation, but beneath are extremely variable, especially in the female. Adult flight behaviour is very similar to that of the common *J. villida*. Both are able to tolerate high temperatures as they settle on hot bare earth or sand in full tropical sunlight.

The larvae feed on *Antirrhinum* (garden snapdragon), *Angelonia salicariifolia*, *Buchnera linearis* (all Scrophulariaceae), *Thunbergia alata*, *Hygrophila salicifolia*, *Asystasia scandens*, *A. gangetica* and *Pseuderanthemum variabile* (all Acanthaceae).

Genus *CETHOSIA* Fabricius, 1807

This genus of more than a dozen species ranges from India and China through south-east Asia to New Guinea and Australia. The adults are large and often brightly coloured, with strongly variegated pattern beneath the wings, and crenulate or dentate termen, especially of the hindwing. They have curious very long simple tarsal claws, and smooth eyes. The larvae, which feed on species of Passifloraceae, are brightly banded, with long tapering spines, including a pair on the head. The pupa (Figs. 17L, M) is grotesquely shaped, with various projections and expansions of the anterior and dorsal surfaces. Two species occur in northern Australia.

78. *Cethosia penthesilea paksha* Fruhstorfer, 1905

Plate 19, figs. 11 (male), 11A (female), orange lacewing.

DISTRIBUTION.   Northern Territory, including Melville Island and Groote Eylandt. It has been recorded from the Kimberley, north-western Australia, but this needs confirmation.

Sometimes the forewing is without the white band near the apex and the white markings beneath may be reduced. Although treated as the Australian subspecies by Stitchel in 1938, *doddi* Oberthur, 1911, was simply a name given to an aberrant specimen of *paksha* from Darwin.

Larvae were reared at Darwin by M. C. Hall and the above descriptions were based on his colour transparencies. He found the larvae fed gregariously on *Adenia heterophylla* ssp. *australis* (Passifloraceae) and D. Bell has also found larvae feeding on this plant.

79. *Cethosia cydippe chrysippe* (Fabricius), 1775

Plate 20, fig. 8 (male), red lacewing.

DISTRIBUTION.   Cape York to Townsville; fairly common in rain-forest clearings.

The striking colours of this fine species always attract attention as it flies or rests with wings outspread in the sunshine, or feeds at *Lantana* and other flowers. Variation in the adult is slight, but Cape York specimens have been known as subspecies *imperialis* Butler, 1876.

The larvae feed gregariously on *Adenia heterophylla* (Passifloraceae).

Genus *VINDULA* Hemming, 1934

The genus, which includes four large, strongly dimorphic species, is distributed from Sri Lanka and India through south-east Asia and the Philippines to New Guinea, Australia and the Solomons. In both sexes there is a short projection on the termen of the hindwing at vein M_3. As in *Cethosia*, the larvae are brightly banded, with long branched spines on the body and a pair of similar spines on the head. They also feed on species of Passifloraceae. The pupae (Figs. 17N, O) are even more grotesquely shaped than in *Cethosia* and resemble a distorted dead leaf. Only one species occurs in northern Australia.

80. *Vindula arsinoe ada* (M. R. Butler), 1874

Plate 20, figs. 7 (male), 7A (female), cruiser.

DISTRIBUTION.   Moa Island, and Cape York to Mackay; males are common, females much less so.

The males often fly in patchy sunlight and shade in rain-forest openings, and feed at *Lantana* and other flowers. The females sometimes fly high and are most readily caught as they feed at flowers. They are taken much less frequently than males.

The larvae feed at night on *Adenia heterophylla* and *Passiflora aurantia* (both Passifloraceae). The pupa is very sensitive to disturbance and responds by a jerky wriggling movement. Both its colour and form must have a strong protective function, but if these devices do not deter an enemy, the sudden movements of the pupa no doubt have that effect.

Genus *VAGRANS* Hemming, 1934

Confined to the Oriental and Australian regions, *Vagrans* contains a single species distributed from India to the south Pacific. It is related to *Phalanta* and *Cupha*, from which it may be distinguished by the short tail at the end of vein M_3 of the hindwing. The larva has numerous long branched spines, including a pair on the head, and the pupa (Fig. 17H) has two long slender anterior spines and three pairs of similar dorsal spines.

81. *Vagrans egista propinqua* (Miskin), 1884

Plate 20, fig. 9 (male), Australian vagrant.

DISTRIBUTION.   Moa and Thursday Islands, Claudie River and Cooktown to Ingham, north Queensland; usually not common, but adults have been taken at Cairns and Kuranda from November to April and in June.

Adults are usually found along the edges of rain forest or on tracks or in small clearings in rain forest, where they settle on low vegetation or on bare rocks or stones.

The larvae feed on *Xylosma ovatum* and *Homalium circumpinnatum* (both Flacourtiaceae) and are said to resemble those of *Cupha*.

Genus *PHALANTA* Horsfield, 1829

This small genus occurs in tropical Africa and Madagascar, and from India and southern Japan through south-east Asia to northern Australia, Aru, New Guinea, the Bismarck Archipelago and the Solomons. The larva has numerous short branched spines, but the head is without prominent spines; the pupa (Fig. 17K) has short dorsal tubercles.

82. *Phalanta phalantha araca* (Waterhouse and Lyell), 1914

Plate 19, fig. 12 (male), leopard.

DISTRIBUTION.   Northern Territory. Once regarded as a rarity, this species has been taken fairly commonly in recent years at Darwin and the Daly River, and also near the South Alligator River and on the Cobourg Peninsula (E. D. Edwards).

The adults feed at flowers and have often been observed flying close to the ground near their food plant. Sometimes, however, they fly rapidly near the tree-tops and settle on the upper parts of the trees (D. Bell).

The larval food plant is apparently not *Petalostigma quadriloculare* (Euphorbiaceae), as previously reported, but *Flacourtia indica* (Flacourtiaceae) (D. Bell). The life history was discovered by D. C. Hewson at Darwin. During May the larvae were very active, fed freely and grew rapidly. Eggs hatched in about two days, larval development was completed in another seven days, and the pupal duration was about four days. The pupae are very beautiful objects, but in their natural surroundings would be very difficult to detect. They are suspended by the cremaster beneath a leaf of the food plant at an angle of about 30 degrees to the leaf surface. The wings become deep reddish brown in colour just before the adult emerges.

Genus *CUPHA* Billberg, 1820

This genus contains about ten species distributed from India, Taiwan and the Philippines, through south-east Asia to Australia, New Guinea and the Solomons. Only one species occurs in Australia. The larva has numerous branched spines and a pair of long branched spines on the head. The pupa (Fig. 17J) has a pair of long slender anterior spines, and three pairs of dorsal spines.

83. *Cupha prosope* (Fabricius), 1775

This species occurs both in New Guinea and eastern Australia. Two subspecies are recorded from within Australian limits.

a. *Cupha prosope prosope* (Fabricius), 1775

Plate 20, fig. 11 (male), Australian rustic.

DISTRIBUTION.   Moa, Thursday and Murray Islands, and Cape York to the Clarence River. It is common north from Rockhampton and Byfield, and near Cairns adults have been collected almost throughout the year.

The adult flies in sunny openings and small clearings in rain forest and fringing forest along creeks, frequently settling on foliage near the ground, with wings expanded.

The larvae feed on *Flacourtia jangomas*, *Scolopia braunii* and *Xylosma ovatum* (all Flacourtiaceae). The striking pupa is rather similar in shape to that of *Vagrans egista*.

b. *Cupha prosope turneri* (Butler), 1876

Plate 20, fig. 11A (male).

DISTRIBUTION. Within Australian limits, only at Darnley Island; this subspecies was originally described from Port Moresby, but has been recorded from other New Guinea localities.

ADULT.   Similar to typical *prosope*, but forewing above usually without the small extensions of the outer margin of the orange band near the tornus.

Some specimens taken at Darnley Island seem to be intermediate between *turneri* and typical *prosope*.

Subfamily ACRAEINAE

(Plates 15, XV)

This subfamily is primarily tropical in distribution, and includes a few species from the Oriental and Australian regions. It also occurs in the tropics of Central and South America, but is best developed in Africa. Only one species occurs in Australia.

The adults are medium-sized with rather narrow and rounded wings, sometimes partly devoid of scales. Like the Danainae they have a rather slow and graceful flight, unless disturbed, and are apparently distasteful to predators.

Structurally the Acraeinae are more closely allied to the Nymphalinae, and like them the larvae have series of long branched spines, and the pupae, which are suspended head downwards by the cremaster, have angular projections anteriorly. The eggs have strong vertical ribs, a flattened base, and their height is greater than the diameter.

Genus *ACRAEA* Fabricius, 1807

The genus includes many African species, but is represented by only one in the Australian region.

84. *Acraea andromacha andromacha* (Fabricius), 1775

Plates 15, fig. 14 (male), XV, glasswing.

DISTRIBUTION. North-western Australia, north from Onslow (Mrs Milne), Northern Territory, including the Alice Springs area, Moa and Prince of Wales Islands, and Cape York to southern New South Wales, and occasionally to Melbourne, Victoria. It is common in northern Australia and eastern Queensland, but usually rare as far south as Sydney, although it is fairly common there in some years. A few specimens were taken in Adelaide in 1973–74 and one in Broken Hill in 1976.

The larvae feed on wild passion vines, *Passiflora* spp., including *P. alba*, and *Adenia heterophylla* (Passifloraceae). Although the eggs are freely laid on the cultivated passion vine *Passiflora edulis* and on granadilla (*Passiflora* sp.) the larvae cannot develop on these plants. However, at Melbourne where adults were taken on several occasions in 1973–74, eggs were laid on *Passiflora mollissima* (banana passion fruit) and larvae were successfully reared on this plant (W. N. B. Quick). South of Broome, W.A. (M. S. Moulds) and in central Australia the larvae have been found feeding on *Hybanthus enneaspermus* (Passifloraceae). The young larvae feed gregariously, but the older larvae usually occur singly. The pupa (Fig. 17I) is cream with black and orange markings.

Family Libytheidae

BEAKS

(Plate 20)

THIS small family contains medium-sized butterflies, all of which are usually referred to the one genus *Libythea*. They are closely related to the Nymphalidae but may be distinguished by the long and porrect, beak-like labial palpi of both sexes and the forelegs of the female.

The adults have a rapid and jerky flight, but rest frequently; males sometimes congregate at damp patches on the ground, but the females are taken much more rarely. In Africa migrations of *Libythea* have been noted. The eggs are bottle-shaped, their height being about twice their diameter, with strong vertical ribs. The larvae are more or less cylindrical, although a little broader in the middle, with a small head, and body covered with minute hairs. Superficially they resemble those of *Catopsilia* (Pieridae) rather than the Nymphalidae. The pupa, however, is suspended only by the cremaster, as in the Nymphalidae, and is smooth and rather angular.

Only one species occurs in Australia, and the whole family includes little more than a dozen species.

Genus *LIBYTHEA* Fabricius, 1807

The genus is widespread but in each region is represented by few species. It occurs in southern Europe, Africa, the Oriental and Australian regions and in America. The adults are said to favour drying river beds and, in dry weather, sometimes congregate at wet spots on roads.

1. *Libythea geoffroy* Godart, 1820

The species has a wide distribution from Burma, Thailand, Borneo and Java eastwards to New Guinea and northern Australia, the Bismarck Archipelago, the Solomons, New Caledonia and the Loyalty Islands. It displays strong sexual dimorphism. Two subspecies have been recognized from Australia.

a. *Libythea geoffroy genia* Waterhouse, 1938

Plate 20, figs. 12 (male), 12A (female), Australian beak.

DISTRIBUTION. North-western Australia and Northern Territory. Specimens have been taken at Cape Leveque (G. Thomas), Windjana Gorge, 130 km east of Derby (M. S. Moulds), Condillac and Cassini Islands, Prince Regent River and Wyndham, W.A., and at Darwin and Cape Wessel (E. D. Edwards), N.T.

The male is somewhat variable. The amount of violet-blue on the forewing may be reduced, when the white spots then appear more distinct. The intensity of the orange band and the extent of the violet-blue of the hindwing also varies, specimens with more violet-blue having a less distinct orange band.

During the heat of the day the adults tend to rest in deep shade: at Wyndham, W.A., in March several were taken at rest beneath the verandah of a house, and at Windjana Gorge, W.A., M. S. Moulds discovered several at rest in the shade of densely foliaged trees, where they were difficult to distinguish from dead leaves.

The early stages have not been described.

b. *Libythea geoffroy nicevillei* Olliff, 1891

DISTRIBUTION. Warraber (Sue) (R. C. Lewis) and Moa Islands, and Cape York to Cairns, Herberton, and Magnetic Island. Only a few specimens have been taken, mainly at Cape York, Coen, Cooktown, Cairns, Chillagoe and Herberton. The subspecies has also been reported from Magnetic Island, near Townsville (D. P. Sands), and from Yeppoon (B. Clark), but these records need confirmation.

ADULT.   Male: similar to *genia* but larger; forewing with smaller less distinct white spots and more prominent dark brown veins, hindwing with orange band less distinct. Beneath hindwing pale grey marbled with purple.

Female: similar to *genia* but larger; forewing with smaller and less distinct white spots; hindwing pale orange at base, with less prominent orange band, its outer margin less irregular. Beneath without purple marbling of male.

Family Lycaenidae

BLUES AND COPPERS
(Plates 13, 14, 20–29, XV–XX; Figs. 18, 19)

THE majority of butterflies in this family are small or very small, but a few are larger, especially in the Australian region. They are noted for their brilliant metallic blue, violet or orange coloration, and sometimes even green, brown or white. The females are usually somewhat duller than the males, but many are equally brightly coloured. The undersides of the wings are often cryptically marked, but sometimes they are patterned in lustrous green or red. The species are most numerous in the tropics, and those of northern Australia usually belong to exotic tropical Oriental and Papuan genera. In addition Australia has some interesting endemic genera, with few or no New Guinea representatives. The adults often fly rapidly and settle, with wings closed, high on the foliage of trees; others have a weak erratic flight close to the ground; the males of some tend to collect on hill-tops where they may exhibit territorial behaviour. With one exception they all fly in sunshine and soon settle when clouds obscure the sun. The exception is *Liphyra brassolis* which sometimes flies at dawn or dusk. Migratory flights have been recorded in only a few species, such as *Lampides boeticus*.

The hindwings in many species bear one or more flimsy slender tails, near the tornus, usually associated with one or more prominent tornal spots. These probably have a protective function, as they tend to direct any attack from a predator towards the posterior angles of the hindwings, parts that can be lost without disadvantage to the insect itself.

The effect is enhanced by the common habit of these lycaenids to settle with their heads directed downwards, and to move the hindwings back and forth against one another.

Most Lycaenidae exhibit sexual dimorphism and in some it is so strongly developed that it is difficult to associate the sexes correctly.

The males of some species have sex-brands or areas of modified scales on the wings, or special sex-scales (androconia) may be scattered amongst the normal scales of the upper wing surface.

The larvae are usually onisciform (shaped like a wood-louse or slater), with head largely hidden beneath the prothorax. The cuticle is relatively much thicker than in other lepidopterous larvae. The hairs or setae are often greatly reduced or inconspicuous, but the larvae of many species are clothed with dense short or minute secondary setae which may be branched or clubbed, or even have expanded stellate tips (Fig. 18). Sometimes there are also paired tubercles. The larvae of most species are associated with ants, some even being found during the day within the ants' nests, from which they crawl to their feeding places each evening attended by the ants. In these larvae there are

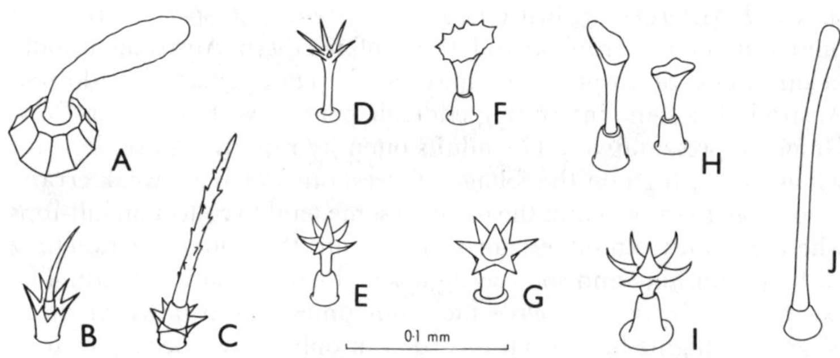

Fig. 18. Secondary setae in larvae of Lycaenidae: A, *Ogyris amaryllis* (p. 279); B, *Theclinesthes onycha* (p. 324); C, *Catopyrops florinda* (p. 319); D, *Hypochrysops byzos* (p. 252); E, *H. digglesii* (p. 258); F, *H. delicia* (p. 243); G, *H. apelles* (p. 254); H, *H. ignitus* (Little Desert, Vic.) (p. 245); I, *H. ignitus* (Yeppoon, Q.) (p. 245); J, *Lucia limbaria* (p. 231).

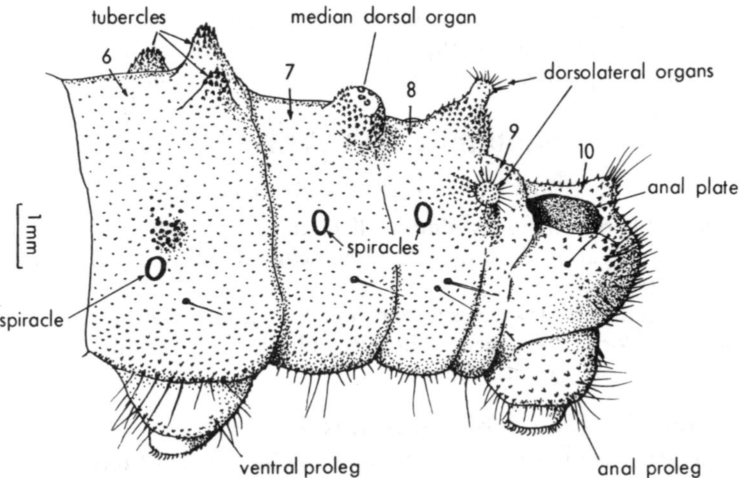

Fig. 19. Dorsal and dorsolateral organs in Lycaenidae: abdominal segments 6–10 in larva of *Jalmenus evagoras*.

scattered epidermal glands which are believed to secrete ant-attractant substances. Most also have a median dorsal organ, sometimes called the "honey gland", on the seventh abdominal segment (Fig. 19) which secretes a fluid eaten by the attendant ants. For long it has been suggested that, in return, the ants provide the larva with some measure of protection, but this may be negligible.

In addition there is usually a pair of dorsolateral eversible organs on the eighth abdominal segment (Fig. 19), sometimes opening at the top of dorsolateral projections.

The larvae of Lycaenidae (Plates XV–XX) usually feed on young terminal foliage, or on buds, or flowers, but others feed within fruits, or on seeds within the pods of legumes. Should their food supply be exhausted the larvae of many species become cannibalistic. A few are normally carnivorous, feeding on aphids, coccids and leaf hoppers; some feed upon the larvae and pupae of ants. Rarely the larvae are fed by the ants within their nests.

The pupae (Plates XV–XX) are usually stout and rounded in cross-section, and may be smooth, rugose, or covered with short erect hairs, or minute tapering, clubbed or even branched spinules.

Subfamily LYCAENINAE

(Plates 13, 14, 21–29, XV–XX; Figs. 18, 19)

The subfamily includes almost all of the Australian species. They range in size from the large *Liphyra brassolis* to the tiny *Freyeria trochylus* and *Zizula hylax*.

Altogether 134 species are recognized from Australia, divided amongst forty-five genera. The subfamily is predominantly tropical and subtropical in distribution.

Genus *LIPHYRA* Westwood, 1864

The genus contains two species, one of which ranges from northern India through south-east Asia to the Moluccas, northern Australia and Papua, whereas the other is confined to New Guinea. The adults are large stout-bodied species with orange and black wings, and fly like a large lycaenid. They are said to be active mainly at dawn and dusk, but have often been observed and collected at other times of the day.

The larvae are oval and flattened, with a hard leathery skin, which protects their soft legs and prolegs, and beneath which the head may be withdrawn to avoid attack from the green tree-ants, *Oecophylla smaragdina*, in the nests of which the larvae live. The stout pupa is formed within the hardened skin of the mature larva.

1. *Liphyra brassolis major* Rothschild, 1898
Plates 25, fig. 20 (male), XVII, XVIII, moth butterfly.

DISTRIBUTION.   Northern Territory, including Groote Eylandt, Moa Island, and Cape York to Mackay; not common. It may occur even as far south as Yeppoon, where the green tree ant *Oecophylla* is plentiful. This subspecies also occurs in Papua and New Britain.

In freshly emerged specimens (Plate XVIII, fig. 2) the legs, antennae, head, thorax and the upperside of the wings are covered with greyish white scales, and the abdomen is clothed with long brown woolly lateral scales. These are lost while the adult is escaping from the ants' nest or during subsequent flight.

230

The eggs hatch in about three weeks. The larvae are found within the nests which *Oecophylla* ants construct amongst the foliage of ever-green trees by sewing many leaves together with silk spun by the ant larvae.

When the larva is due to moult it attaches the crochets of its prolegs to a pad of silk and about three or four days later its old skin splits downwards under the front edge and in each direction around the margin. The cast skin largely retains the shape of the larva and might easily be mistaken for a live larva. When it reaches maturity the larva again attaches its crochets to a pad of silk and, instead of its skin splitting, it becomes strongly convex above and beneath and acquires a hard and brittle texture. The larval skin is completely moulted and merely has the function of protecting the enclosed thin-skinned pupa from ant attack.

The adult emerges in about three weeks. After the pupal skin has been shed, the anterior dorsal segment of the protective puparium breaks open to release the very moth-like adult (Plate XVIII, fig. 2). At first its wings are more abbreviated than is usual in butterflies and are covered with greyish white deciduous scales. The abdomen appears greatly enlarged and the antennae very thick, owing to their covering of long woolly deciduous scales. These scales become entangled in the mandibles of any attacking ants allowing the butterfly to escape without injury.

Genus *LUCIA* Swainson, 1833

The genus contains a single Australian species, which is widely distributed in eastern Queensland and New South Wales, and in Victoria and south-eastern South Australia.

2. *Lucia limbaria* Swainson, 1833

Plates 27, figs. 7 (male), 7A (female), XV, small copper.

DISTRIBUTION. Mackay, Queensland, to Victoria and south-eastern South Australia; local. It occurs near the coast, but is more common inland. In southern Queensland it has been taken as far west as Augathella (M. De Baar) and in New South Wales as far as the Warrumbungle National Park and Narrandera (E. D. Edwards). In South Australia it has been recorded from Brinkworth, near Clare, to

Adelaide and the Mt Lofty Ranges. A specimen was recorded last century from Lord Howe Island, but there are no more recent records.

The adults are inconspicuous as they fly close to the ground and settle frequently on the surface or on herbage. They also feed at the flowers of low-growing plants such as *Helipterum* and *Helichrysum*.

The larvae usually feed during the day, but sometimes in drier more inland localities at night, on *Oxalis corniculata* (yellow wood sorrel, Oxalidaceae) and both larvae and pupae are attended by numerous small black ants *Iridomyrmex* sp. (*gracilis* group), which excavate tunnels up to five or six centimetres in depth in the soil at the base of the *Oxalis* plant. The larvae are found during the day in the ant galleries, or sometimes under stones or logs if the food plant happens to be growing beside them. The pupae, which are rather delicately formed and easily damaged, are found in the same situations.

Genus *ACRODIPSAS* Sands, 1980

The genus contains seven small, usually rather dull-coloured species referred until recently to *Pseudodipsas*.

All the species except *A. illidgei* are known to hill-top in areas of dry eucalypt forest. The larvae of three of the species are believed to be predatory on the larvae and pupae of ants.

3. *Acrodipsas brisbanensis* (Miskin), 1884

This rare species, known from south-eastern Queensland, New South Wales, the Australian Capital Territory, and eastern and central Victoria, was once confused with the closely related *A. cuprea*. Two subspecies are recognized.

a. *Acrodipsas brisbanensis brisbanensis* (Miskin), 1884

Plate 23, fig. 6 (male), large ant-blue.

DISTRIBUTION.   Southern Queensland and New South Wales; usually rare. Specimens have been taken at Theodore (C. G. Miller), Bunya Mts (M. De Baar), Mt Tibrogargan (B. Cantrell) and Millmerran (J. Macqueen); the original specimen, a female, came from Brisbane, where the species must be rare. Most specimens have been collected in the Sydney area and the Blue Mts. Specimens have also been taken near Armidale (C. W. Frazier), near Grafton (C. G. Miller), near Queanbeyan (W. Graham, S. Brown), and at Canberra (M. Fletcher).

Superficially this species is very similar to *A. cuprea*. In the forewing of males, however, the basal half is never coppery, the termen is usually nearly straight and the postmedian band beneath is broken at M_3. The markings beneath both wings are usually narrower and more obscure in *A. brisbanensis*. The genitalia are also distinct. Very few females of either species have been taken, but the markings beneath the wings of female *A. brisbanensis* are narrower and less conspicuous than in *A. cuprea*. The hindwing of both species, especially in males, is squarer than in the other species of the genus. Males of *A. brisbanensis* have usually been taken flying very rapidly around eucalypts on the tops of small steep hills, settling on the foliage. Females have also been taken visiting the same hill-tops. A male and a female from Millmerran were taken on flowers, and a female was taken near Canberra feeding at *Bursaria* flowers (S. Brown). Despite careful search the early stages have not yet been discovered.

b. *Acrodipsas brisbanensis cyrilus* (**Anderson and Spry**), 1897

DISTRIBUTION. Victoria; usually very rare. Specimens are known from near Moe, Springvale, Cranbourne, near Broadford, and Flinders Peak in the You Yangs. Most specimens have been collected in December and January.

ADULT. The subspecies was distinguished from typical *brisbanensis* by the spots and bands beneath the wings being very narrow. However, these slight differences may prove to be so variable that the subspecies cannot continue to be recognized.

4. *Acrodipsas cuprea* (**Sands**), 1965

Plate 22, figs. 5 (male), 5A (female), cuprea ant-blue.

DISTRIBUTION. Central and southern Queensland to central Victoria; not uncommon locally. Adults have been collected at the Expedition Range, Isla Gorge, Mt Tinbeerwah, near Noosa, Toowoomba, at several localities in New South Wales including Grafton, Boambee, Port Macquarie, Newcastle, Toronto, West Head, National Park, Lawson, Blackheath and Pambula, near Canberra in the Australian Capital Territory, and in the Wellington River valley, north of Heyfield, near Moe, and at Launching Place and Yarra Junction, Victoria. A female has been taken at 1,750 m on Mt Ginini in the Brindabella Range, A.C.T. (K. Omoto).

Males of this species may be separated from the closely related *A. brisbanensis* by the more convex termen of the forewing, the broader bands beneath the wings, rendered more prominent by the white edging, the more or less straight postmedian band beneath the forewing, and the genitalia. Males taken south of Newcastle all have the basal half of the forewing above distinctly coppery, but in those from farther north the forewing is usually uniformly dark brown. However, a series taken at Toowoomba by J. Macqueen includes some with the upperside more or less suffused with violet. Females are much less commonly taken, but they too have the markings beneath the wing broader and more conspicuous than in *A. brisbanensis*, and the blue areas above are usually more violet in colour and more restricted in *A. cuprea*. Adult males have usually been taken flying around eucalypts on the summits of small steep hills or ridges, and females have also been taken visiting these hill-tops.

Eggs are laid in groups of two or three on tree trunks infested with small black ants (apparently *Crematogaster* sp.) and are probably collected by the ants and taken to their nests in galleries made in the tree by long-horned beetles. At first the larvae are believed to be fed by the ants, and later to attack and eat the ant larvae and pupae. C. G. L. Gooding reared larvae and pupae from borer holes in the trunk of an *Acacia baileyana*, which also carried larvae of *Hypochrysops delicia* attended by *Crematogaster* ants. On another occasion he found freshly emerged adults on a tree stump infested with small black ants.

5. *Acrodipsas myrmecophila* (Waterhouse and Lyell), 1913

Plate 22, figs. 6 (male), 6A (female), small ant-blue.

DISTRIBUTION. Central Queensland to Victoria; known from Theodore, Expedition Range, near Kumbia, Mitchell, Millmerran and Brisbane, Q., near Grafton, Dungog, Sydney, Stanwell Park, and near Deniliquin (J. C. Le Souef), N.S.W., and Broadford, Ringwood and Ocean Grove, Victoria; rare. It probably occurs at intervening localities.

At Ocean Grove the eggs are laid in batches of up to fifty on dead stumps of *Acacia pycnantha* containing colonies of a small black ant stated to be *Iridomyrmex nitidus*. Until the eggs hatch about four weeks later, they are attended by the ants. The larvae are found within the ant nests and their movements are said to be more rapid than in other

234

lycaenid larvae. The pupae are attached in groups of eight or ten to the walls of the ant galleries or in smaller groups in cracks of the dead tree stumps. Ants were observed running over the freshly emerged adults without disturbing them in any way. Adults have also been reared at Mitchell from pupae found under stones at the foot of "bindee" trees. Larvae were present beneath the stones and in cracks at the foot of the trees, attended by large numbers of small black ants.

Little is known about the feeding behaviour of the larvae but, from the following very interesting observations made by J. Macqueen at Millmerran, they are predators of the ant larvae and pupae. *A. myrmecophila* larvae were introduced to a captive ant colony containing ant larvae and pupae. There was no evidence that the ants fed the butterfly larvae but, by the time the latter pupated, the ants had heaped up many collapsed ant pupae. At one stage when ant larvae and pupae were in short supply, the butterfly larvae were given slices of apple. There was no sign of feeding on the apple, although the larvae made small hollows from which they apparently sucked the apple juice. This and the collapsed ant pupae suggest that the butterfly larvae feed by sucking fluids from their prey. Clearly much could be learnt about the life history of *A. myrmecophila* and *A. illidgei* by studying their behaviour in artificial ant colonies.

6. *Acrodipsas illidgei* Waterhouse and Lyell, 1914

Plate 23, fig. 7 (female), Plate 29, fig. 5 (male), Illidge's ant-blue.

DISTRIBUTION.   Brisbane to Burleigh Heads, Queensland; usually rare but many taken recently near Burleigh (S. Johnson) and Redland Bay (C. Hagan). Adults fly September to December and in February.

This species was originally regarded as a subspecies of *A. myrmecophila*. However it is larger, has much broader markings beneath the wings and very different male genitalia. The larvae and pupae are also associated with a different species of ant.

The larvae and pupae have been found within the nests of the small black ant *Crematogaster laeviceps*. The ant nests were in a hollow branch of a mangrove at Burleigh and under patches of bark on the trunk of a bloodwood (*Eucalyptus*) at Southport. Only three or four larvae or pupae were present in each colony and they were usually surrounded by the larvae and pupae of the ants. The larvae were sluggish and, when rearing was attempted in the absence of ants, died. In September, 1979 S. Johnson found a pupa lying unattached in a hollow twig of a mangrove near Redland Bay.

The food requirements of the larvae, as in *A. myrmecophila*, are still not understood, but it is very probable that they are predacious on the immature stages of the ants.

7. *Acrodipsas arcana* (Miller and Edwards), 1978

Plate 14, fig. 5 (male), Plate 29, fig. 12 (female).

DISTRIBUTION. Known so far from only one hill-top about 24 km west of Grafton, New South Wales, and from Theodore, Queensland (C. G. Miller). Adults are on the wing from September to April.

With *A. hirtipes* and *A. melania*, *A. arcana* belongs to the *A. illidgei* group of species, all of which have swollen fore and hind femora and very short mid tibiae. Both *A. illidgei* and *A. arcana* differ from the northern *A. hirtipes* and *A. melania* by the ground colour beneath being brown rather than greyish. The male forewing in *A. arcana* is more pointed and its termen less convex than in *A. illidgei*. The females of *A. arcana* are without the broad blue central area of both wings found in *A. illidgei*. In both sexes the bands beneath the wings are broader and darker in *A. illidgei* and show less displacement at the veins.

The males of *A. arcana* show a strong tendency to hill-top, and have been taken flying in company with *A. brisbanensis*, *A. cuprea* and *A. myrmecophila*. The early stages are unknown.

8. *Acrodipsas hirtipes* Sands, 1980

Plate 29, figs. 3 (male), 4 (female).

DISTRIBUTION. Known so far only from hill-tops near Coen, northern Queensland. Adults were taken in November.

ADULT. Male: above forewing brown, basal area reaching end of cell and tornus coppery brown, veins in outer half of wing and narrow terminal line dark brown; hindwing coppery brown, basal area dark brown, termen narrowly black, a narrow light blue subterminal line and a pair of black subtornal spots. Beneath light grey-brown, bands narrowly brown, hindwing with pale blue subterminal line and two orange subtornal spots centred black.

Female: above dull brown-black, termen of both wings black, hindwing with two black subtornal spots inwardly edged with blue scales, a narrow blue subterminal line near tornus, fringe chequered light grey and black. Beneath a lighter grey than in male, bands darker brown; hindwing with pale blue subterminal line near tornus and three orange subtornal spots, two centred with black.

This is a small species, with rounded wings as in *A. myrmecophila*. The club of the antenna is shorter and relatively broader than in other species of the genus. The coppery brown colour of the males is rather similar to that of *A. cuprea* from southern New South Wales, and beneath the greyish ground colour is similar to that of *A. melania*. The dull black coloration of the female above is unlike that of any other species.

Specimens have been taken on a hill-top in company with *A. melania*.

9. *Acrodipsas melania* Sands, 1980

Plate 29, figs. 13 (male), 14 (female).

DISTRIBUTION.   Known from near Coen, northern Queensland, and one specimen from Cape York. Most specimens have been taken in November.

ADULT.   Male: above grey-brown, both wings with bar at end of cell and termen narrowly dark grey; hindwing with a narrow light blue line near termen and a pair of obscure black subtornal spots. Beneath light grey-brown, bands broad, light brown edged darker brown and obscurely light grey; hindwing with pale blue subterminal line and two orange subtornal spots centred with black.

Female: above forewing dark grey, a broad central area reaching base and inner margin light blue, a large dark grey spot at end of cell and a dull black terminal line; hindwing light blue, costa and termen broadly dark grey, a dark grey spot at end of cell, a dark grey postmedian band, and a dull black terminal line, a narrow blue subterminal line and three dark grey subterminal spots near tornus. Beneath light grey, bands dark brown, postmedian band of both wings broader than in male.

The forewing in the male is pointed and the termen is nearly straight. The greyish ground colour of the wings beneath is similar to that of *A. hirtipes*, but differs from other species. The female is rather similar to *A. illidgei* above, but is a paler blue; the postmedian band on the hindwing above is absent in *A. illidgei*. The swollen fore and hind femora, and short mid tibiae, show that the species belongs to the *A. illidgei* species group.

The species has been taken hill-topping in company with *A. hirtipes*.

Genus *PARALUCIA* Waterhouse and Turner, 1905

The genus contains three species, which are confined to Australia.

10. *Paralucia pyrodiscus* (Rosenstock), 1885

This species is distributed from Queensland to Victoria. It is usually more common than *P. aurifera* away from the coast, and its distribution extends farther north. It does not occur in Tasmania.

a. *Paralucia pyrodiscus pyrodiscus* (Rosenstock), 1885

Plate 27, fig. 6 (male), dull copper.

DISTRIBUTION. From Duaringa and the Expedition Range, central Queensland, to Lakes Entrance, Victoria; sometimes common locally. In New South Wales it has been recorded inland as far as the Warrumbungle National Park, and recently in the Australian Capital Territory (W. Graham).

Both sexes resemble *P. aurifera* in shape, but are larger, with orange-yellow central areas and broader postmedian band beneath the forewing. The adults have a rather jerky flight, mainly in the vicinity of the larval food plant *Bursaria*, growing in moister situations in areas of relatively low rainfall.

The larvae feed at night on the foliage of *Bursaria spinosa* (blackthorn, Pittosporaceae), and are attended by small black ants, *Notoncus* sp. They shelter during the day in the ants' nest at the base of the plant, being found resting chiefly on the stem or main roots. Pupation usually occurs in the ants' nest.

b. *Paralucia pyrodiscus lucida* Crosby, 1951

DISTRIBUTION. Central and western Victoria. The type locality is Melbourne, but specimens from Kiata and Dimboola, western Victoria, approach this subspecies rather than *pyrodiscus*. At Eltham adults are on the wing from early December to early February, but are most numerous in late December.

ADULT. Male: similar to *pyrodiscus*, but both wings with a shining orange-yellow area, in hindwing triangular, occupying tornal half of wing.

238

Female: similar to *pyrodiscus*, but central area of forewing brighter, hindwing with variable orange-yellow central area, rarely reaching almost to base, and sometimes reaching termen, subterminal dark brown spots prominent.

The immature stages and life history are very similar to those of *pyrodiscus*. The larvae likewise feed on *Bursaria spinosa*.

11. *Paralucia aurifera* (Blanchard), 1848

Plate 27, figs. 5 (male), 5A (female), bright copper.

DISTRIBUTION.   Southern Queensland to central Victoria and northern, eastern and southern Tasmania; common but local. Specimens have been taken as far north as Blackbutt and the Bunya Mts (M. De Baar), Queensland, some 400 km farther south than Duaringa, the known northern limit of *P. pyrodiscus*. In the A.C.T. the species occurs at altitudes up to 1,200 m.

The orange areas in both sexes above are variable in size and shape, especially in the female, and the markings beneath the wings are also variable. Adults have a rapid jerky flight, as in *P. pyrodiscus*, but settle frequently, often on small shrubs, twigs, stones or even on the ground. They are usually found flying near *Bursaria*.

In spring the eggs hatch in from eight to twelve days. The larvae are attended by small black ants (*Iridomyrmex ?nitidiceps*) and feed at night usually on the foliage of *Bursaria spinosa* (blackthorn, Pittosporaceae). E. D. Edwards recently discovered larvae feeding on *Citriobatus pauciflorus* (also Pittosporaceae) near Sydney. During the day the larvae are usually found in the ants' nests at the base of the food plant. Pupae may also be found in the ants' nests or under dead leaves on the ground. Larvae reach maturity in from six to seven weeks in early summer and the pupal duration is about three weeks. At least two generations are completed annually.

12. *Paralucia spinifera* Edwards and Common, 1978

Plate 14, fig. 14 (male), Plate 29, fig. 7 (female).

DISTRIBUTION.   Known so far from only a small area about 20 km east of Bathurst, New South Wales, at 1,000 m. Adults were collected in October.

Adults of both sexes are smaller than the other species of *Paralucia* and differ in having a long apical spine on the fore tibia. The species is also distinguished by the purple central areas and the shape of the wings in the male, and the brown or purple basal areas of the wings in the female.

Adults have been observed flying close to the ground near the larval food plant. At rest they bear a close resemblance to *Neolucia agricola*.

The larvae feed on *Bursaria spinosa* (Pittosporaceae) growing in eucalypt woodland. During the day young larvae occur on the food plant but older larvae and pupae are found in the nests of small black ants at the base of the *Bursaria* shrubs (J. R. Turner). In captivity a larva matured in about forty-eight days, but the pupal duration is not known. There may be only one generation annually.

Genus *PSEUDODIPSAS* C. and R. Felder, 1860

The genus contains small, fast-flying species restricted to rain forest.

13. *Pseudodipsas eone iole* Waterhouse and Lyell, 1914

Plate 22, figs. 7 (male), 7A (female), eone blue.

DISTRIBUTION.   Prince of Wales Island, and Cape York to the Atherton Tableland and Paluma; uncommon. At Kuranda adults have been taken from September to January and in March and April.

Mature larvae were found during the day on the lower surface of the large juvenile leaves of young *Clerodendrum cunninghamii* (Verbenaceae) near Paluma, north Queensland. They ate irregular holes in the leaves and were attended by a few small black ants, *Iridomyrmex gilberti*. Pupation occurs in curled dead leaves on the ground near the foot of the food plant. At Cairns larvae have been recorded on *Faradaya splendida* (Verbenaceae); at the Claudie River larvae fed on the leaves and young shoots of *Smilax* (Smilacaceae) and were attended by small black ants (M. De Baar).

14. *Pseudodipsas cephenes* Hewitson, 1874

Plate 22, figs. 8 (male), 8A (female), cephenes blue.

DISTRIBUTION. Kuranda and the Atherton Tableland, Queensland, to near Iluka, New South Wales; usually rare near the southern limits of its range, but it has been taken in fair numbers at Broken Head, N.S.W., in February and March, and also in June and July.

The male has very pointed forewings and the tornal area in both sexes is produced. The extent of the blue areas in both sexes varies greatly; males are known with the basal areas bearing only scattered blue scales. This is a rain-forest species.

The eggs are laid singly or in groups of two or three on the twigs and leaf petioles of the food plant. The larvae vary somewhat in the brightness of their colours. They are found during the day on the lower surface of the leaves of the food plant, or in sheltered positions along the twigs or branches, or in holes in the trunk, but pupae have been found in a hollow branch. The larvae and pupae are attended by the dull black ant *Iridomyrmex gilberti*. On the south coast of Queensland food plants include *Diospyros fasciculosa* (Ebenaceae) and *Smilax australis* (Smilacaceae). On the latter they were found associated with the larvae of *Hypochrysops miskini*, both sheltering in the same situations and attended by the same species of ant. A larva of *P. cephenes*, which was taken to Sydney, accepted leaves of the mistletoe *Dendrophthoe vitellina* (Loranthaceae) and was reared successfully.

Genus *HYPOCHRYSOPS* C. and R. Felder, 1860

The genus is distributed from the Moluccas, Lesser Sunda Islands and Timor to the Aru Islands, Australia, New Guinea, the Bismarck Archipelago and the Solomons, but also includes a single species in Thailand, Malaya and Borneo. It is well developed in New Guinea, but almost equally so in Australia where eighteen species are now known, and others will almost certainly be discovered. It is absent from Tasmania.

Some of the most brilliant and beautiful Lycaenidae belong here. The wings above are various shades of shining blue, purple, orange or green, and beneath have metallic green markings or bands of spots that are often bright red in colour and outlined with metallic green or blue. When the adults are seen in flight or at rest little idea is obtained of their brilliance. The species are usually extremely local in their occurrence.

The larvae are usually attended by ants and during the day are often found in the ants' nests at the foot of the food plant, in cracks or crevices, in holes produced by boring insects, or in curled or joined leaves. The pupae occur in similar places, attached by anal hooks and a central girdle. Many plant genera are used by the larvae as food, including a fern (Pteridophyta), the only known food plant of an Australian butterfly outside of the Gymnospermae and Angiospermae. The life history of some species is still unknown.

15. *Hypochrysops theon medocus* (Fruhstorfer), 1908

Plate 23, figs. 3 (male), 3A (female).

DISTRIBUTION.   Prince of Wales Island, and Cape York to the Claudie River; not common.

This striking species can be readily identified in both sexes by the pattern beneath the wings. The adults fly at rain-forest margins, settling frequently on foliage well above the ground.

The larvae feed on the epiphytic or terrestrial fern *Drynaria quercifolia* (Polypodiaceae), which has large deeply dissected fronds (G. Daniels). The younger larvae are thought to feed within galleries in the rhizome of the fern, occupied by colonies of the small ant *Iridomyrmex cordatus*. Older larvae feed at night on the fronds attended by numerous ants, and shelter during the day amongst debris at the base of the fern. Larvae reach maturity in about twenty-four days. Pupae have been found amongst the debris or partly concealed in the rhizome of the fern.

16. *Hypochrysops hippuris* Hewitson, 1874

Plate 23, fig. 4 (male), Plate 29, fig. 9 (female).

DISTRIBUTION.   Claudie River, Cape York Peninsula.

This is another brilliant but poorly known species easily distinguished by the pattern of the wings beneath. The female is smaller than *H. theon*, with a larger white patch on the forewing above, a more extensive white costal area of the hindwing, and a larger metallic blue basal area. With more persistent collecting at Cape York, the distribution of this fine species in far northern Australia will probably be extended.

According to J. F. R. Kerr, who has collected many of the known specimens, the males are very sluggish and appear to spend most of their time settled on the foliage of trees in the rain forest. They flew only when the tree on which they rested was violently shaken and even then returned promptly to settle once again on the tree. The life history has not been discovered, but G. Daniels collected a female that had been settling on and near a bird's nest fern, and this may possibly be the food plant.

17. *Hypochrysops halyaetus* Hewitson, 1874

Plate 21, figs. 12 (male), 12A (female), western jewel.

DISTRIBUTION. South-western Australia, from about 160 km north of the Murchison River to Perth. A few specimens are known from the Peron Peninsula and from as far inland as Wongan Hills. Adults have been taken from August to November, but mainly during November, and may be fairly common locally.

This is an easily recognized species, with brilliant colour both above and beneath the wings. Females from the Geraldton district have an orange-red subterminal line on the forewing above, whereas a long series from Chittering, about 65 km from Perth, lack this line. Adults have usually been captured flying around wattles (*Acacia*), or feeding at the small yellow flowers of *Verreauxia reinwardtii* (Goodeniaceae).

Larvae were found at Geraldton, W.A., in August 1972 by R. Grund. They were feeding on *Acacia xanthina* (Mimosaceae), and were attended by small brown and black ants, *Crematogaster* sp. Larvae were also discovered by R. Grund on *Jacksonia* sp. (Fabaceae). The pupa has not been described, but is probably very similar to *H. delicia*.

18. *Hypochrysops delicia* Hewitson, 1875

This species occurs widely in coastal and subcoastal areas from northern Queensland to Victoria. Three subspecies are recognized, differing principally in the width of the black marginal areas of the wings; these tend to be wider in specimens from southern localities.

a. *Hypochrysops delicia delicia* Hewitson, 1875

Plate 21, figs. 8 (male), 8A (female), blue jewel.

DISTRIBUTION. Southern Queensland to the Illawarra district, New South Wales, on the coast and tablelands; not common and extremely local. In Queensland adults have been taken as far north as Maryborough. At Sydney there are two generations annually; the adults fly mainly from late October to December, and again from February to mid-April (D. P. Sands).

This subspecies has the black wing margins of intermediate width, although this and the shade of the blue central areas vary to some extent. Specimens from Millmerran (J. Macqueen) have paler blue areas, especially in the female, but are nearer typical *delicia* than *duaringae*. There is also some variation in the size and extent of the orange-red markings above, especially in the female. Occasionally females have almost a complete orange-red subterminal line on the hindwing, and sometimes this line is also present on the forewing. The adults are difficult to capture and indeed are very inconspicuous as they fly rapidly around the tree-tops, usually settling well out of reach. However, many specimens have been reared from larvae and pupae. The males fly on hill-tops.

The larvae feed on several species of *Acacia* (Mimosaceae). In coastal southern Queensland and on Stradbroke Island they have been reared from *A. cunninghamii*, at Toowoomba from *A. mearnsii*, at Stanthorpe from *A. irrorata*, and at Sydney from *A. implexa*. The eggs are laid on the trunks of the wattles, usually upon large trees already infested with wood borers. The larvae are attended by numerous small black ants, *Crematogaster* sp., which have their nests in cracks, under loose bark, or more especially in the tunnels made by wood-boring insects in the trunk, branches and twigs. During the day the larvae shelter in these places and at night move out to the foliage to feed. They pupate in the same situations, several pupae sometimes being found together in the one shelter.

In cool areas there is an annual life cycle, eggs being laid from October to December, the larvae growing slowly through the summer, autumn and winter, and pupating in late winter and early spring. In such areas larval growth is very slow, or is suspended altogether, during the coldest months, but is resumed as soon as temperatures increase or if larvae are taken indoors. Breeding may be confined to only a few trees in any one area, but the larvae may be found there year after year.

b. *Hypochrysops delicia delos* (Waterhouse and Lyell), 1914
Plate 21, fig. 9 (male).

DISTRIBUTION. Southern tablelands of New South Wales, the Australian Capital Territory and Victoria; very local. Adults have been taken near Braidwood, at Canberra, and several localities in Victoria, as far west as the You Yangs, north of Geelong (W. N. B. Quick) and the Grampians (K. L. Dunn). They fly mainly from November to January, but occasionally later.

ADULT.   Similar to typical *delicia* in both sexes, but above with broader black wing margins and the orange-red subterminal markings sometimes reduced.

As in typical *delicia* adults are extremely inconspicuous in flight, but a few have been taken in eastern Victoria feeding at buddleia flowers, and some have even been captured in a light trap at Canberra. As in typical *delicia* the males tend to fly on hill-tops.

The life history and early stages are similar to those of typical *delicia*. Larval feeding and growth are almost suspended during the winter. In Victoria the food plants include *Acacia mearnsii*, and *A. melanoxylon*. Near Canberra adults have been reared from larvae and pupae collected on *A. mearnsii* and attended by *Crematogaster* sp., but a larva has been taken on *A. melanoxylon*.

c. *Hypochrysops delicia duaringae* (**Waterhouse**), 1903
Plate XV.

DISTRIBUTION.   Northern and central Queensland; so far known from near Mt Garnet, and from Duaringa, Westwood, Stanwell, Bluff, Expedition Range, Spectacle Creek and Isla Gorge.

ADULT.   Similar to typical *delicia* in both sexes, but with much narrower black margins to the wings and paler, more brilliant, central areas.

The life history and larvae and pupae (Plate XV, figs. 8, 9) resemble those of typical *delicia*. Near Mt Garnet they were discovered on *Acacia flavescens* during April. The young larvae sheltered during the day in curled phyllodes on the tree. One pupa was found between two phyllodes joined with silk, presumably by some other larva. Older and fully mature larvae and pupae were taken from borer holes and cracks in the trunks near the base of the same trees. They were attended by numerous small black ants, *Crematogaster* sp. Adults emerged from the pupae in May. In central Queensland the larvae were found feeding on *Alphitonia excelsa* (Rhamnaceae) by A. F. Atkins, who noted that they were greener than those of typical *delicia*.

19. *Hypochrysops ignitus* (**Leach**), 1814
This species has the widest distribution, occurring in north-western, eastern, southern and south-western Australia, and in southern New Guinea. Five subspecies have been recognized, four of which occur in

Australia and one in New Guinea. The subspecific status of a single male in the Australian National Insect Collection from the Edgar Range, south-east of Broome, W.A., has still to be determined. The larval food preferences are much more diverse than in the other species.

a. *Hypochrysops ignitus ignitus* (Leach), 1814

Plate 21, figs. 1 (male), 1A (female), fiery jewel.

DISTRIBUTION. From the Broadsound Range, 40 km west of Marlborough, Yeppoon and the southern Expedition Range, central Queensland, to Victoria and south-eastern South Australia. In New South Wales most specimens have been caught or reared in the Sydney area, where the adults fly in spring and autumn. Victorian localities include the Dandenong Ranges, Ocean Grove, Redesdale, Dimboola and the Big and Little Deserts, where there is apparently only a single generation annually, with adults on the wing from December to early March. In South Australia this subspecies is rare, having been recorded at Victor Harbor, the Mt Lofty Ranges, Melrose (A. F. Atkins) and the Yorke Peninsula. It probably has two generations each year in South Australia.

The subspecies varies greatly both above and beneath. The extent and depth of orange on the costa of the forewing varies, and is especially noticeable in specimens from central Victoria; in some specimens the upperside is lightly suffused with orange. Occasionally the number of spots beneath the wings may be reduced. The rapid flight of the adult makes it very inconspicuous, although it is not difficult to catch when settled on a twig. Males tend to collect on hill-tops.

The larvae and pupae are attended by numerous small black ants, *Iridomyrmex* sp. (*nitidus* group). If there is not already a nest of this ant at the base of the plant where the butterfly deposits its eggs, the ants soon find the eggs and establish a colony. They first build an ant-byre over the eggs and, when the larvae hatch, are said to either carry the young larvae to the young terminal shoots or perhaps to feed them. As the larvae grow they are constantly attended by the ants as they crawl upwards to their food each evening and back to the ant-byre, or ants' nest just beneath the soil surface, for shelter during the day. If the larvae are numerous and the food plant small, the leaves may be completely stripped and even the green bark consumed. As many as thirty larvae have been found at the base of one small tree. Sometimes larvae shelter among dead leaves around the base of the food plant, and occasionally feed during the day. Pupation usually occurs in the ants' nest or in the ant-byre, and the wings of the newly emerged adult do not begin to expand until it has crawled towards the light and out of the nest.

At Yeppoon and Westwood the larvae feed on *Alphitonia excelsa* (red ash, Rhamnaceae) and their presence is indicated if the leaves have freshly eaten irregular patches on the upper or lower surface. North of Rockhampton they have been found on *Dodonaea lanceolata* (Sapindaceae), and in the Expedition Range on *Acacia juncifolia* (Mimosaceae). Near Brisbane they have been reared from *Exocarpos latifolius* (Santalaceae), *A. excelsa* and *Banksia* sp. (Proteaceae), but at Burleigh Heads most larvae have been found on *Eucalyptus* spp. (Myrtaceae), usually blue gum or iron bark, but a few on *A. excelsa*, *Cupaniopsis anacardioides* (Sapindaceae), *Elaeocarpus obovatus* (Elaeocarpaceae), and *Acacia aulacocarpa* (Mimosaceae). Near Millmerran the larvae have been reared from *Dodonaea attenuata* (hop bush) and *Heterodendrum diversifolium* (both Sapindaceae). Additional food plants recorded at Stanthorpe are *Exocarpos cupressiformis* (Santalaceae), *Pomaderris* sp. (Rhamnaceae), *Cassinia quinquefaria* (Compositae), *Jacksonia scoparia* (Papilionaceae), and *Acacia neriifolia* (Mimosaceae). Near Sydney the larvae usually feed on *Acacia decurrens* (Mimosaceae), but also on *Dodonaea triquetra* (Sapindaceae). In Sydney gardens they have been recorded damaging *Camellia* (Theaceae) and a plum tree (*Prunus*, Rosaceae). They have also been found feeding on the foliage of *Rubus fruticosus* (blackberry, Rosaceae). Near Geelong, Victoria, the usual food plants are *Acacia pycnantha* (golden wattle) and *A. decurrens*, but in the Little Desert, western Victoria, the larvae feed on *Brachyloma daphnoides* (Epacridaceae). In South Australia the recorded food plants are *Acacia pycnantha* (Mimosaceae) and *Choretrum glomeratum* (Santalaceae).

b. *Hypochrysops ignitus chrysonotus* Grose-Smith, 1899

DISTRIBUTION. Cape York to Paluma, north-eastern Queensland; in coastal areas and as far inland as Mareeba and Mt Garnet.

ADULT. Male: similar to typical *ignitus* but central areas above usually much brighter bluish purple, markings beneath brighter.

Female: similar to typical *ignitus* but central areas above variable in size and colour, and usually brighter bluish purple, beneath as in male.

The shape of the hindwing varies considerably in males taken west of Paluma. In some the hindwing is rounded as in typical *ignitus*, whereas in others the tornus of the hindwing is more acute as in specimens of *chrysonotus* from farther north.

c. *Hypochrysops ignitus olliffi* Miskin, 1888

Plate 21, fig. 2 (male), dingy jewel.

DISTRIBUTION.   South-western Australia; known only from the Esperance (C. G. Miller), Albany and Denmark areas. Adults have been collected in November and December.

ADULT.   Male: similar to typical *ignitus*, but spots and bands beneath much narrower and paler, with less conspicuous metallic bluish green.

Female: similar to typical *ignitus* but purple above; beneath as in male.

Adults have been observed flying around *Acacia*, upon which the larvae probably feed. At Esperance C. G. Miller found adults feeding at flowers growing amongst the sand dunes.

d. *Hypochrysops ignitus erythrinus* (Waterhouse and Lyell), 1909

DISTRIBUTION.   Northern Territory; known from only a few specimens captured at Darwin in February and March.

ADULT.   Male: similar to typical *ignitus* but with central areas above duller, brownish purple, with narrow dull brown margins to both wings; beneath light grey-brown, with spots and bands narrower, dull red, edges broader metallic green.

Female: similar to typical *ignitus* but central areas above duller, light bluish purple, margins brown; beneath as in male.

20. *Hypochrysops epicurus* Miskin, 1876

Plates 21, fig. 4 (male), XVI, dull jewel.

DISTRIBUTION.   Maryborough to Port Macquarie. Specimens have been taken from September to November, and in January and February, at Brisbane, Stradbroke Island, Burleigh Heads, Brunswick

Heads, Ballina, Evans Head, Coffs Harbour, and Port Macquarie (C. W. Frazier). Former habitats at Burleigh Heads and Currumbin have been destroyed or threatened by coastal development.

This is a very dull species with the orange markings beneath scarcely as broad as their metallic green margins. In the female the extent of the purplish blue central areas and orange suffusion above is very variable. The adults settle on the tops of the taller mangroves and seldom fly unless disturbed. They appear to fly more readily during the afternoon.

The larvae are always attended by the small dull black ant *Iridomyrmex ?nitidiceps*, and feed on the foliage of the mangrove *Avicennia marina* (Verbenaceae). They eat only the lower surface of the leaves and shelter and pupate either in rolled leaves or in hollows or holes in the branches or trunk. In early spring pupae are usually found in the less accessible holes and crevices in the branches, but in summer a greater number occur in the leaf shelters.

21. *Hypochrysops cyane* (Waterhouse and Lyell) 1914

Plate 21, figs. 3 (male), 3A (female), cyane jewel.

DISTRIBUTION. Coen to Sydney, rare. Specimens have been taken at Coen (G. B. Monteith), Cooktown (D. Drake), Cairns (F. P. Dodd), Kuranda, Mackay, Bunya Mts, Toowoomba, Goondiwindi, Millmerran, Brisbane, and Sandgate in Queensland, and at Grafton, Toronto, Woy Woy and Casula in New South Wales.

This species is larger than *H. ignitus* and, in the male, the hindwing is more produced at the tornus. The brown-black termen of the wings above, and the red spots and bands beneath are narrower. In the female the central areas above are bluer than in coastal specimens of *H. ignitus* and, as in the male, the markings beneath are much narrower.

The larvae and pupae are always attended by small black ants, stated to be *Iridomyrmex itinerans*, which form colonies in hollow branches and borer holes in *Angophora costata* (Myrtaceae), the larval food plant. While very young the larvae remain in groups on the foliage, skeletonizing the leaves as they feed. Later they shelter during the day in holes and hollows in the branches occupied by the ants, and emerge at night to feed. Their feeding is then along the leaf margins and, if larvae are numerous, the foliage soon acquires a frayed or fretted appearance. Pupation occurs in the ant galleries where the larvae shelter.

22. *Hypochrysops piceatus* Kerr, Macqueen and Sands, 1969

Plate 23, fig. 5 (male).

DISTRIBUTION.   Known only from the Leyburn district and near Millmerran on the Darling Downs, southern Queensland. Adults are on the wing in October and November and again in January and February. The species appears to be rare near Millmerran but more common near Leyburn.

This is one of the smallest species in the genus, recalling *H. ignitus* but in the male with much broader dark margins and with the hindwing rather more produced at the tornus, as in *H. cyane*. The markings beneath the wings in both sexes appear very dark owing to the black edges of the spots and bands, and the reduced metallic coloration except on the outer side of the subterminal band. The pattern beneath the wings provides a reliable means of distinguishing *H. piceatus* from *H. ignitus*, *H. cyane* and *H. epicurus*, but the male genitalia also differ in these four species.

The adults of *H. piceatus* have a slower, more fluttering mode of flight than these closely related species. They are very inconspicuous on the wing and spend most of their time settled near the tops of the *Casuarina* trees on which the larvae feed. They have never been seen flying around any other species of tree.

Larvae feed on *Casuarina luehmannii* (bull oak, Casuarinaceae), and were found sheltering in borer holes in the upper branches during the day, attended by small black ants, *Iridomyrmex* sp. (*itinerans* group). This is the same ant that attends the larvae of *Ogyris aenone* which feed on mistletoe growing on the same *Casuarina* trees. Pupae probably also occur in borer holes used by the larvae.

23. *Hypochrysops miskini* (Waterhouse), 1903

Plate 21, figs. 7 (male), 7A (female), Miskin's jewel.

DISTRIBUTION.   Kuranda to Paluma and Maryborough to Southport, Queensland; rare. It is known mainly from Kuranda and from Southport, but it has also been taken at Herberton, Innisfail, Kirama State Forest, Paluma, Maryborough (M. J. Manski), Cooloola (M. De Baar), and Brisbane. Females are taken more commonly than males.

The female is easily recognized by its pale blue coloration above, the extent of which is very variable. The male has the costa and termen of the forewing more arched than in other similarly coloured species. The black-edged bar at the end of the cell beneath the hindwing is a useful point of distinction in both sexes. The original specimen came from Coomera in southern Queensland. Males from northern Queensland have the apex and termen of the forewing more broadly black. The species is probably confined to rain forest or rain-forest margins.

The eggs are laid singly or in groups of up to five on the undersurface of the leaves, leaf petioles, and stems of the food plant, and sometimes even on the twigs and branches of the tree over which the food plant scrambles. The larvae feed on *Smilax australis* (Smilacaceae), a common climbing or scrambling plant with large green leaves and thorny stems. They are always attended by the small black ant *Iridomyrmex gilberti* and may be associated with the larvae of *Pseudodipsas cephenes*. The pupae are found in hollow twigs, curled dead leaves, or in other convenient shelters near by. During the day the older larvae are probably found in the same shelters.

24. *Hypochrysops cleon* Grose-Smith, 1900

Plate 29, fig. 11 (male).

DISTRIBUTION. Claudie River, north-eastern Queensland; in Australia known only from a series of males collected by M. De Baar and S. J. Johnson in June, 1978.

ADULT. Male: above shining violet-blue; forewing with apex and termen black; hindwing with termen black. Beneath forewing grey towards tornus, almost black towards termen, remainder dull orange with light metallic green lines and spots; hindwing light grey towards base, whitish in outer half, spots and bands orange-red edged with light metallic green, band at end of cell edged black.

Female: above pale blue; fore- and hindwings with costa and termen broadly black. Beneath similar to male but forewing paler; hindwing with band at end of cell edged more broadly black.

The male is a much brighter purplish blue above than in *H. miskini*, and the bands beneath are brighter, more continuous and more distinctly edged with metallic green. The female is similar in shape and general appearance to *H. miskini*. Beneath the hindwing of both sexes the band at the end of the cell is more distinctly margined with black than in *H. miskini*.

The males from the Claudie River were taken as they settled on a small tree growing in open woodland on a ridge adjacent to rain forest. The early stages are not known.

25. *Hypochrysops pythias euclides* Miskin, 1888
Plate 21, figs. 5 (male), 5A (female), peacock jewel.

DISTRIBUTION.   North-eastern Queensland, from Port Douglas to Paluma; on the coast and up to 900 m; usually rare, but more common near Tully.

The metallic green markings beneath the wings are the broadest of any Australian species. The species is probably confined to rain forest.

The larvae have been found in early December (M. De Baar) feeding on the leaves of *Triumfetta rhomboidea* (Tiliaceae) at Paluma, and when the supply of food plant was exhausted, the larvae accepted *Commersonia bartramia* (Sterculiaceae). Larvae rested during the day on the lower surface of the leaves between the main veins and did not appear to be attended by ants. Pupation occurred on the foliage of the food plant in a crude shelter formed by the larva webbing two leaves together or folding over a leaf and webbing it with silk. The pupal duration in early January was ten to twelve days.

26. *Hypochrysops byzos* (Boisduval), 1832
This species is south-eastern in distribution, ranging from southern Queensland to Victoria. Two subspecies are recognized.

a. *Hypochrysops byzos byzos* (Boisduval), 1832
Plates 21, figs. 6 (male), 6A (female), XVI, yellow spot jewel.

DISTRIBUTION.   Southern Queensland and New South Wales; in some years fairly common near Sydney, but more usually rare. In Queensland it is known only from near Stanthorpe on the tablelands, and from a single female labelled "Cunnamulla" in the Australian Museum, but in New South Wales it occurs both on the coast and on the tablelands, from near Armidale, Mt Victoria, Mittagong, Sydney and Tathra.

This is a very distinctive species, especially in the female. The extent of the orange patches above and especially the size and number of the spots beneath the hindwing are variable. Of the purple species, the male of *H. byzos* has the richest purple above and the markings beneath are also broad and bright.

The larvae are not usually attended by ants, but occasionally one or two may be present. They feed at night or in the late afternoon and early morning on the dark green upper surface of the leaves of the food plant, eating irregular patches which indicate the presence of larvae. They rest on a silk membrane spun beside the midrib on the lower surface of the leaf, their colour and texture matching very closely that of the pale hairy undersurface of the leaf. Near Sydney the larvae feed on *Pomaderris lanigera* (Rhamnaceae), a shrub growing on the Hawkesbury sandstone and, less frequently, on *P. sieberana*, *P. ferruginea* and *P. andromedifolia* (G. Daniels). Near Stanthorpe they are usually found on *P. lanigera* but occasionally on another *Pomaderris*. The latter is the usual food plant near Armidale (C. W. Frazier). The pupae are usually attached to the undersurface of the leaves, but near Sydney are more often found in a curled dead leaf on the ground. Between December and March the pupal duration at Sydney varies from thirteen to fifteen days, but from September to November ranges from twenty-three to twenty-eight days.

b. *Hypochrysops byzos hecalius* Miskin, 1884

DISTRIBUTION. Australian Capital Territory, south-eastern New South Wales and eastern and central Victoria; local, fairly common at times, but usually rare. It is known from Mt Tidbinbilla and the Brindabella Range, A.C.T. (700 to 1,000 m), Brown Mt, N.S.W. (D. Binns), and the Wellington River valley, Mt Buller, Falls Creek, Mt Donna Buang, Toora, Narracan, the Dandenong Ranges, Taggerty and Gisborne, Victoria. The adults are on the wing as early as October in Victoria, and all stages may be present during the late spring and summer.

ADULT. Male: similar to typical *byzos*, but more shining purple above, and orange-red subtornal band of hindwing usually extended into a thin terminal line, beneath pale yellow, spots and bands smaller.

Female: similar to typical *byzos*, but central patches above lighter orange, hindwing with termen broadly orange, beneath much duller.

The most interesting feature of this subspecies is the yellow ground colour of the male beneath, resembling the female rather than the male of typical *byzos*.

The life history is similar to that of typical *byzos*. At Mt Tidbinbilla and in Victoria the larvae feed on *Pomaderris aspera*, but they have also been found on *P. eriocephala* in the Brindabella Range and on *P. lanigera* in Victoria.

27. *Hypochrysops apelles apelles* (Fabricius), 1775

Plates 22, figs. 14 (male), 14A (female), XVI, copper jewel.

DISTRIBUTION.   Darwin, Northern Territory, islands of Torres Strait, and Cape York to the Richmond River; fairly common at Yeppoon and farther north. A male in the Australian Museum was collected last century near Newcastle, N.S.W.

This species is only slightly larger than *H. ignitus* and *H. byzos*, and is easily identified by its orange upperside and the markings beneath. Adults from Cape York are brighter, with the apex of the forewing and the tornus of the hindwing more pointed. The adults are usually found flying around and settling on the mangroves on which the larvae feed, but away from the coast they probably also frequent their food plants. At the Claudie River they have been observed flying in rain forest.

The larvae eat irregular patches from either surface of the leaves of the food plant, and hide during the day in a curled leaf or between leaves spun together by themselves or by other insects. At Yeppoon they are often found between leaves previously joined by the larva of a tortricid moth, *Procalyptis parooptera*. The larvae are always attended by small black ants, *Crematogaster* sp. In coastal areas they usually feed on mangroves, including *Rhizophora stylosa* at the Tweed River, *R. stylosa*, *Bruguiera gymnorrhiza* and *Ceriops tagal* (all Rhizophoraceae) at Burleigh Heads, *C. tagal* and *Avicennia marina* var. *resinifera* (Verbenaceae) at Maryborough and Urangan, and *C. tagal* at Yeppoon. On Moa Island they were found on *Alphitonia* (Rhamnaceae), on Prince of Wales Island on *Barringtonia* (Barringtoniaceae), at Cooktown on *Terminalia* (Combretaceae), and at Townsville on *Planchonia careya* (Lecythidaceae). Away from the coast they have been reared from *Angophora floribunda* (Myrtaceae), and from *Acacia flavescens* (Mimosaceae) near Mt Garnet; near Cape York they were found feeding on *Alphitonia excelsa* (red ash, Rhamnaceae).

28. *Hypochrysops apollo* **Miskin, 1891**

This striking species occurs on Prince of Wales Island, and in coastal areas from Cape York to Ingham.

a. *Hypochrysops apollo apollo* **Miskin, 1891**

Plates 23, figs. 13 (male), 13A (female), XVI.

DISTRIBUTION.   Cairns to Ingham; fairly common but local.

This fine species can be readily identified by its size, its brilliant reddish orange wings above and intricate pattern beneath. Adults are seldom caught because they fly high and fast. Sometimes both males and females are seen walking over the outside of the bulbous food plants. The species is found in the coastal paper bark (*Melaleuca*) swamps, where *Ogyris aenone* and *Hypochrysops narcissus* also occur.

The larvae live in the galleries that are found within the bulbous epiphytic plant *Myrmecodia beccarii* (ant plant, Rubiaceae), which grows commonly on *Melaleuca* (paper bark), *Tristania suaveolens* (swamp mahogany), and other coastal trees. They are associated with small ants which live in large colonies in the same galleries. The ant normally associated with *Myrmecodia* is the native *Iridomyrmex cordatus*, but it has often been replaced by the introduced pest species *Pheidole megacephala*, which appears to be largely indifferent to the larvae of *H. apollo*. The larvae feed on the internal tissue of the plant, and sometimes on the *Myrmecodia* leaves; they gradually enlarge the galleries and make an irregular hole to the outside. Pupation occurs within the enlarged gallery and the adult escapes through the hole after emergence. The presence of larvae and pupae can often be detected by the pale freshly eaten internal surfaces of the galleries, whereas surfaces that have not been damaged recently are dark brown in colour. Generally not more than one larva is found in any one plant.

b. *Hypochrysops apollo phoebus* **(Waterhouse), 1928**

DISTRIBUTION.   Prince of Wales Island and Cape York to the Claudie River; fairly common. Adults have been taken at Cape York in most months. This subspecies has been recorded recently from near Port Moresby, Papua New Guinea.

ADULT.   Male: similar to typical *apollo* but forewing less pointed, above much deeper orange-red; beneath forewing orange-red, costa and apex darker reddish brown, markings more obscure, hindwing darker, markings darker, less contrasted.

255

Female: above similar to typical *apollo*; beneath as in male of *phoebus* but paler.

The life history is probably similar to that of typical *apollo*. At Cape York a pupa was found on one occasion in a hole in an unidentified bulbous epiphyte, which from the description must have been a *Myrmecodia*.

29. *Hypochrysops elgneri* (Waterhouse and Lyell), 1909

This species is known from New Guinea, Prince of Wales Island and Cape York. Two subspecies are recognized from Australia, one of which is shared with Papua New Guinea.

a. *Hypochrysops elgneri elgneri* (Waterhouse and Lyell), 1909

Plate 23, fig. 14A (female).

DISTRIBUTION.   Prince of Wales Island, Torres Strait; rare. Adults have been collected from May to July. This subspecies is also found in Papua New Guinea.

b. *Hypochrysops elgneri barnardi* Waterhouse, 1934

Plate 23, fig. 14 (male).

DISTRIBUTION.   Cape York to the Claudie River and the McIlwraith Range; rare. Adults have been taken in June, September to November and in January.

This subspecies is larger and brighter than the other. Nothing is known about the life history of either. At the Claudie River and McIlwraith Range adults rested on *Melaleuca* trees (M. De Baar and D. P. Sands), and were induced to fly only when the trees were shaken. G. Daniels found that adults of both sexes could be shaken from *Tristania suaveolens* at the Claudie River; once disturbed the males rarely returned, whereas he took females only on these trees and they readily returned to them after disturbance.

30. *Hypochrysops polycletus* (Linnaeus), 1758

This species ranges from the Moluccas to New Guinea, the Aru Islands and north-eastern Australia. Typical *polycletus* occurs in the Moluccas. Two subspecies are known from within Australian limits.

a. *Hypochrysops polycletus rovena* **Druce, 1891**

Plate 21, figs. 11 (male), 11A (female), rovena jewel.

DISTRIBUTION. North-eastern Queensland, from Weipa, Batavia Downs, and the Claudie River to the McIlwraith Range, Coen, Cooktown, Kuranda and Mackay; not common.

This is a handsome species that cannot be confused with any other. It occurs in rain forest, but on Cape York Peninsula has been taken more frequently in the drier vine thickets. The life history is not known.

b. *Hypochrysops polycletus rex* **(Boisduval), 1832**

DISTRIBUTION. Within Australian limits only one female has been taken at Darnley Island. The subspecies also occurs in New Guinea.

ADULT. Male: similar to *rovena* but larger, above with narrower black margins, beneath darker.

Female: similar to *rovena*, but above with green at base instead of blue.

31. *Hypochrysops narcissus narcissus* **(Fabricius), 1775**

Plate 21, figs. 10 (male), 10A (female), narcissus jewel.

DISTRIBUTION. Islands of Torres Strait, and Cape York to Ingham; sometimes common.

The adults are very inconspicuous in flight, and settle frequently. The species is found mainly in swampy areas where *Melaleuca* (paper bark) and *Tristania suaveolens* (swamp mahogany) grow. Both *Hypochrysops apollo* and *Ogyris aenone* occur in the same situations. Adults of *H. narcissus* also fly around mangroves.

The larvae have been found feeding on the upper surface of the leaves of the mistletoe *Dendrophthoe vitellina* (Loranthaceae) growing on *Tristania*. As a result of the larval feeding the leaves tended to roll into hollow tubes in which the larvae sheltered during the day. They were attended by small black ants. Near Cooktown larvae were found feeding on the lower surface of the leaves of *Terminalia* (Combretaceae), attended by small ants; pupal skins were found in the ants' nest in the hollow tree trunk. Larvae were also collected there and at Hartleys Creek, north of Cairns, on mangroves. They were sheltering in the curled brown leaves along with larvae of *Hypochrysops apelles*. Although no pupae of *H. narcissus* were found in the curled leaves, these

257

would provide the only pupation sites for larvae on the young mangroves, whereas hollow branches would have been available for pupation on the larger mangroves.

32. *Hypochrysops digglesii* (Hewitson), 1874

Plates 22, figs. 9 (male), 9A (female), XV, Diggles' blue.

DISTRIBUTION. Cape York, Queensland, to near Byron Bay, New South Wales; uncommon. It has recently been reared in fair numbers from near Broken Head, N.S.W. The species has also been taken in New Guinea.

As in other species of *Hypochrysops*, the adults are very inconspicuous on the wing.

The larvae feed at night on the leaves of mistletoes, sheltering during the day in a curled dead leaf of the food plant, in hollow branches, under loose bark on the host tree, or on a nearby tree, in curled dead leaves at the base of the host tree, or in the ants' nests. The pupae are found in the same situations. Both larvae and pupae are attended by numerous small black ants, *Crematogaster* sp. The larvae have been recorded feeding on the mistletoes (Loranthaceae) *Amyema conspicuum* growing on *Alphitonia excelsa* (red ash), *Dendrophthoe vitellina* and *Muellerina celastroides*.

Genus *PHILIRIS* Röber, 1891

The genus includes numerous species distributed from the Moluccas, the Kai and Aru Islands, through New Guinea to the D'Entrecasteaux Islands, the Bismarck and Louisiade Archipelagoes and northern and eastern Australia.

33. *Philiris diana* Waterhouse and Lyell, 1914

This species is known from the Milne Bay area of New Guinea, from the Claudie River, Cape York Peninsula, and from Kuranda and Paluma. It is the largest species occurring in Australia. Two subspecies are recognized.

a. *Philiris diana diana* Waterhouse and Lyell, 1914

Plate 22, figs. 4 (male), 4A (female), diana moonbeam.

DISTRIBUTION. Kuranda and Paluma, north Queensland; rare. Adults occur in rain forest.

b. *Philiris diana papuana* **Wind and Clench, 1947**

DISTRIBUTION.   Bamaga, Captain Billy Creek, Shelburne Bay, and the Claudie River, Cape York Peninsula. It also occurs in the Milne Bay area, New Guinea.

ADULT.   Male: similar to typical *diana*, but forewing lacks the small white central patch, hindwing beneath with a black terminal line joining black terminal dots on veins near tornus.

Female: similar to typical *diana*, but forewing with white area reduced to white central patch suffused with blue scales; hindwing with white apical area much enlarged, contrasting with a broad black terminal area, veins black. Beneath as in male.

Adults have been taken only in rain forest.

34. *Philiris nitens* **(Grose-Smith), 1898**

This species occurs in north-eastern Queensland and Papua New Guinea.

a. *Philiris nitens nitens* **(Grose-Smith), 1898**

Plate 22, figs. 2 (male), 2A (female), blue moonbeam.

DISTRIBUTION.   McIvor River, north of Cooktown, to Ingham; fairly common.

Adults are found in rain forest.

Near Cairns larvae feed on *Glochidion philippicum* (Euphorbiaceae) (D. P. Sands).

b. *Philiris nitens lucina* **Waterhouse and Lyell, 1914**

DISTRIBUTION.   Cape York to the Claudie River and McIlwraith Range, north-east of Coen; rare at Cape York, more common at the Claudie River.

ADULT.   Male: similar to typical *nitens*, but margins more broadly black.

Female: similar to typical *nitens*.

Females from the Claudie River are variable, and resemble Cairns specimens. The males of *lucina* are, however, consistently distinct.

35. *Philiris fulgens kurandae* **Waterhouse, 1903**

Plate 22, figs. 3 (male), 3A (female), purple moonbeam.

DISTRIBUTION.   Cape York to Paluma, north Queensland; not common.

In the male the contrasting purple forewing and blue hindwing make recognition easy. The blue areas in the female are variable in extent. The species is restricted to rain forest.

36. *Philiris ziska titeus* **D'Abrera, 1971**

Plate 23, fig. 11 (male).

DISTRIBUTION.   Claudie River, Cape York Peninsula.

The female of this subspecies has not so far been recognized with certainty, and only eleven males have been taken in Queensland, all in rain forest.

37. *Philiris innotata* **(Miskin), 1874**

This species occurs in north and south Queensland and in northern New South Wales.

a. *Philiris innotata innotata* **(Miskin), 1874**

Plates 22, figs. 1 (male), 1A (female), XIX, common moonbeam.

DISTRIBUTION.   From Rockhampton and Yeppoon, Queensland, to Port Macquarie, New South Wales; fairly common. It has been taken inland as far as Carnarvon Gorge, Bulburin State Forest, Monto and Gayndah. In central Queensland it has been collected in August and from December to April.

The adults are usually found flying around *Ficus* (native figs, Moraceae) upon which the larvae feed. Unlike the other Australian species of the genus, this is not restricted to rain forest, but also occurs along creeks and rivers where the food plant grows.

Near Brisbane the larvae feed on *Ficus* sp. (black or water fig), and have also been reported on *F. opposita* (rough-leaved fig). The hairy larvae and pupae match the underside of the leaves where they are found.

b. *Philiris innotata evinculis* **Wind and Clench, 1947**

DISTRIBUTION.   Cape York to Mackay; common at times.

ADULT.   Male: similar to typical *innotata* but above purplish blue, with narrower brown-black terminal areas.

Female: similar to typical *innotata* but above with central blue areas more extensive.

It is doubtful if this can be maintained as a separate subspecies, and further collecting may well show that the ranges of this and typical *innotata* are continuous.

The life history is similar to typical *innotata*. Near Cairns the larvae feed on *Ficus opposita* (rough-leaved fig, Moraceae).

Genus *ARHOPALA* Boisduval, 1832

The genus contains many species distributed from Afghanistan, Sri Lanka and India to China, Korea and Japan, and through south-east Asia to New Guinea, northern Australia and the Solomons. It is richly developed in India and Burma, in Malaya and in New Guinea. Australia has only four species, all of which occur north of the Tropic. The wings are often a brilliant blue or purple above, but beneath are usually protectively patterned in various shades of brown. The brilliant colours of the adults are seen only in flashes during their rapid flight, but for much of their time they remain settled on trees with their wings closed. The adults are robust insects, with a short curved tail to the hindwing at the end of vein CuA_2.

38. *Arhopala wildei wildei* Miskin, 1891

Plate 25, figs. 4 (male), 4A (female), small oakblue.

DISTRIBUTION.   Moa Island and Cape York to Etty Bay, near Innisfail, and Tully, Queensland; not common.

The species is easily recognized by its smaller size, the colour of the wings beneath, and the white central areas of the female above. It seems to be restricted to the rain forest and flies less rapidly than the other Australian species. The early stages are not known.

39. *Arhopala centaurus* (Fabricius), 1775

The species includes nine subspecies distributed from Sumatra and Java to New Guinea, northern Australia, the Bismarck Archipelago and the Solomons.

a. *Arhopala centaurus centaurus* (Fabricius), 1775

Plates 25, figs. 1 (male), 1A (female), XIX, dull oakblue.

DISTRIBUTION.   Moa, Thursday, Horn, and Prince of Wales Islands, and Cape York to Yeppoon; common. There are two males in the Australian Museum from as far inland as Chillagoe, north Queensland.

Although this is the least brilliant of the three larger Australian species, it is nevertheless a handsome insect. The species is mainly coastal in distribution, occurring near mangroves or around dwarfed or larger rain-forest trees growing on sand dunes or behind beaches. A migration of this species near the Claudie River, Cape York Peninsula, was observed by M. S. Moulds.

The pupae are found in small shelters made from the leaves of the food plant, and both larvae and pupae are attended by *Oecophylla smaragdina* (green tree-ants). Near Cardwell the larvae were found on the young foliage of *Eucalyptus intermedia* and on *Melaleuca quinquenervia* (both Myrtaceae) (W. N. B. Quick).

b. *Arhopala centaurus asopus* Waterhouse and Lyell, 1914

DISTRIBUTION.   North-western Australia and Northern Territory. Adults have been taken at Koolan Island in Yampi Sound and at the Drysdale River, W.A., and at Darwin, Brocks Creek, Katherine, Daly River Crossing and Groote Eylandt, N.T.

ADULT.   Male: similar to typical *centaurus* but duller purple above.
Female: similar to typical *centaurus* but duller purple above, and base shining blue.

At the Drysdale River, W.A., larvae were found on *Eucalyptus* sp. (Myrtaceae), attended by green tree-ants, *Oecophylla smaragdina*.

40. *Arhopala madytus* Fruhstorfer, 1914

Plate 25, figs. 2 (male), 2A (female), bright oakblue.

DISTRIBUTION.   Cape York to Yeppoon; fairly common.

The male is darker above than *A. centaurus* or *A. micale*. The female is similar above to *A. centaurus* but rather brighter. In both sexes the underside is darker than in *A. centaurus*. All three species are found in the same situations.

41. *Arhopala micale* Boisduval, 1853

The species ranges from the Moluccas, through New Guinea to the Louisiade Archipelago, the Kai and Aru Islands, and northern Australia. Sixteen subspecies have been distinguished, including three in Australia. This is the brightest Australian species, with the wings beneath richly patterned in dark purplish browns.

a. *Arhopala micale amphis* **Waterhouse, 1942**

Plate 25, figs. 3 (male), 3A (female), common oakblue.

DISTRIBUTION. Coastal Queensland from 50 km north of Cooktown to Yeppoon; common.

Both larvae and pupae are attended by green tree-ants, *Oecophylla smaragdina*. At Cairns the larvae have been recorded feeding on several families of native trees including *Terminalia muelleri* (Combretaceae), *Cryptocarya hypospodia* (Lauraceae), *Acmena* sp. (Myrtaceae), *Faradaya splendida* (Verbenaceae), *Heritiera littoralis* (Sterculiaceae), *Glochidion ferdinandi* (Euphorbiaceae), *Cordia dichotoma* (Boraginaceae), *Hibiscus tiliaceus* (Malvaceae), and *Cupaniopsis anacardioides* (Sapindaceae).

b. *Arhopala micale amytis* **(Hewitson), 1862**

DISTRIBUTION. Moa, Thursday, and Prince of Wales Islands, and Cape York Peninsula as far south as the Claudie River, McIlwraith Range and Port Stewart.

ADULT. Similar to *amphis* in both sexes, but less greenish above, with narrower black margins. Beneath variable, similar to *amphis*, but with more extensive metallic green scaling near tornus.

This is an even more striking subspecies than the one found farther south. However as specimens of *A. micale* from localities between Cooktown and just north of Coen show intermediate characters, the differences between *amytis* and *amphis* may prove to be clinal.

c. *Arhopala micale amydon* **Waterhouse, 1942**

DISTRIBUTION. Northern Territory, including Groote Eylandt, and Murray and Darnley Islands, Torres Strait.

ADULT. Male: above rich purple, with very narrow black margins. Beneath variable, similar to *amytis* with prominent area of metallic green scaling near tornus of hindwing.

Female: above dull violet-blue, with black margins broader than in male. Beneath similar to male.

This is a much less brilliant subspecies than either *amphis* or *amytis*, and may be distinguished by the colour above, especially in the female.

Near Darwin the larvae have been found on *Cupaniopsis anacardioides* (Sapindaceae), attended by green tree-ants, *Oecophylla smaragdina*.

Genus *OGYRIS* Westwood, 1851

The genus contains robust, fast-flying species, distributed throughout mainland Australia and Kangaroo Island. Twelve species are known from the Australian mainland, and three from New Guinea; the genus is absent from Tasmania. The wings above are often brilliantly coloured, but beneath they exhibit protective patterns and coloration. The adults usually fly high around the tree-tops, or near the parasitic mistletoes upon which the larvae feed.

The egg is spherical or mandarin-shaped, either smooth or pitted. The larvae nearly always feed at night and are usually attended by ants. Most feed on mistletoes (Loranthaceae), but the food plant of at least one species is *Choretrum* (Santalaceae). They shelter during the day beneath loose bark, in crevices, or in the attendant ants' nest, usually at the foot of the tree upon which the mistletoe is growing. The pupae, which are always attached by the anal hooks and a central girdle, are found in the places where the larvae shelter.

42. *Ogyris genoveva* Hewitson, 1853

This species ranges from north Queensland to the Flinders Ranges, South Australia, both in coastal areas and in the drier inland. Six subspecies have been recognized, but the status of these is doubtful. The males differ only slightly from one another, mainly in the shade of the violet-purple and the width of the black margins above. The females vary to some extent in the size of the cream patch of the forewing and the extent and colour of the metallic central areas above. At one time confused with *O. zosine*, *O. genoveva* may usually be distinguished by the rich violet tint to the purple coloration of the male above, and in the female by the metallic bluish-green or bluish central areas of both wings, the larger cream patch on the forewing, and the broader hindwings.

The larvae are nearly always found within permanent or temporary nests of sugar ants (*Camponotus*), usually at the foot of the trees on which the mistletoe food plant is growing. Often several larvae, and sometimes as many as one hundred, are found together in the one ants' nest. Pupation also occurs within the ants' nest and, after the adults emerge, the ants destroy the pupal skins.

a. *Ogyris genoveva genoveva* Hewitson, 1853

Plate 24, figs. 1 (male), 1A (female).

DISTRIBUTION. Brisbane to north-eastern New South Wales;

usually not common. Two males in the Australian Museum from Kuranda and Little Mulgrave River, north Queensland, seem to belong here also.

The larvae feed at night on several species of mistletoe, usually attended by the sugar ant *Camponotus nigriceps*. It has been suggested that the ants may collect young larvae from several trees and bring them to the one nest, which would account for the large numbers of larvae sometimes found together. The ants may even feed the first instar larvae. The larvae usually follow the same track to and from the food plant and on smooth-barked trees this is clearly visible when the larvae are numerous.

When searching for larvae of this species, the ants' nests can usually be located by giving the tree carrying the mistletoe a sharp tap or by scratching around the base of the tree with a stick. If a nest is present, the ants will usually swarm out. By carefully digging around the base of the tree, the ants' nest can be opened and any larvae or pupae uncovered. They are usually found in a chamber just beneath the surface of the soil or under a stone. As the larvae are difficult to rear in captivity it is wise to leave them, covering up the nest carefully with pieces of bark or stones, and then covering these with leaves. A few weeks later the nest can again be examined and the pupae collected.

b. *Ogyris genoveva duaringa* **Bethune-Baker, 1905**

DISTRIBUTION.   Inland central and southern Queensland, from Springsure, Duaringa and Mt Morgan to Millmerran, Stanthorpe and Charleville (J. C. Le Souef), from many localities in inland New South Wales, including Gunnedah, Coonabarabran, Baradine, Cobar, 88 km west of Nyngan, Mt Hope, Condobolin, Lake Cargelligo, West Wyalong and Leeton, and from Mildura, Victoria. A specimen has also been reared from Yeppoon, in coastal central Queensland, by J. C. Le Souef.

ADULT.   Male: similar to typical *genoveva* but usually a little smaller and slightly paler violet above.

Female: similar to typical *genoveva* but basal areas above usually much bluer, in hindwing sometimes produced along veins M_3, CuA_1 and CuA_2, almost reaching small spots on either side of these veins near termen.

At Millmerran and Toowoomba females with greenish central areas above have been found in the same locality as those with bluish central areas. Western New South Wales females have much more blue above than *gela* and some specimens do not differ greatly from less extreme examples of the Flinders Ranges subspecies *splendida*.

Near Stanthorpe the larvae have been taken feeding on *Amyema pendulum* growing on *Eucalyptus*. Both larvae and pupae were found in the nest of the sugar ant *Camponotus nigriceps*, rarely under loose bark on the tree trunk. J. Macqueen has recorded them at Millmerran on *A. miquelii, A. bifurcatum, A. quandang, A. miraculosum* and *Dendrophthoe glabrescens*. At West Wyalong the larvae feed on *A. miquelii*. Colonies of from two to more than one hundred larvae have been found in the one ants' nest. There are two generations annually.

c. *Ogyris genoveva gela* Waterhouse, 1941

DISTRIBUTION.   Eastern New South Wales; local. The original specimens came from near Sydney, but the subspecies has been taken near Armidale on the Northern Tablelands, near Canberra in the Australian Capital Territory, and at Scone, and Murrurundi.

ADULT.   Male: similar to typical *genoveva* but smaller, brighter in colour, with broader black margins.

Female: similar to typical *genoveva* but basal areas above greener, spots near termen of hindwing rarely present, cream patch on forewing broad and often extending to vein CuA_2.

The larvae feed at night on several species of mistletoe, including *Amyema miquelii* and *Dendrophthoe vitellina*, and are usually attended by the large yellow-brown or dark brown *Camponotus* ants. Near Canberra and Armidale the ants are probably *C. consobrinus*. Though nearly always in the ants' nests at the foot of the host tree, the larvae are occasionally found with the ants near the mistletoe food plant, under a large slab of loose bark or in a partly rotted branch of the tree. Very occasionally the larvae are attended by small black ants. Near Sydney there are two generations annually.

d. *Ogyris genoveva araxes* Waterhouse and Lyell, 1914

DISTRIBUTION.   Central and western Victoria, including Broadford, Melton, Bacchus Marsh, Dimboola, Horsham and Lake Hattah.

Adults are on the wing during November and December.

ADULT. Male: similar to *gela*, but smaller, above paler violet-purple, black margins broader.

Female: similar to *gela*, but smaller, metallic bluish green central areas more restricted, cream patch on forewing above usually extending broadly to CuA_2, with smaller indentation on its inner edge.

This subspecies has an annual life cycle.

e. *Ogyris genoveva genua* Waterhouse, 1941

DISTRIBUTION. Mt Lofty Ranges, South Australia; very local. There may be two generations annually.

ADULT. Male: similar to *araxes*, but larger, darker violet-purple above, and black margins broader.

Female: similar to *araxes*, but central bluish green areas more restricted, cream patch on forewing above very broad, usually extending broadly to CuA_2, with inner edge nearly straight, with only a slight indentation, beneath extending to costa where it is white, hindwing beneath with more light brown suffusion than in other subspecies.

The life history is similar to that of typical *genoveva*.

The larvae feed on *Amyema miquelii* and *A. pendulum* growing on *Eucalyptus*, and are attended by sugar ants, *Camponotus* sp. (*consobrinus* group).

f. *Ogyris genoveva splendida* Tindale, 1923

DISTRIBUTION. The Flinders Ranges area, South Australia; rare.

ADULT. Male: similar to *araxes* but a little brighter bluish purple above, with whitish costal streaks near apex more conspicuous.

Female: similar to typical *genoveva*, but with central areas light metallic blue and much more extensive, in hindwing reaching termen, enclosing three subterminal black spots between veins M_1 and CuA_1, forewing with light metallic blue streak near apex.

Some adults are very similar to *duaringa*, although the central blue areas of the female are usually more extensive.

The life history is similar to that of typical *genoveva*.

Both sexes have been reared by M. W. Mules from pupae he collected near Cradock, S.A. The pupae were found in the hollow butts of mallee. Apparently the larvae enter the butts through the attendant *Camponotus* ant nests at the base of the tree. Larvae have recently been collected at the same site by J. C. Le Souef, who found them in a hollow of a mistletoe-bearing mallee, about two metres from the ground. They were also attended by *Camponotus* ants.

43. *Ogyris zosine* Hewitson, 1853

This species is known from Western Australia, the Northern Territory and from localities in eastern Australia from Cooktown to Ballina. Most records are from coastal areas, but the species also occurs near Tom Price, W.A., near Alice Springs, N.T., and at Mt Isa and other inland localities in Queensland.

The northern subspecies *typhon* is remarkable for the great variation found in the females, which range in colour from dull dark purple, through bright bluish purple, pale purple, and pale purplish blue to pale blue; and more than one form may occur in the one colony of larvae. The males of *O. zosine* are a duller and darker purple than *O. genoveva*, with the tornus of the hindwing more produced; the hindwings of the female are distinctly narrower than in *O. genoveva*. Three subspecies have been distinguished.

The larvae and pupae are attended by sugar ants (*Camponotus*) and especially in southern areas are usually found in small colonies under loose bark or in cracks on the trunk of the host tree, or under stones, or in curled leaves on the ground near the base of the tree. In northern Queensland they often occur in temporary sugar ant nests in the soil at the foot of the tree, but also under loose bark on the trunk. Those in underground ant galleries usually pupate attached to the tree trunk well below the soil surface. The ants do not destroy the empty pupal skins after the adults emerge and these may often be found when searching for live larvae and pupae.

a. *Ogyris zosine zosine* Hewitson, 1853

Plate 24, fig. 1B (female), purple azure.

DISTRIBUTION. Maryborough and Gayndah to Ballina and Evans Head; local. A specimen is known from Toowoomba (J. Macqueen).

The larvae are not as broad as *O. genoveva*, with more prominent lateral projections of the eighth abdominal segment, more conspicuous hairs, and darker and more clearly defined markings. At Burleigh Heads they are attended by *Camponotus claripes* (sugar ant) and at Evans Head by one of the two species of *Camponotus* attending the larvae of *typhon* in north Queensland.

b. *Ogyris zosine typhon* **Waterhouse and Lyell, 1914**

DISTRIBUTION.   Darwin, Elliott (J. C. Le Souef) and the Alice Springs area (E. D. Edwards, G. Griffin), Northern Territory, and northern and central Queensland, from the Claudie River and Coen to Mt Isa (J. C. Le Souef), Hughenden, Bogantungen, Rockhampton and Yeppoon. Specimens taken near Parraburdoo, 70 km south of Tom Price, Western Australia, by D. F. Lascelles, and on Moa Island, Torres Strait, by G. B. Monteith and R. Lachlan probably belong here. In central Queensland adults have been recorded from August to October and from February to May.

ADULT.   Male: similar to typical *zosine*, above variably purple but usually duller. Beneath paler, ground colour suffused with brown.

Female: similar to typical *zosine* but with basal areas of both wings very variable, ranging in colour from dull to bright purple, purplish blue and pale blue; forewing with cream patch variable in size. Beneath paler with brown suffusion.

The various female forms occur together in the one locality.

Larvae of *typhon* are attended by two species of *Camponotus* ants near Townsville. Near Clermont the food plant is *Amyema quandang* growing on *Acacia harpophylla* (brigalow) (E. J. Dumigan), and near Alice Springs it is *Amyema maidenii* growing on mulga (E. D. Edwards).

c. *Ogyris zosine zolivia* **Waterhouse, 1941**

DISTRIBUTION.   Hayman, Whitsunday and Lindeman Islands, Queensland.

ADULT.   Male: similar to typical *zosine* but larger, above slightly paler purple, with broader black margins; beneath paler.

Female: similar to typical *zosine* but larger, purple areas more restricted.

The male is a brighter purple above than *typhon* and is not as pale beneath. As in typical *zosine* only a purple form of female is known.

44. *Ogyris idmo* Hewitson, 1862

This species occurs in western New South Wales and Victoria, South Australia, and south-western Australia. Little is known of the behaviour of the adults, and neither the early stages nor the larval food plant has been reported.

a. *Ogyris idmo idmo* Hewitson, 1862

Plate 24, figs. 3 (male), 3A (female), large brown azure.

DISTRIBUTION. South-western Australia; rare. Adults have been taken from October to December.

The adults fly within about one metre of the ground and often settle with closed wings on the ground or on stones or sticks, the protective pattern beneath the wings making them difficult to see. They have also been taken feeding at flowers such as *Pimelea*, and on more than one occasion flying around a low-growing creeper.

Near Bunbury A. F. Atkins noticed females settling on the ground and retreating beneath low bushes, but although some of the bushes sheltered *Camponotus* ant nests no oviposition was observed and no larvae or pupae were found in the ant nests. Nearby a few males were noticed hill-topping on the higher sand dunes. There is probably only a single generation annually.

b. *Ogyris idmo halmaturia* Tepper, 1890

DISTRIBUTION. Western New South Wales, western Victoria and South Australia, including Kangaroo Island; very rare. Specimens have been recorded from Mildura (B. Vardy), the Little Desert, Dimboola, the Grampians and Portland in Victoria, and from Kangaroo Island, from near Ceduna, Blackwood, Victor Harbor and Brimbago, South Australia. Adults have been collected in October and November. There is a specimen in the National Museum of Victoria from Broken Hill, New South Wales.

ADULT. Male: similar to typical *idmo* but forewing with more rounded apex and more convex termen, brown-black margins of both wings much broader.

Female: similar to typical *idmo* but basal areas of wings a brighter bluish purple and more extensive.

As in typical *idmo* there is probably a single generation annually.

An adult female was reared in 1945 by M. W. Mules from a pupa collected near Kiata, Victoria. The pupa was found in the nest of a sugar ant, *Camponotus nigriceps*, established at the base of a small tree, *Eucalyptus viridis*, apparently without any mistletoe. He also collected a fresh female which had apparently emerged from a pupa in the same ants' nest.

45. *Ogyris otanes* C. and R. Felder, 1865

Plate 24, figs. 10 (male), 10A (female), small brown azure.

DISTRIBUTION. Western New South Wales, north-western Victoria, South Australia, including Kangaroo Island, and south-western Australia; rare. It was once taken at Moonta, Yorke Peninsula, and Nuriootpa and Goolwa, S.A., and occurred more commonly on Kangaroo Island, where its habitats have now been greatly disturbed or destroyed; there is a male in the National Museum of Victoria from Broken Hill, and a few specimens have been taken in the Stirling Range, W.A. Recently the species has been found breeding in the Big Desert, north-western Victoria, and has been recorded at Red Bluff (A. D. Bishop) and one or two other localities in the north-west of the State. In the Big Desert it has been caught or reared from October to December and in March.

The adults fly about one metre above the ground and often settle with wings closed on the ground or on sticks and stones, where their protectively coloured wings harmonize perfectly with the background.

On Kangaroo Island the larvae feed on the bark of the shoots and stems of the small shrub *Choretrum glomeratum* (common sour bush, Santalaceae) growing in moist sandy soil. The presence of larvae is indicated by the plants presenting a scorched appearance. Larvae and pupae are found in the nests of a sugar ant, *Myrmophyma ferruginipes*, at the base of the food plants. They occur in galleries excavated by the ants at a depth of five or six centimetres, usually on the undersides of roots. As in *O. genoveva* the empty pupal skins are destroyed by the ants after the adults emerge. The food plant of the larvae in the Big Desert is also thought to be *Choretrum* sp.

46. *Ogyris abrota* Westwood, 1851

Plate 24, figs. 9 (male), 9A (female), dark purple azure.

DISTRIBUTION. Southern Queensland to south-eastern South Australia; in Queensland rare on the coast but fairly common at Toowoomba and the Bunya Mountains; fairly common in New South Wales and Victoria, especially at lower altitudes; recently discovered near Canberra, A.C.T. (E. D. Edwards); only one specimen is known from South Australia. Adults are on the wing mainly during October and November, and again from February to April.

The species is easily recognized by the rich purple male, somewhat similar in colour to *O. genoveva*, and the dark brown female with a very large central pale patch on the forewing. Though usually cream in colour, several females reared by D. P. Sands have a white patch. The pattern beneath both wings is also distinctive. The species shows little variation throughout its extensive range.

The larvae are attended by small ants, usually *Crematogaster* sp., but sometimes *Froggattella kirbyi* or *Technomyrmex ?albipes*. They have been recorded feeding on mistletoes growing on *Eucalyptus* and *Banksia*, including *Dendrophthoe vitellina*, *Muellerina eucalyptoides* and *M. celastroides*. *M. eucalyptoides* grows commonly on *Eucalyptus camaldulensis* (red gum) and on introduced ornamentals even in the inner suburbs of Melbourne, where *O. abrota* may still be found breeding (W. N. B. Quick). The young larvae are found mainly under loose bark on the food plant itself, and mature larvae and pupae are usually found under loose bark not very far from the base of the mistletoe. Should there be no convenient loose bark for shelter, larvae will wander much farther in search of a suitable pupation site. There are apparently two generations annually, but the seasonal history deserves further study. The larvae are sometimes heavily parasitized, particularly the second or summer generation.

47. *Ogyris olane* Hewitson, 1862

This species ranges from northern Queensland to South Australia. In southern Queensland and New South Wales it is more common on the tablelands than on the coast, and in Queensland occurs even 480 km inland, at Mitchell. Two subspecies are recognized, typical *olane* having the more inland distribution.

a. *Ogyris olane ocela* **Waterhouse, 1934**

Plates 24, figs. 11 (male), 11A (female), XVII, olane azure.

DISTRIBUTION.   Northern and central Queensland to Victoria and South Australia; from sea level to about 1,060 m in Queensland and New South Wales, and 600 m in Victoria. It is generally more common on the tablelands. The most northern records are from 24 km west of Kuranda, from near Townsville and at Clermont, Dululu, Rockhampton, and the Carnarvon Range. Specimens from South Australia may be better placed here than in typical *olane*. In Victoria adults are on the wing from September to May, but in central Queensland they fly as early as July.

Specimens from Clermont are a brighter purple above than those from other localities, but nevertheless resemble this subspecies rather than the next.

The larvae are sometimes attended by one or two small black ants, but more often are without ants. Near Canberra the attendant ant is *Crematogaster* sp. The larvae feed at night on mistletoes growing on *Eucalyptus*, especially *Amyema pendulum* and *A. miquelii*. The latter is also the food plant at Toowoomba and Millmerran (J. Macqueen). When young they are usually found in ones or twos under loose bark close to the food plant, but later occur singly under small pieces of loose bark on the trunk of the eucalypt near the ground. Pupae are found singly in the same situations as the mature larvae. There are apparently two generations annually.

b. *Ogyris olane olane* **Hewitson, 1862**

DISTRIBUTION.   Inland southern Queensland and New South Wales. It is known from Dalby and Mitchell, Q., and from many localities in inland N.S.W., including Lightning Ridge, Bourke, Cobar, Walgett, Brewarrina, Coonabarabran and Lake Cargelligo. Some adults reared from pupae collected at Wombeyan Caves, N.S.W., only about 72 km from the coast (E. D. Edwards), and others reared from larvae and pupae taken near Inglewood, Victoria (J. C. Le Souef), seem to be nearer this subspecies than to *ocela*.

ADULT.   Male: similar to *ocela* but much smaller, basal areas brighter purple, extending into cell of forewing and much nearer termen of hindwing.

Female: similar to *ocela* but smaller, basal areas more extensive and much bluer.

There is considerable variation in the shade of the basal areas of the wings, and the distinction between the two subspecies is probably not well defined.

The larvae feed at night on *Amyema miquelii*, sheltering during the day under loose bark near the food plant or in borer holes. They are usually attended by small black ants. The pupae are found under loose bark some distance from the food plant, often near the foot of the *Eucalyptus* host tree.

48. *Ogyris barnardi* Miskin, 1890

This species ranges from Clermont and Lotus Creek, 97 km north-west of Marlborough, Queensland, to Spencer Gulf, South Australia.

a. *Ogyris barnardi barnardi* Miskin, 1890

Plate 24, figs. 12 (male), 12A (female), Barnard's azure.

DISTRIBUTION.   Inland central and southern Queensland as far west as Cunnamulla (J. C. Le Souef), Stradbroke Island, and near Boggabilla (M. De Baar), Lightning Ridge, Brewarrina, Cobar, Bourke, Dubbo, and the Murrurundi district, New South Wales. The original specimen came from the Dawson River, Queensland.

The larvae frequently have one or two small black or brown ants in attendance. They feed on *Amyema quandang* (Loranthaceae) growing on *Acacia* (J. Macqueen), and have also been recorded from *A. miraculosum* (T. H. Guthrie).

b. *Ogyris barnardi delphis* Tindale, 1952

DISTRIBUTION.   The area extending from about 75 km north-west of Port Augusta to Iron Knob and Whyalla, South Australia; common near Iron Knob. Adults are on the wing in October and November.

ADULT.   Male: similar to typical *barnardi* but usually smaller, darker purple above, with broader black margins; beneath with larger bluish white cell bars in forewing.

Female: similar to typical *barnardi* but usually smaller, dull bluish purple above, with rather narrower black margins; beneath as in male.

The subspecies is adapted to much more arid conditions than typical *barnardi*, and might be expected to have an even wider distribution in mulga-saltbush plant communities than is known at present. Adults have been taken flying around *Acacia* trees carrying the food plant.

Both larvae and pupae occur under loose bark on the host tree or in borer holes in the butt or the branches of the mistletoe food plant. The larvae feed on *Amyema quandang* growing on *Acacia sowdenii* (myall).

49. *Ogyris ianthis* Waterhouse, 1900

Plate 24, figs. 7 (male), 7A (female), Sydney azure.

DISTRIBUTION.   Dalby and Millmerran, Queensland, to Sydney, New South Wales; rare. Most specimens have been reared from near Sydney, Millmerran and Leyburn but one female has been taken near Dalby, a male at Glen Aplin, near Stanthorpe, and a female at Gunnedah, N.S.W. No doubt intermediate localities will be discovered. Near Sydney and at Millmerran there appear to be two generations annually, for adults fly mainly from late October to mid November and from late January to late February.

This species is remarkable for its strong sexual dimorphism. The black apical area of the forewing is broader in Sydney males, which fly very rapidly on certain hill-tops, but are seldom caught.

The larvae feed on several species of mistletoe. At Sydney they have been reared from *Muellerina eucalyptoides* and *Dendrophthoe vitellina* growing on *Eucalyptus* and *Angophora*. In the Millmerran district J. Macqueen has found the larvae on *D. glabrescens* and *Amyema miquelii* on *Eucalyptus*, *A. quandang* on *Acacia*, and *A. linophyllum* on *Casuarina*. They are always attended by the small black and brown ant *Froggattella kirbyi*. The larvae, usually several together, are found under loose bark or in borer holes and cracks in the host tree, both on the trunk and on the smaller branches close to the mistletoe. Near Sydney D. P. Sands has noticed that the larvae are relatively more slender and the pupae more elongate than in other species except *O. iphis*. They are thus well adapted to sheltering and pupating in borer holes, where they are more usually found. The pupal duration in January ranges from fifteen to twenty days.

50. *Ogyris iphis* Waterhouse and Lyell, 1914

This species is known only from the Darwin area, Northern Territory, and from north-eastern Queensland, west of Kuranda, Paluma, and Charters Towers. Two subspecies are recognized.

a. *Ogyris iphis iphis* **Waterhouse and Lyell, 1914**

Plate 24, figs. 8 (male), 8A (female), Dodd's azure.

DISTRIBUTION. North-eastern Queensland, from west of Kuranda and Paluma, and from the Burra Range, about 135 km south-west of Charters Towers.

This species is closely allied to *O. ianthis,* but sexual dimorphism is less pronounced. Males have been taken flying on hill-tops in open *Eucalyptus* forest.

The life history and early stages are very similar to *O. ianthis.* Eggs are laid on the butt or on bare twigs of the mistletoe (Loranthaceae) food plant. Younger larvae usually remain close to the mistletoe; older larvae and pupae are found mainly in borer holes and hollow branches of the host tree. Both larvae and pupae are attended by the small brown and black ant *Froggattella kirbyi,* which also attends the larvae of *O. ianthis.* West of Kuranda larvae have been found feeding on *Amyema miquelii* growing on *Eucalyptus* and *Angophora.* There and west of Paluma they occurred on *Eucalyptus* infested with *Dendrophthoe vitellina,* and at Burra Range they have been reported on *Amyema quandang.*

b. *Ogyris iphis doddi* **Waterhouse and Lyell, 1914**

DISTRIBUTION. Known only from Darwin, Northern Territory; very rare.

ADULT. Male: similar to typical *iphis,* but above with narrower black margins, ground colour beneath forewing suffused with dull orange.

Female: similar to typical *iphis* but above with narrower black margins, beneath paler, forewing with larger orange patch below cell.

51. *Ogyris aenone* **Waterhouse, 1902**

Plates 24, figs. 2 (male), 2A (female), XVI, Cooktown azure.

DISTRIBUTION. Thursday and Horn Islands, from Cooktown to Ayr, northern Queensland, and on the western Darling Downs at Millmerran and Leyburn, southern Queensland; fairly common at

Leyburn, rarer in the Millmerran district. A single pupa has been taken near Ayr, and no doubt other intermediate localities will be discovered between Ayr and southern Queensland. At Leyburn adults have been seen or reared by J. Macqueen throughout the year, except in May, June and December.

This is the largest of the light metallic blue species, from the remainder of which it may easily be distinguished by the prominent lobe of the hindwing, the pale blue, whitish or orange subapical patch of the forewing above, and the distinctive pattern of the underside. Adults of both sexes taken during the winter near Leyburn and Millmerran are much darker beneath than those taken at other times of the year.

The known distribution of *O. aenone* is remarkable. In north Queensland it occurs in the moist *Melaleuca* (paperbark) coastal swamps, whereas J. Macqueen and J. F. R. Kerr have found it on the western fringe of the Darling Downs in dry savannah woodland, where the adults are usually seen flying around *Casuarina*.

Near Cairns larvae have been reported on the mistletoe *Dendrophthoe vitellina* (Loranthaceae) sheltering during the day beneath loose bark on the host tree. The pupae were found in the same situations. Both larvae and pupae were attended by the small brown *Pheidole* ants which colonize the epiphytic *Myrmecodia* (ant plant) growing on the same trees. Near Millmerran, however, the larvae feed on the mistletoes *Lysiana exocarpi* and *Amyema linophyllum*, both parasitizing *Casuarina luehmannii*, and occasionally on *A. miquelii* growing on eucalypts. There the larvae and pupae are always attended by a small black ant, *Iridomyrmex* sp. (*itinerans* group).

52. *Ogyris oroetes* Hewitson, 1862

This species is represented by two subspecies, one of which occurs widely in Queensland and New South Wales and the other in South Australia and south-western Australia. Elsewhere the species is known from two adults from Derby and one from the Hamersley Range, north-western Australia, and others from near Alice Springs, Northern Territory, but further specimens are needed from these areas to determine their subspecific status. The adults can be recognized by the relatively straight postmedian band beneath the forewing in both sexes, and the absence of orange-red in the cell beneath the forewing of the female.

a. *Ogyris oroetes oroetes* **Hewitson, 1862**

Plate 24, figs. 4 (male), 4A (female), silky azure.

DISTRIBUTION.   From near Cairns and Kuranda, Queensland to Scone and Cessnock, New South Wales. It has been taken as far inland as Mt Isa (J. C. Le Souef), Clermont, Rolleston, Carnarvon Range, Mitchell and Coonabarabran. The original specimen came from Brisbane. In central Queensland adults have been recorded from June to August and in April.

In the male the lilac tint to the blue above the wings is characteristic. Males from near Kuranda and Cairns have narrower black margins than those from southern Queensland.

The larvae usually feed at night on mistletoes (Loranthaceae) growing on *Eucalyptus*, the recorded species being *Amyema pendulum*, *A. miquelii* and *A. bifurcatum*. The larvae and pupae are found, rarely more than two or three together, under loose bark, often near the food plant.

b. *Ogyris oroetes apiculata* **Quick, 1972**

DISTRIBUTION.   From 70 km north (R. H. Fisher) and 50 km south-east (E. D. Edwards) of Wentworth, New South Wales; from Mt Hope, north-western Victoria (J. C. Le Souef); from Waikerie (M. Moore), Canopus (R. H. Fisher) and the Flinders Ranges, South Australia; and from an area extending from 140 km east of Balladonia to Kellerberrin and north-west to Three Springs and Geraldton (N. McFarland), Western Australia. This subspecies probably has a much wider distribution in the drier areas of western New South Wales and Victoria, and South and Western Australia.

ADULT.   Similar to typical *oroetes*, but above with broader black margins, especially at the apex of the forewing (W. N. B. Quick). Males from south-western Australia also differ by the colour of the wings above, which are without the lilac suffusion found in typical *oroetes* or in specimens of *apiculata* from the Flinders Ranges. Adults collected near Alice Springs by E. D. Edwards are similar to *apiculata*, but the males are bright purple above. A male from the Hamersley Range, W. A. (M. S. Moulds) is also purple above, but with narrower black margins.

Mature larvae are pinkish or greyish brown and are almost devoid of markings. They feed on *Amyema miquelii* (Loranthaceae) growing on *Eucalyptus*.

53. *Ogyris amaryllis* Hewitson, 1862

This is the most widely distributed *Ogyris*, occurring in all the mainland States, both on the coast and far inland.

a. *Ogyris amaryllis amaryllis* Hewitson, 1862

Plate 24, figs. 5 (male), 5A (female), amaryllis azure.

DISTRIBUTION. Nambour, Queensland, to Mittagong, New South Wales, both on the coast and tablelands. Coastal areas from which the species has been reared include the Maroochy River (N. Stockton), Brisbane, the Richmond River, Tuggerah, Minto, Careel Bay, Penrith, and Menangle Park. In addition it has been reared at Stanthorpe and Armidale. The original specimen came from Brisbane. Specimens from Wombeyan Caves, N.S.W., may belong here but additional specimens are needed from this locality.

This subspecies varies somewhat in the width of the black outer margins.

The larvae and pupae are usually attended by small black ants and are found under loose bark, in crevices or cracks on the trunk of the host tree, usually near the base, or in borer holes in the branches carrying the mistletoe food plant. The larvae feed on *Amyema cambagei* growing on *Casuarina*.

b. *Ogyris amaryllis amata* Waterhouse, 1934

DISTRIBUTION. Australian Capital Territory: along the Molonglo, Murrumbidgee and Cotter Rivers and their tributaries. Adults are on the wing from October to March.

ADULT. Male: similar to typical *amaryllis*, but smaller, a deeper shining blue above, and much broader black outer margins.

Female: similar to male, but with broader black margins and costa of forewing broadly black.

Although adults can often be seen flying around *Casuarina*, they usually fly too high and too rapidly to permit capture.

The life history is similar to typical *amaryllis*. The larvae likewise feed on *Amyema cambagei*, infesting *Casuarina cunninghamiana* fringing the rivers. They are attended by small black ants, *Iridomyrmex ?nitidiceps*, and the larvae and pupae occur singly under bark or in crevices near the base of the host tree, or more commonly in hollows and tunnels produced by boring insects in the branches carrying the mistletoe. There are probably two generations annually.

c. *Ogyris amaryllis meridionalis* **Bethune-Baker, 1905**
Plate XVI.

DISTRIBUTION.   Throughout much of inland eastern and southern Australia, where it is sometimes very common. It ranges from inland central and southern Queensland, New South Wales west of the tablelands, western Victoria, South Australia, and Western Australia south from Exmouth (M. S. Moulds). Specimens from Camooweal, Winton, Clermont, Millmerran, Leyburn, Toowoomba, including those from the foot of the Dividing Range near Toowoomba, and from Mitchell, Charleville and Cunnamulla are included here. In addition, specimens from 30 km west of Fairview, Cape York Peninsula, and from Petford, north Queensland, are tentatively assigned here. Bethune-Baker first gave the name to specimens from Victoria. Adults are on the wing from September to May.

ADULT.   Male: above shining bright blue, outer margins narrowly black. Beneath dark grey, apex and termen paler, cell bars whitish edged metallic blue, postmedian band dark brown; hindwing grey, markings brown edged darker brown.

Female: above shining bright blue; forewing with costa and termen broadly black, a broad black bar at end of cell, sometimes merged with broader black subapical area, veins beyond cell black, whitish costal streaks near apex; hindwing costa and termen black. Beneath similar to male, forewing with two scarlet spots in cell, hindwing rich brown centrally and on termen.

The larvae are usually attended by small black ants: those from southern inland New South Wales proved to be *Iridomyrmex* sp. (*rufoniger* group). The young green larvae feed openly during the day on the young shoots of the mistletoe, but by the third instar they become slate-grey in colour and shelter during the day under loose bark or in cracks near the food plant. Older larvae and pupae are found singly under loose bark or in cracks, usually near the mistletoe but sometimes near the base of the host tree.

The larvae have been reared from a wide range of mistletoes (Loranthaceae), including *Amyema linophyllum* growing on *Casuarina luehmannii* (bull oak), *A. miquelii* and *Muellerina eucalyptoides* on *Eucalyptus*, *A. miraculosum* on *Santalum lanceolatum* (sandalwood), *Amyema quandang* on *Acacia pendula* (boree), *Acacia harpophylla* (brigalow), and other wattles, *Amyema maidenii* on *Acacia aneura* (mulga) and *Acacia harpophylla*, *Amyema preissii* on *Acacia*, *Amyema cambagei* on *Casuarina cunninghamiana* (river oak), *A. congener* on *Geijera parviflora* (wilga), *A. bifurcatum* on *Angophora* and other trees, and from *Amyema sanguineum* (E. D. Edwards). G. Daniels has observed females ovipositing on *A. miraculosum boormanii* growing on *Geijera parviflora* (wilga). Although *Lysiana exocarpi* has also been recorded as a food plant, P. Atsatt found that the larvae of *meridionalis* died when fed the foliage of *Lysiana* spp.

d. *Ogyris amaryllis hewitsoni* Waterhouse, 1902

Plate 24, figs. 6 (male), 6A (female).

DISTRIBUTION. Coastal and subcoastal Queensland, from Cairns to Tin Can Bay, south of Maryborough, and Gayndah.

The male is usually paler beneath than in *meridionalis*.

The life history and early stages are said to resemble those of typical *amaryllis* or *meridionalis*. At Mackay and Yeppoon the adults fly commonly around the mistletoe *Amyema mackayense* growing on mangroves; this mistletoe is no doubt the food plant.

e. *Ogyris amaryllis parsonsi* Angel, 1951

DISTRIBUTION. North-western Australia, Northern Territory and northern South Australia; the type specimens came from Aileron, N.T., but others are known from Brocks Creek, Central Mount Stuart, Hermannsburg and many localities in the Alice Springs area (E. D. Edwards), N.T., from the north-west of South Australia, and from the Hamersley Range and Broome (M. S. Moulds), and the Edgar Range, 186 km south-east of Broome, W.A. There appear to be at least two generations annually.

ADULT. Male: similar to *hewitsoni* but above a richer shade of blue, with narrower black margins.

Female: similar to *hewitsoni* but above a richer shade of blue, a narrower black bar at end of cell in forewing, cell and central area suffused white.

Variation has been noted in the shade of blue above, and in the strength of the markings both above and beneath the wings.

Near Aileron, N.T., the larvae feed on a greyish mistletoe (Loranthaceae) growing on *Casuarina decaisneana* (desert oak), and are attended by small black ants. During the day they shelter beneath loose bark or in other crevices on the trunk of the host tree or, when suitable shelter is not available there, in the ant galleries in the soil. From larvae collected in May, adults were reared the following August and September. In the Northern Territory adults were collected in April and May, and south-east of Broome in August. In the Northern Territory and South Australia (R. Manskie) adults flew around mulga (*Acacia aneura*) carrying the mistletoe *Amyema maidenii*.

Genus *JALMENUS* Hübner, 1818

This genus contains nine species confined to eastern, southern and western Australia, in the east south from Cape York, and in the west south from Port Hedland. It is absent from Tasmania.

The eggs are nearly spherical, slightly flattened, with a micropylar depression above and with a reticulated pattern of ridges from which arise numerous pointed spines. They are laid in clusters. With one exception, the mature larvae have paired blunt tubercles at least on some segments; they feed during the day singly or in groups, usually on *Acacia* but sometimes on *Cassia*, *Heterodendrum*, and even *Eucalyptus*. Both larvae and pupae are attended by ants. The pupae are usually shiny, ranging in colour from pale grey to brown, usually with darker markings, or black, and are attached by the anal hooks and central girdle to twigs or small branches of the food plant, or sometimes to a communal web previously spun by the larvae.

54. *Jalmenus evagoras* (Donovan), 1805

This species ranges from central Queensland to central Victoria, both in coastal areas and well inland. The adult has a long sinuous tail at vein CuA_2 of the hindwing, and the termen is also strongly toothed at CuA_1 and at $1A + 2A$, with a smaller projection at M_3. Two subspecies are recognized.

a. *Jalmenus evagoras evagoras* (Donovan), 1805
Plates 25, fig. 5 (male), XVII, XVIII, common imperial blue.

DISTRIBUTION. South-eastern Queensland, eastern New South

Wales, the Australian Capital Territory, and eastern and central Victoria; common but local. It occurs on the tablelands as well as near the coast, and in coastal Queensland extends as far north as Kroombit Tops, west of Gladstone.

Specimens from southern New South Wales and Victoria are nearly always darker above than those from southern coastal Queensland and have broader black markings beneath.

The eggs have a more coarsely reticulated pattern of ridges, with fewer spines, than in *J. ictinus*. The larvae feed in groups during the day on *Acacia* (Mimosaceae), including *A. decurrens*, *A. dealbata*, *A. rubida*, *A. melanoxylon*, *A. irrorata*, *A. polybotrya*, *A. cunninghamii*, *A. falcata*, *A. spectabilis*, *A. binervata*, *A. ingramii* and *A. neriifolia*; at Sydney often on *A. mearnsii*, but also on other species such as *A. terminalis*. They are usually found on plants up to one or two metres in height, and often strip them of foliage. The presence of larvae can often be recognized at a distance by the damage they cause. They are attended by swarms of small black ants, belonging to two or three species of *Iridomyrmex*, and pupate in groups on communal webs spun by the larvae amongst the foliage or twigs. Despite the presence of ants larval parasitism by small braconid wasps is often heavy.

Near Sydney at least two generations are passed each year and the immature stages are most common in October and early February. Eggs laid in the late summer do not hatch until the following spring (J. Macqueen and E. D. Edwards) and the larvae produce adults by November and December. Eggs laid by these adults hatch in a relatively short time and produce the next generation of adults in about two months.

b. *Jalmenus evagoras eubulus* **Miskin, 1876**

Plate 25, fig. 6 (male).

DISTRIBUTION.   Central and southern inland Queensland and far northern New South Wales. It is known from Eungella (D. E. A. Morton), 97 km north-west of Marlborough, Gogango Range, Duaringa, Edungalba, Eidsvold and Millmerran, Q., and 35 km south of Boggabilla, N.S.W. (M. De Baar).

This subspecies is normally distinguished from typical *evagoras* by the very pale whitish blue- or green-tinted central areas above and the narrower black markings beneath the wings.

Normal *eubulus* is probably confined to the brigalow areas of central and southern Queensland and northern New South Wales.

The early stages are similar to typical *evagoras*. The larval food plants include *Acacia harpophylla* (brigalow) and *A. penninervis* (Mimosaceae); larvae have been known to eat the waxy covering of scale insects. J. Macqueen has noticed that this subspecies, in contrast to typical *evagoras* and other species such as *J. ictinus* and *J. daemeli*, apparently cannot adapt itself to cleared brigalow areas, and larvae are never found on the brigalow regrowth in such areas.

55. *Jalmenus eichhorni* Staudinger, 1888

Plates 25, fig. 7 (male), XIX, northern imperial blue.

DISTRIBUTION.   Cape York and Weipa to Trevethan Creek, 27 km south of Cooktown. Adults are fairly common and have been taken from March to November.

This species may be distinguished from all others by the presence of a narrow black subterminal band beneath the forewing.

The larvae feed on *Acacia* (Mimosaceae) including *A. crassicarpa* at Weipa, and are attended by meat ants (*Iridomyrmex* sp.). Unlike *J. evagoras*, the tubercles bear black bristles and dorsolateral tubercles are absent. The larvae are not as deeply indented as *J. ictinus* and the prothoracic plate bears black bristles. They pupate on the food plant, and in June the pupal duration was about twelve days.

56. *Jalmenus ictinus* Hewitson, 1865

Plates 25, fig. 11 (male), XVIII, ictinus blue.

DISTRIBUTION.   Northern, central and southern Queensland to near Melbourne, Victoria; usually more common in the drier sub-coastal and inland areas. In Queensland it is known to occur as far north as an area 96 km west of Bowen; similar specimens from Kuranda and the Atherton Tableland belong to *J. pseudictinus*.

Superficially the adults are almost identical with those of *J. pseudictinus* from the same locality, but the forewing costa is slightly less arched near the base and the male valvae have a strong rounded prominence at the base.

The reticulated pattern of ridges on the egg is finer than in *J. evagoras*, and the spines are more numerous. The spines are more prominent than in *J. icilius*. In coastal areas larvae vary considerably in depth of coloration but are never green. Inland, however, green

284

larvae with darker markings are common. The larvae are apparently always attended by the common mound or meat ant *Iridomyrmex* sp. (*purpureus* group). They feed singly or in groups, usually on *Acacia* (Mimosaceae), including *A. decurrens, A. dealbata, A. mearnsii, A. melanoxylon, A. cunninghamii, A. pendula* (boree), *A. harpophylla* (brigalow), *A. rubida* and *A. bidwillii*. Near Canberra they have been noticed attacking and hollowing out galls growing on *A. rubida* (D. Ferguson). At Millmerran they also feed on *Heterodendrum diversifolium* (Sapindaceae). They pupate singly or in twos or threes on the food plant, usually on the leaves, twigs, or branches, but sometimes under loose bark, in cracks in the trunk, or in cracks in the soil at the base of the tree. Unlike *J. icilius* they have not been noticed in any form of shelter constructed by the larvae. In inland areas the pupae, even on one tree, vary greatly in coloration, and much more so than near the coast. Inland they are sometimes very pale, the extreme being pale yellowish brown with the darker coloration restricted to the wing margins, antennae and eyes. In coastal districts pupae are normally deep brown to black, and very pale pupae seldom if ever occur.

57. *Jalmenus pseudictinus* Kerr and Macqueen, 1967
Plate XVIII.

DISTRIBUTION. Known from Kuranda, Herberton, 65 km south of Mt Garnet (W. R. Hindson), Cardwell, 16 km west of Paluma, and Mt Surprise (M. S. Moulds), north Queensland, the Gogango district, central Queensland, and the Millmerran district, south Queensland; very local. In the north adults and larvae have been collected in April and October; near Millmerran adults have been reared in November and December, and in February and March.

ADULT. Similar to *J. ictinus* in both sexes, but forewing costa slightly more arched, male valvae rather narrower and more elongate, without a strong rounded prominence at base of costa.

The adults may very easily be mistaken for Queensland specimens of *J. ictinus*, but the very distinct larvae, attended by two widely different species of ants, show that the two species are distinct.

The larvae are much greener than *J. ictinus* from the same areas, with much less prominent tubercles, and without the black bristly hairs that are a feature of *J. ictinus*. In the Millmerran district they nearly always feed on *Acacia harpophylla* (brigalow, Mimosaceae), but are occasionally found on *Heterodendrum diversifolium* (Sapindaceae). Larvae found recently by J. F. R. Kerr and J. Macqueen about 16 km west of Paluma, north Queensland, were feeding on *A. flavescens*. Both larvae and pupae are invariably attended by numerous small reddish ants with black abdomens, *Froggattella kirbyi*. The pupae are rather more elongate than *J. ictinus*, with more pointed abdomen. Near Millmerran several pupae often occur together in cracks and crevices in the trunk of the food tree, or under loose bark of neighbouring dead trees, or in other convenient crevices near by. There pupae are occasionally found on the twigs and phyllodes of the food plant, but these are generally parasitized. Near Paluma pupae were found amongst debris at the foot of the tree and on phyllodes; the latter were not parasitized.

58. *Jalmenus daemeli* Semper, 1879
Plate 25, fig. 8 (male), Dämel's blue.

DISTRIBUTION.   Cairns to Brisbane, and inland at Hughenden (S. Brown), 97 km north-west of Marlborough, several localities between Rockhampton and Anakie, Eidsvold, Gayndah, Toowoomba, Stanthorpe, Millmerran, and 72 km north of Narrabri, N. S. W. (R. P. Field); very local, common at the more inland localities. At Millmerran adults have been reared or taken from November to April.

The pale markings beneath the wings distinguish this from *J. ictinus* and allied species.

The larvae feed singly on *Acacia* (Mimosaceae), especially *A. cunninghamii* and *A. harpophylla* (brigalow), but also *A. neriifolia*, *A. bidwillii*, *A. bancroftii*, *A. macradenia*, *A. decurrens* and *A. irrorata*. Larvae sometimes occur on the young foliage of *Heterodendrum diversifolium* (Sapindaceae). At Toowoomba they have been found on *Eucalyptus melanophloia* (silver-leaf ironbark) and *Eucalyptus* sp. (bloodwood, Myrtaceae), as well as *Acacia*. They are attended by numerous small black ants, *Iridomyrmex* sp. (*rufoniger* group), and pupate singly on small branches of the food plant.

59. *Jalmenus lithochroa* Waterhouse, 1903
Plate 25, fig. 12 (male), lithochroa blue.

DISTRIBUTION.   Restricted to south-eastern South Australia, where it has been taken at Melrose, Yunta and near Adelaide, from October to early April, but especially in January and February. The species must be very rare near Adelaide nowadays.

This is a relatively dull species, with a shorter tail than in previous species. Specimens from near Adelaide usually have obscure brown markings beneath, but those collected and reared by R. H. Fisher from near Melrose were very distinctly marked.

The larvae feed on *Acacia pycnantha* (Mimosaceae) near Adelaide, and are attended by numerous small black ants. At Melrose the larvae were found on *A. victoriae* and were attended by a large black ant. Those taken back to Adelaide would not accept the foliage of *A. pycnantha*. Younger larvae sheltered during the day, often two or more together, beneath or between small branches of the food plant, and fed at night, whereas larger larvae fed during the day on the young leaves and flower buds. Pupae were attached near the base of the trunks of *A. victoriae*, or were sometimes attached to debris and dead leaves caught in the foliage.

60. *Jalmenus inous* Hewitson, 1865

Plate 25, fig. 9 (male), inous blue.

DISTRIBUTION. South-western Australia, south from Carnarvon.

As in *J. icilius*, the hindwing in this species is without a tail, but it is usually larger than *J. icilius* and the broad dark brown bands with white margins beneath the wings usually distinguish it. In some specimens, however, the markings beneath are very obscure. The significance of variation in this and other Western Australian species deserves close study.

The larvae feed at night on *Acacia cyanophylla* (Mimosaceae) from October to December and perhaps later, resting during the day singly or in twos or threes along the stems of the food plant just below ground level. They are attended by numerous small black ants, identified as *Iridomyrmex ?gracilis*. The pupae are attached to the stem of the food plant also just below ground level, or to fallen leaves or debris on the ground, and are attended by the ants. At Bunbury, during the first week in November, all stages were present. Larvae and pupae were also plentiful in sandy areas near Mandurah from the second week in November to early December.

61. *Jalmenus icilius* Hewitson, 1865

Plates 25, fig. 10 (male), XVIII, icilius blue.

DISTRIBUTION.   Southern inland Queensland and inland New South Wales, the Australian Capital Territory, and from north-central and north-western Victoria to the Mt Lofty Ranges, Yorke Peninsula, Eyre Peninsula, and Flinders and Musgrave Ranges, South Australia, and in Western Australia south from Kalbarri and inland as far as Meekatharra (M. Hutchison); it has also been recorded from near Alice Springs, central Australia; local.

This is one of the smallest species of the genus, characterized by the very short projection to the hindwing and the very pale markings beneath the wings. Adults vary considerably in size and in the brightness of the markings above and beneath. Specimens from south-western Queensland, the Yorke Peninsula, S.A., and from the Geraldton and Carnarvon areas of Western Australia tend to be smaller and duller than those from elsewhere.

The larvae feed on *Cassia nemophila*, *C. artemisioides* (Caesalpiniaceae) and *Acacia* (Mimosaceae), including *A. pendula* (boree), *A. harpophylla* (brigalow), *A. aneura* (mulga), *A. pycnantha*, *A. rubida*, *A. victoriae*, *A. calamifolia*, *A. parramattensis* (D. Ferguson) and *A. anceps*. They are attended by variable numbers of small black ants, *Iridomyrmex* sp. Those attending larvae on *A. pendula* near Yanco, N.S.W., were *Iridomyrmex* sp. (*rufoniger* group). The larvae sometimes pupate in a crude shelter formed by drawing together two or three phyllodes with a few strands of silk, but more often without shelter on a phyllode, a twig, or in a fork or crack of the tree trunk, or under fallen leaves, bark or debris at the base of the tree; sometimes pupae are found in the ant galleries.

62. *Jalmenus clementi* Druce, 1902

Plate 14, fig. 3.

DISTRIBUTION.   North-western Australia; known from an area between Exmouth (M. S. Moulds), the Yule River and the Hamersley Range; very local.

As in *J. inous* and *J. icilius* the hindwing is without a tail, and the terminal projections are even less noticeable. The species is smaller and duller than *J. inous*, with the central greenish or bluish areas more brassy. The valva of the male genitalia is shorter and broader than in either of the other two species.

Larvae of *J. clementi* differ from other *Jalmenus* larvae in not having prominent dorsal tubercles on any segment. They are very variable in colour and pattern, but mature larvae seem to be more commonly brown in colour. Young larvae and occasionally older larvae and pupae are attended by small *Iridomyrmex* ants. The pupae, which vary in size, are usually attached to the broad phyllodes of the food plant, sometimes between two phyllodes loosely drawn together with a few strands of silk, or in the hollows of deformed or curled phyllodes.

Adults, larvae and pupae were collected at Tom Price in the Hamersley Range in late December and early January by J. Olive, and in November by M. S. Moulds. The larvae were feeding on *Acacia inaequilatera* (Mimosaceae), a wattle with broad leathery glaucous phyllodes, each of which has a stout terminal spine and a pair of basal spines. They showed a strong preference for the young growth.

Genus *PSEUDALMENUS* Druce, 1902

The genus contains only a single endemic species, so far known only from New South Wales, the Australian Capital Territory, Victoria and Tasmania.

The egg is mandarin-shaped, with a flat base and dense pattern of irregular raised ridges on the surface, with intervening shallow pits. The larvae are without noticeable projections or protuberances. They are slightly flattened, with the seventh and eighth abdominal segments strongly depressed, the depressed area containing the spiracles of these two segments, the median dorsal organ, and the pair of dorsolateral eversible organs. The anal plate is also depressed. The skin is covered with minute spinules, each arising from a stellate base. The head is also clothed with very fine short hairs, and the main dorsal and lateral setae are minutely branched.

The larvae feed exposed during the day attended by numerous small black ants, probably *Iridomyrmex foetans*. The dark brown or dull black pupae have a finely rugose surface. They are found under loose bark, or in crevices or holes, or under stones, attached by the anal hooks and a central girdle.

63. *Pseudalmenus chlorinda* (Blanchard), 1848

This species displays a strong geographic variation and seven subspecies have been recognized. A remarkable situation occurs in Tasmania where four subspecies are found. There populations separated by only a few kilometres have widely different markings. Both in Tasmania and on the mainland *P. chlorinda* is one of the earliest species to fly in the spring. The larvae complete their development during the early summer, and the pupae remain dormant during the remainder of the summer and throughout the winter.

a. *Pseudalmenus chlorinda zephyrus* Waterhouse and Lyell, 1914

Plate 25, fig. 13 (male).

DISTRIBUTION. The Southern Tablelands and south coast of New South Wales, the Australian Capital Territory, eastern Victoria as far west as Gisborne, and coastal north-eastern Tasmania; very local. Specimens have been taken in the Brindabella Range, A.C.T. (900 to 1,200 m) and in the Tinderry Mts, the Budawang Range, 20 km west of Ulladulla, and Monga, southern N.S.W.

This is a variable subspecies; specimens from northern Tasmania usually lack an orange spot on the inner side of the black bar at the end of the cell in the forewing.

The larvae feed openly on the young leaves of *Acacia* (Mimosaceae), usually *A. melanoxylon* (blackwood) or *A. dealbata*, and apparently nearly always on *A. dealbata* in Tasmania. At 900 m in northern Victoria larvae are found nearly always on *A. dealbata* but in the A.C.T. more frequently on *A. melanoxylon* (S. Brown). They occur on large trees or on small bushes less than a metre in height. In the Budawang Range (500 m) larvae have been found on *A. obtusifolia* (E. D. Edwards), and in the Monga State Forest on *A. trachyphloia*. Larvae feeding on *A. melanoxylon* often pupate under loose bark or in cracks or borer holes in the trunk of the food tree. Those feeding on *A. dealbata* often pupate under loose bark on the trunk of a neighbouring *Eucalyptus*, usually *E. viminalis*, with which the attendant ants, *Iridomyrmex foetans*, are associated. Both larvae and pupae are usually attended by numerous ants. At lower altitudes in Victoria larvae mature in about one month, from mid-November to mid-December, but at 900 m above sea level have been found well into January. Pupae remain dormant until after the following winter and the adults emerge from the end of August until November or December; they are rarely taken in January even at the higher altitudes.

b. *Pseudalmenus chlorinda chlorinda* (Blanchard), 1848

DISTRIBUTION.   Eastern Tasmania, from south of Hobart, the eastern shores of the Derwent River, and the east coast near Swansea inland to the South Esk and upper Tamar valleys; very local. The original specimens probably came from Richmond.

ADULT.   Male: above similar to *zephyrus*, but forewing with more restricted orange markings, often only two obscure orange spots, one beyond black bar at end of cell and other towards tornus, and with a larger area of scattered orange scales between them; hindwing without central orange patch, red subterminal band narrow and restricted to three spots near tornus. Beneath with ground colour dull grey.

Female: similar to male, but forewing with a prominent orange patch beyond black bar at end of cell; hindwing with a small irregular orange central area, subterminal red band slightly more extended than in male, but less than in female *zephyrus*. Beneath similar to male.

Specimens of this subspecies may be reliably identified by the distinctly dull grey tint of the ground colour beneath.

The larvae of this and other Tasmanian subspecies usually feed on *Acacia dealbata* or, more rarely, *A. mearnsii*, but have not been recorded on *A. melanoxylon*. They pupate beneath loose bark on the trunks of neighbouring *Eucalyptus* trees, especially *E. viminalis*, sometimes more than three metres away from the food plant, which harbour the numerous small black ants, reported to be *Iridomyrmex foetans*, that attend the larvae and pupae. Larvae may be found on *A. dealbata* plants ranging in height from only a few centimetres to four or five metres.

c. *Pseudalmenus chlorinda myrsilus* (Westwood), 1851

DISTRIBUTION.   Tasman Peninsula and the coast opposite Maria Island, south-eastern Tasmania.

ADULT.   Male: above similar to *zephyrus*, but forewing with a prominent orange area, crossed by black veins, and nearly surrounding broad black bar at end of cell; hindwing with a small irregular orange patch at end of cell, subterminal red band broad, extending nearly to vein M_1 and enclosing two black spots near tail, the one nearer tornus very small. Beneath whitish grey, forewing markings as in *zephyrus*, hindwing with a narrow median band of black spots from costa to vein M_3, a uniform red subterminal band.

Female: similar to male but forewing with more extensive orange patch, veins not black; hindwing with much larger orange patch at end of cell, often connected to broad subterminal red band. Beneath similar to male.

In size, brilliance of markings and the whitish ground colour beneath, this subspecies approaches *chloris*.

d. *Pseudalmenus chlorinda conara* Couchman, 1965

DISTRIBUTION.   North-central midlands, Tasmania. The original specimens came from Conara; others are known from the area between Conara, Bothwell and Kempton.

ADULT.   Male: above similar to typical *chlorinda*, but forewing with orange area beyond cell almost completely overlaid with black scales, base sprinkled with blue scales; hindwing with central red band obscured by black scales. Beneath dull grey, subterminal red band absent or reduced to a thin line between veins M_2 and CuA_1, and above this replaced by black subterminal spots.

Female: similar to typical *chlorinda*, but orange area of forewing completely overlaid with black scales and only very faintly indicated; hindwing black except for subterminal red band. Beneath as in male, but hindwing with a broad complete postmedian band of black spots, subterminal red band absent or reduced to three dull red spots near tail and above this replaced by black subterminal spots.

This is a melanic subspecies which appears to have a very restricted distribution.

e. *Pseudalmenus chlorinda fisheri* Tindale, 1953

DISTRIBUTION.   Grampians, western Victoria. Specimens have been taken at Mt Victory (365 m) and Mt Rosea (700 m).

ADULT.   Male: above similar to *zephyrus*, but orange patch of forewing prominent, enclosing on three sides black bar at end of cell, and crossed by black veins; hindwing with a small orange patch at end of cell, a broad red subterminal band extending to vein M_1 and partly enclosing two black spots near tail. Beneath similar to *zephyrus*, hindwing with broad red subterminal band partly or largely enclosing a series of black spots.

Female: similar to male, but hindwing with much larger orange patch at end of cell, and broader red subterminal band.

This subspecies has more prominent orange and red markings than *zephyrus*, with the black subterminal spots of the hindwing more prominent.

The larvae feed on shrubs of *Acacia melanoxylon* (blackwood) and are attended by small black ants. Pupae are found in crevices in the bark of the *Acacia*, and are apparently not attended by ants. Adults have been collected in November, and mature larvae and pupae in late December. A few adults emerged from these the next month, but most of the pupae remained dormant until the following September. Thus under some conditions a partial second generation of this subspecies may occur.

f. *Pseudalmenus chlorinda chloris* Waterhouse and Lyell, 1914

Plate 25, fig. 14 (male).

DISTRIBUTION. Katoomba, Blackheath, Oberon, Mittagong, Bellmore Falls (S. Brown) and Mt Kembla districts, New South Wales; rare and extremely local, occurring principally in gullies at altitudes of from 730–1,200 m.

This is one of the brightest subspecies; it can be distinguished from all but *myrsilus* and *barringtonensis* by the silky white ground colour beneath. Unlike *myrsilus* the orange patch of the forewing extends well into the cell, almost as much before the black cell bar as after it. In the male the veins crossing this patch are not black. In both sexes the red subterminal band of the hindwing is not quite as broad as in *myrsilus*, and the black spots near the tail are not enclosed by red. It can be separated from *barringtonensis* by the separate central and subterminal orange and red areas of the hindwing.

The larvae feed on *Acacia elata* (cedar wattle), *A. dealbata*, *A. melano-xylon* and *A. terminalis*, and consume relatively large quantities of the foliage; they sometimes also eat galls which grow on the food plant. They pupate under stones or in curled leaves on the ground, or in borer holes, when they have been feeding on small wattles, or probably in borer holes in the branches and trunks when feeding on larger trees.

They feed during November and December, and remain as pupae until the following spring. Adults are on the wing from September to November.

g. *Pseudalmenus chlorinda barringtonensis* **Waterhouse, 1928**

DISTRIBUTION. Barrington Tops, New South Wales, from 1,200–1,400 m. A few specimens taken by C. W. Frazier at Point Lookout, New England National Park (1,580 m), apparently belong to this subspecies.

ADULT. Male: above similar to *chloris*, but hindwing with red subterminal band enlarged and merging with central orange patch, and enclosing two black spots near tail, the one nearer tornus very small. Beneath similar to *chloris*.

Female: similar to male, but orange area of forewing and orange and red areas of hindwing larger.

This is the brightest subspecies and can be recognized by the single large orange-red area on the hindwing above and the silky white ground colour of both wings beneath.

The larvae feed during the day on young foliage of *Acacia dealbata*, and occasionally of *A. melanoxylon*, in small groups when young but singly later. They may be found from November to January. Both larvae and pupae are attended by numerous small black ants. Pupae are present from December until the following spring; they are found under loose bark near the base, or in borer holes in the trunk or branches of the food tree, under sticks, leaves or stones on the ground near by, or under loose bark on the trunks of neighbouring eucalypts. Adults are on the wing from October to January.

Genus *HYPOLYCAENA* C. and R. Felder, 1862

The genus is well developed in tropical Africa and is also distributed through the Oriental and Australian regions as far as the Bismarck Archipelago.

64. *Hypolycaena phorbas* (Fabricius), 1793

This species occurs in New Guinea, the Bismarck Archipelago and northern Australia. Two subspecies are known from Australia.

a. *Hypolycaena phorbas phorbas* (Fabricius), 1793

Plates 25, figs. 15 (male), 15A (female), XVIII, common tit.

DISTRIBUTION. Islands of Torres Strait, and Cape York to Yeppoon; common.

The adults are easily recognized. They fly rapidly and settle on the ends of twigs.

The larvae are very variable in colour and markings. They are attended by numerous green tree-ants *(Oecophylla smaragdina)* and feed on many different plants, including *Cupaniopsis anacardioides* (Sapindaceae), *Faradaya splendida, Clerodendrum floribundum* (Verbenaceae), *Cassia alata, C. fistula* (P. S. Valentine) (Caesalpiniaceae), *Planchonia careya* (Lecythidaceae), *Flagellaria indica* (Flagellariaceae), *Acmena* (Myrtaceae), and mistletoe (Loranthaceae). They are usually found beneath the leaves during the day and pupate head downwards, sometimes several together, on twigs or stems of the food plant.

b. *Hypolycaena phorbas ingura* Tindale, 1923

DISTRIBUTION. Northern Territory, including Groote Eylandt; fairly common.

ADULT. Male: similar to typical *phorbas*, but forewing usually a duller blue, hindwing usually more purplish. Beneath pale grey, markings as in typical *phorbas*.

Female: similar to typical *phorbas*, but central white area of forewing much less prominent; beneath pale grey or greyish white, markings as in male.

Males are variable and can scarcely be distinguished from typical *phorbas*. However, the less prominent white area in the forewing of the female appears to be a constant feature of this subspecies.

65. *Hypolycaena danis turneri* (Waterhouse), 1903

Plates 25, fig. 16 (male), XIX, black and white tit.

DISTRIBUTION. Islands of Torres Strait, and Cape York to Innisfail and Kuranda; common at Cairns.

This is a very distinctive, easily recognized species.

The larvae feed on the flowers of orchids (Orchidaceae) and are particularly troublesome in Cairns where they attack the exotic *Vanda* orchids and the native *Dendrobium bigibbum* (Cooktown orchid) grown in gardens. If buds and flowers are not available the larvae feed on the seed pods, leaves, or even the younger stems.

Genus *DEUDORIX* Hewitson, 1863

The genus ranges from India and Sri Lanka to Taiwan, through south-east Asia and the Philippines to the Moluccas, New Guinea, Australia, the Solomons, the New Hebrides and Samoa. Only two species reach Australia.

66. *Deudorix epijarbas* (Moore), 1857

This is the most widely distributed species in the genus, its range encompassing that of all the other species, from India and Taiwan to New Guinea and the islands of the south-west Pacific. Two subspecies have been distinguished in Australia.

a. *Deudorix epijarbas diovis* Hewitson, 1863

Plate 25, figs. 17 (male), 17A (female), cornelian.

DISTRIBUTION.   Mackay to Gosford; rare south of the Manning River, usually uncommon farther north. Away from the coast it has been taken at the Expedition Range, Toowoomba, Stanthorpe and Millmerran.

The males are very bright insects and cannot be mistaken. The females are dull, but the shape of the hindwing and the markings of the underside should assist in placing them. Adults of both sexes fly rapidly and settle on the tips of twigs, often high above the ground.

The larvae feed on the seeds of *Harpullia pendula* (tulipwood, Sapindaceae), which are contained in a bright orange capsule. J. F. R. Kerr has noticed that, as in *Virachola democles*, the larva makes a small round hole in the capsule through which it pushes its faecal pellets with its head and which it then blocks with its flattened anal plate. When the seeds are devoured, the larva can then move to another capsule, entering it by chewing a hole in the rind. Pupation occurs within the empty seed capsule and the newly emerged adult escapes through the hole made by the larva. In Queensland the larvae have also been reported feeding on *Macadamia* nuts, and a pupa was found in a seed pod of *Buckinghamia celsissima* (both Proteaceae) (M. De Baar). On another occasion, larvae which had eaten all the *Harpullia* seeds supplied accepted loquat *(Eriobotrya japonica)* seeds.

296

b. *Deudorix epijarbas dido* **Waterhouse, 1934**

DISTRIBUTION.   Cape York to Kuranda, Cairns, El Arish and Tully; not common.

ADULT.   Male: above similar to *diovis*, but with central areas above dull red; forewing with terminal fringe brown; hindwing veins black. Beneath similar to *diovis*, but tinged with purple, hindwing with less metallic bluish green scaling near tornus.

Female: similar to *diovis*, but brown above rather than grey. Beneath as in male but less purplish.

This subspecies is generally duller than *diovis*.

At the Claudie River, Cape York Peninsula, the pupae have been found in hollowed-out fruit of a palm with bipinnate fronds, *Caryota rumphiana* (Arecaceae). Not more than one pupa occurred in any one fruit and, as many eaten-out fruit contained no pupae or pupal skins, larvae probably attack more than one fruit during their development. Each fruit attacked was secured to its stalk with silk.

67. *Deudorix epirus agimar* **Fruhstorfer, 1908**

Plate 23, fig. 1 (male).

DISTRIBUTION.   Islands of Torres Strait, Cape York, the Claudie River, and Rocky River, 20 km north of Silver Plains; not common. This subspecies also occurs in Papua New Guinea, and others are known from the Moluccas, the Aru Islands and New Guinea.

The termen of the forewing in the female is more rounded than in the male. Few Australian specimens of this handsome species are represented in collections, but the species is more abundant in New Guinea. Both sexes resemble *Virachola democles* above, but they are larger, the male is without a sex-brand, and the female has a larger area of white on the forewing and also a white patch on the hindwing.

At the Claudie River adults have been reared by M. De Baar from larvae feeding on the seeds of *Harpullia angustifolia* (Sapindaceae). In Papua New Guinea the larvae also feed on the seeds of *Harpullia*.

Genus *VIRACHOLA* **Moore, 1881**

The genus is well represented in tropical Africa, but only a few species occur in the Oriental and Australian regions. One species is known from north-eastern Queensland and another from the Northern Territory.

68. *Virachola democles* (Miskin), 1884

Plate 23, fig. 2 (female).

DISTRIBUTION.   Prince of Wales Island and Batavia Downs to the Claudie River, Coen, Port Stewart, Cooktown, Cairns, and the Basilisk Range near Innisfail, Queensland; not common in collections.

The female has much broader wings than the male. Both sexes are somewhat like *Deudorix epirus agimar* above, but in the male the blue areas are deeper in colour, and in the female the white area of the forewing is smaller and there is no white area on the hindwing. The pattern beneath the wings readily distinguishes both sexes of the two species. Adults of *V. democles* are much more similar to *V. smilis*, but beneath lack the prominent spot in the cell of the forewing, the spot above the cell of the hindwing, and the postmedian band is more or less continuous. The white patch of the female forewing is much larger than in *V. smilis*. The adults fly swiftly and are not often collected, but once the food plant is recognized adults are not difficult to rear from the larvae and pupae.

Near Cairns the larvae feed on the seeds within the fruits of a climbing plant, *Strychnos minor* (Loganiaceae). At several localities in Cape York Peninsula adults have been reared from larvae and pupae found within the fruits of *S. lucida*. The larva eats out the contents of a fruit, ejecting its faecal pellets through a hole about 4 mm in diameter in the rind. J. F. R. Kerr noticed that when the larva is not feeding it blocks this hole with its anal plate. Before pupating in the hollow shell of the fruit, the larva attaches the fruit to its stem with silk and covers the hole with webbing. The newly emerged adult leaves the fruit through the hole used for ejecting frass after it has ruptured the webbing.

69. *Virachola smilis dalyensis* Le Souef and Tindale, 1970

Plates 13, fig. 9 (female), XIX.

DISTRIBUTION.   Northern Territory; adults have been taken or reared at Darwin, Daly River Crossing, Point Stuart, and east to the South Alligator River.

This species is rather similar to *V. democles* but both sexes may be distinguished by the underside, especially by the prominent red-brown spot in the cell of the forewing, another above the cell in the hindwing, and the broken postmedian band of the forewing. In the female the white patch above the forewing is much smaller.

Adults were taken on flowers of cashew nut in the Northern Territory, in company with *Hypolycaena phorbas ingura*. In flight the two species were difficult to distinguish.

The larvae feed in the small orange-coloured fruits of *Strychnos lucida* (Loganiaceae), a small tree growing commonly near Darwin. They hollow out the inside of the fruit, usually devouring the seed; they pupate within the hollow fruit, usually only one to a fruit, but sometimes two and rarely three. The adult escapes through a round hole in the hard rind previously made by the larva.

Genus *RAPALA* Moore, 1881

The genus is distributed from India and Sri Lanka through south-east Asia and the Philippines to New Guinea and Australia. Only one species is found in eastern Queensland.

70. *Rapala varuna simsoni* (Miskin), 1874

Plate 25, fig. 18 (male), indigo flash.

DISTRIBUTION.   Islands of Torres Strait, and Cape York to Yeppoon; fairly common. This subspecies has also been recorded at Cooran Tableland (G. B. Monteith) and at Brisbane, and there are specimens in the Australian National Insect Collection from New Guinea and Normanby Island.

The life history and early stages have not been described from Australia. However, adults have been reared at Mareeba (R. I. Storey) from larvae feeding on the flowers of litchi, *Litchi chinensis* (Sapindaceae). At Stanwell, near Rockhampton, adults have been noticed laying eggs on the flower buds of *Alphitonia excelsa* (Rhamnaceae); the young larvae eat out the buds and later feed on the young foliage (W. N. B. Quick).

Genus *BINDAHARA* Moore, 1881

The genus contains three species ranging from India through south-east Asia and the Philippines to the Moluccas, New Guinea, Australia and the Solomons. One subspecies occurs in north-eastern Queensland.

71. *Bindahara phocides yurgama* Couchman, 1965

Plate 25, figs 19 (male), 19A (female), Australian plane.

DISTRIBUTION. Islands of Torres Strait, and Cape York to Townsville; rare. At Kuranda adults have been taken from November to May.

The colour and pattern seem to be fairly constant in both sexes, but males from Cape York Peninsula appear to be consistently paler beneath than those from Cairns (J. F. R. Kerr). In particular, the hindwing beneath is cream in colour instead of brownish yellow as in specimens from Cairns.

This species cannot be mistaken for any other, with its long tails and distinctive pattern. It appears to be confined to the rain forest and the adults fly rapidly, somewhat like a skipper, and often settle on the end of a twig high above the ground.

The life history and early stages have not been described from Australia.

In Queensland the larvae were discovered by M. J. Manski feeding on the seeds within the berries of *Salacia chinensis* (Hippocrateaceae).

Genus *ANTHENE* Doubleday, 1847

The genus occurs in Africa and from southern India, through south-east Asia to New Guinea, Australia and the Solomons.

72. *Anthene seltuttus affinis* (Waterhouse and Turner), 1905

Plates 28, figs. 11 (male), 11A (female), XIX, dark ciliate blue.

DISTRIBUTION. Northern Territory, including Groote Eylandt, islands of Torres Strait, and Cape York to Yeppoon; common. A single specimen has been taken as far south as Ballina, New South Wales (D. Yeates).

This species has often been confused with the Oriental *A. emolus* (Godt.), but the male genitalia show them to be distinct. From *A. lycaenoides* it can be separated by the broken or irregular postmedian band beneath the forewing.

300

During the day the young larvae are found beneath the young terminal leaves of the food plant, and the older larvae here and on older leaves and twigs. They feed on *Cupaniopsis anacardioides* (Sapindaceae), *Cassia fistula* (Caesalpiniaceae) (P. S. Valentine) and other plants, and are always attended by numerous green tree-ants, *Oecophylla smaragdina*. The flattened ventral surface of the pupal abdomen is somewhat like that found in *Arhopala*.

73. *Anthene lycaenoides godeffroyi* (Semper), 1879

Plate 28, figs. 12 (male), 12A (female), pale ciliate blue.

DISTRIBUTION.   Northern Territory, Darnley and Moa Islands and Cape York to Bingil Bay, near Tully (W. R. Hindson), Cardwell (J. F. R. Kerr), Townsville, and Herberton; much rarer than *A. seltuttus*.

The broad unbroken postmedian band beneath the forewing, and the whitish central patch above the forewing of the female, will distinguish this species from *A. seltuttus*. Two pairs collected at Darwin by J. F. R. Kerr are much darker than Queensland specimens, and the whitish central patch of the female forewing is largely obscured.

Near Cairns larvae have been reared from *Caesalpinia crista*, *Cassia alata* and *C. fistula* (all Caesalpiniaceae), *Pongamia pinnata* (Fabaceae), *Cupaniopsis anacardioides* (Sapindaceae), and *Faradaya splendida* and *Clerodendrum* (both Verbenaceae). Pupae are found on the stem or leaves of the food plant.

Genus *CANDALIDES* Hübner, 1819

This genus ranges from the Lesser Sunda Islands to the Aru Islands, New Guinea, the Louisiade Archipelago and D'Entrecasteaux Islands and Australia. Twelve species are known from Australia.

The pupae are roughly triangular in cross-section, with a strong dorsal ridge and abdomen expanded laterally into a pair of flanges (Plate XX, figs. 1D, 1E). These flanges are sometimes curled upwards at the sides, and the pupa also has a flattened anterior lobe, which may be strongly indented in the middle.

74. *Candalides gilberti* Waterhouse, 1903

Plate 26, figs. 3 (male), 3A (female), Gilbert's blue.

DISTRIBUTION.   North-western Australia, Northern Territory, including Melville Island; rare. Adults have been taken at the Prince Regent River, W.A., Darwin, Howard Springs, Adelaide River, Cobourg Peninsula, South Alligator River (E. D. Edwards), Mt Bundey (J. F. R. Kerr), Daly River Crossing and 80 km south-west of Daly River Crossing, N.T.

Both sexes have the apex of the forewing rather more acute than in the next four allied species, and the female is without the large white central area on each wing.

The life history and early stages have not been recorded, but the shape of the pupa is said to be typical of the genus. At Mt Bundey J. F. R. Kerr found a pupal skin of this type under loose bark on a *Eucalyptus* carrying mistletoe and suggested that the larvae may feed on mistletoe as in *C. margarita*.

75. *Candalides margarita margarita* (Semper), 1879

Plate 23, fig. 9 (male).

DISTRIBUTION.   Thursday, Horn and Prince of Wales Islands, and Cape York to Port Macquarie; fairly common. Another subspecies occurs in New Guinea and the Aru Islands.

The male is distinguished by the clearly visible trident sex-mark on a dull purplish blue forewing, and by the valva of the genitalia, which is broad at the base and tapered to a bent rounded apex. The female is very similar to *C. helenita*, with the white patch of the hindwing reaching the apex, but the bases of both wings are dull blue. The white patches are much larger in females from north Queensland than in those from farther south. The adults are on the wing throughout the year in southern coastal Queensland. Compared with *C. absimilis*, which occurs in the same localities, they are more robust and fly more vigorously.

The larvae feed on the young shoots and flower-buds of several species of mistletoe (Loranthaceae). Those feeding on the flower-buds are often pinkish brown in colour. In coastal southern Queensland the usual food plants are *Amyema congener*, *Amylotheca dictyophleba*, *Benthamina alyxifolia*, *Dendrophthoe vitellina* and *Muellerina celastroides*. Near Yeppoon a pupa was found on *Amyema conspicuum* growing on *Alphitonia excelsa* (red ash). Sometimes the larvae are each attended by two or three small black ants, *Technomyrmex sophiae*. They may be found throughout the year, but are more abundant during April. The pupal duration varies from about three weeks in summer to about five weeks in winter. In the Cairns district the larvae have also been recorded on *Cupaniopsis anacardioides* (Sapindaceae).

76. *Candalides helenita helenita* (Semper), 1879

Plate 26, figs. 2 (male), 2A (female), helenita blue.

DISTRIBUTION. Prince of Wales Island, and Cape York to Kuranda, the Atherton Tableland, Bingil Bay, and Paluma; fairly common. At Kuranda it has been taken throughout the year. Another subspecies, *dimorpha* Röber, occurs in New Guinea.

The males may be identified by the greenish blue colour above, and the distinct trident sex-mark on the forewing.

Near Cairns the larvae feed on *Cryptocarya hypospodia* (Lauraceae) and *Brachychiton acerifolium* (Sterculiaceae).

77. *Candalides absimilis* (Felder), 1862

Plate 26, figs. 1 (male), 1A (female), pencilled blue.

DISTRIBUTION. Claudie River, Coen and Cooktown to eastern Victoria; common between Cairns and Sydney. In Victoria it has been taken as far west as the Dandenong Ranges.

In the male the colour of the wings above is rather variable but is usually richer than in either *C. margarita* or *C. consimilis* and, if viewed in ordinary light, the forewing appears to be without a sex-mark. However, if the wing is viewed against a strong light, a trident sex-mark is visible. The valva of the male genitalia is broad at the base, but the narrower apex curves inwards and is divided into a pair of sharp projections. The white central patch in the hindwing of the female does not reach the apex, but is usually larger than that of *C. consimilis*. A close study of *C. persimilis* Waterhouse suggests that it is simply a form of *C. absimilis*.

The larvae feed on the buds and young terminal leaves of plants in several families, including *Cupaniopsis* (Sapindaceae), and *Castanospermum australe* (Moreton Bay chestnut), garden *Wisteria* and *Milettia megasperma* (all Fabaceae). In south coastal Queensland they are said to be most common on the young shoots of *Cupaniopsis* from October to December. There they also feed on the young leaves of *Cassia fistula* (Caesalpiniaceae), and on the flowers of *Alectryon coriaceus* (Sapindaceae) and *Flagellaria* (Flagellariaceae). In Queensland they have been reared from eggs laid in the flowers of *Macadamia integrifolia* (Queensland nut, Proteaceae), and at Sydney they have been recorded destroying the young foliage of *Macadamia* (C. N. Smithers). They are also common in New South Wales on *Brachychiton acerifolium* (flame tree, Sterculiaceae) and have been found on *Erythrina* (coral tree, Fabaceae) (J. R. Turner). The pupal duration is very variable; some of the pupae formed in December produce adults in only fifteen days, whereas others remain quiescent until the following spring.

78. *Candalides consimilis* Waterhouse, 1942

This species ranges from the Claudie River to eastern Victoria, three subspecies being recognized.

The forewing in the male bears a diffuse dull sex-mark; the termen of both wings is less convex and the tornus of the hindwing is more produced than in *C. absimilis*. In the female the central white patches are more or less suffused with purplish blue.

a. *Candalides consimilis consimilis* Waterhouse, 1942

Plate 23, fig. 10 (male).

DISTRIBUTION.   Kuranda to coastal southern New South Wales; fairly common north from Sydney.

The male has narrower black margins than in *goodingi*, and less conspicuous black terminal spots than in either *goodingi* or *toza*. The white patches in the female are more distinct than in *goodingi*.

In coastal south Queensland the larvae feed on the flowers of a small rain-forest tree *Alectryon coriaceus* (Sapindaceae). Larvae that pupated in the latter half of December produced adults in February and March, but some pupae remained dormant until the following September. Near Sydney D. P. Sands has found larvae commonly on *Hedera helix* (ivy, Araliaceae); he has also found them feeding on flowers of *Polyscias sambucifolius* (Araliaceae) and on the young terminal foliage of *P. elegans*. Some of the latter were reared at Canberra on young apricot leaves. At Sydney it has also been reared from *Ceratopetalum gummiferum* (Cunoniaceae) (D. McAlpine).

b. *Candalides consimilis toza* (Kerr), 1967

DISTRIBUTION.   Claudie River, Cape York Peninsula.

ADULT.   Male: similar to typical *consimilis* but forewing with termen less convex, above less purplish, black terminal spots larger in forewing and present between all veins in hindwing. Beneath with markings smaller and less distinct.

Female: unknown.

c. *Candalides consimilis goodingi* (Tindale), 1965

DISTRIBUTION. Australian Capital Territory, the far south of New South Wales, and the eastern half of Victoria. Two specimens have been taken at Tom Groggin and Eden, N.S.W., and one in the Brindabella Range, A.C.T. (J. M. Walker). In recent years several specimens have been taken in the Dandenong Ranges and other ranges to the east of Melbourne.

ADULT. Male: similar to typical *consimilis* but with costa of forewing black, termen in both wings more broadly black, inner edge of terminal area with a series of black triangular inward projections between veins.

Female: similar to typical *consimilis* but hindwing without any trace of a white patch.

Though described as a separate species, the male genitalia of *goodingi* are essentially the same as in *consimilis*; it is therefore treated here as a dark subspecies of *consimilis*. It has been suggested that this subspecies may have become more common in the Dandenong Ranges recently because introduced ivy (*Hedera helix*), a known food plant of *C. consimilis* at Sydney, has become so abundant.

79. *Candalides cyprotus* (Olliff), 1886

Two subspecies are recognized, one from Queensland and the other with a wide distribution from New South Wales to Western Australia. Adults are similar in size to the previous five species. In the male the forewing bears a somewhat diffuse trident sex-mark, and the tornus of the hindwing is even more produced than in *C. consimilis*.

a. *Candalides cyprotus cyprotus* (Olliff), 1886

Plate 26, figs. 8 (male), 8A (female), cyprotus blue.

DISTRIBUTION. New South Wales, western Victoria, South Australia and south-western Australia; sometimes fairly common. In New South Wales it occurs on the northern and central tablelands as well as on the coast. In Victoria it is known from Lake Hattah and the Big and Little Deserts. In South Australia it has been taken rarely from the

Victorian border to the Mt Lofty Ranges. In Western Australia it is known as far north as a little beyond Geraldton. In N.S.W. adults are usually on the wing from September to November, but in South Australia as late as January.

In New South Wales the larvae feed on *Jacksonia scoparia* (dogwood, Fabaceae), but near Geraldton, W.A., N. McFarland and N. B. Tindale found larvae on *Grevillea bracteosa* (Proteaceae).

b. *Candalides cyprotus pallescens* (Tite), 1963

DISTRIBUTION.   Yeppoon and Duaringa, Queensland, to the Casino and Grafton districts, north-eastern New South Wales; west of the Divide it is known from Millmerran and Stanthorpe; sometimes common. Adults are on the wing near Brisbane in spring and autumn.

ADULT.   Male: similar to typical *cyprotus* but larger, above more pinkish, beneath with ground colour paler, markings more distinct.

Female: similar to typical *cyprotus* but larger. Beneath as in male.

The larvae feed on *Jacksonia scoparia* (Fabaceae). The pupa is rather similar in shape to *C. hyacinthinus* (Plate XX, figs. 1D, 1E) but the abdomen is not as flat. The pupal duration is extremely variable, some pupae remaining dormant for up to eighteen months.

80. *Candalides erinus erinus* (Fabricius), 1775

Plate 26, figs. 10 (male), 10A (female), small dusky blue.

DISTRIBUTION.   North-western Australia north from Onslow, Northern Territory, including Groote Eylandt, the islands of Torres Strait, and Cape York to Port Macquarie and Wallis Lake; common.

As in the next three species *C. erinus* has a pair of dark subtornal spots beneath the forewing, and the male is without a sex-mark. In *C. erinus* the subtornal spots are very distinct and the apex of the male forewing is more rounded than in any of the other three species. The adults may be looked for wherever the larval food plant *Cassytha* (Devil's twine or dodder laurel) grows.

The larvae feed on *Cassytha filiformis* (Lauraceae) in southern Queensland, a dodder laurel which grows plentifully near beaches or on coastal sand dunes. They are attended by one or two small black ants. In shape they are similar to *C. hyacinthinus* (Plate XX, figs. 1A–1C) but lack reddish brown dorsal spots.

81. *Candalides geminus* Edwards and Kerr, 1978

Plate 14, fig. 4 (male).

DISTRIBUTION.   Alligator Rivers area, Northern Territory (G. B. Monteith) and Cape York, Queensland, to Barryrenie, near Cowra, New South Wales; south of Cooktown, Q., it has not been recorded from coastal areas, and in N.S.W. is known only from areas west of the Divide.

The ground colour of the wings beneath in both sexes is pale grey and the markings are less conspicuous than in *C. hyacinthinus*. There are marked differences in the genitalia of both sexes between *C. geminus*, *C. hyacinthinus*, *C. acastus* and *C. erinus*. In general appearance specimens from the Northern Territory are remarkably like *C. erinus*.

Near Coonabarabran, N.S.W., the larval food plant is *Cassytha paniculata* (Lauraceae), and the larvae feed openly during the day. Just before pupation they change in colour from green to brown or reddish brown, and pupation probably occurs in sheltered situations near the food plant. Eggs collected by E. D. Edwards in October hatched in about ten days, the larvae matured in twenty-two days, and the pupal duration was thirteen days. Several generations are probably completed during the warm months. The larva is very similar in shape to *C. erinus* and *C. hyacinthinus* (Plate XX, figs. 1B, 1C), but is more brightly coloured. The pupa resembles *C. erinus*. On the Cape York Peninsula adults of *C. geminus* have also been observed flying around *Cassytha* sp., which is no doubt the larval food plant there.

82. *Candalides acastus* (Cox), 1873

Plate 26, figs. 11 (male), 11A (female), blotched blue.

DISTRIBUTION.   Southern Australia, south from Burrum Heads, Q., on the east coast and Carnarvon, W. A., in the west, and northern and eastern Tasmania from sea level to about 300 m; common. In South Australia it occurs as far north as Wilpena Pound, Flinders Ranges. At Sydney adults are on the wing almost throughout the year.

The adults may be recognized by the diffuse dark grey-brown areas near the termen beneath both wings. In shape and in the colour above they resemble *C. hyacinthinus* more closely than *C. erinus* but are smaller.

The larvae feed at night on *Cassytha* (Lauraceae). In coastal south Queensland they have been found only on a very fine, low-growing species *C. glabella*. During the day they rested close to the ground on leaves of *Lomandra* and *Patersonia* parasitized by the *Cassytha*. In South Australia they have been recorded on *C. pubescens* as well as *C. glabella*, eating out the flower buds. The pupa is smaller and narrower than *C. erinus* or *C. hyacinthinus* (Plate XX, figs. 1D, 1E) with the abdomen scarcely flattened.

83. *Candalides hyacinthinus* (Semper), 1879

The species is widely distributed throughout eastern, southern and south-western mainland Australia. Three subspecies are here recognized.

This is the largest of the *erinus* group of species, with the apex of the forewing in the male more pointed, and the termen of both wings less convex, than in the other three.

a. *Candalides hyacinthinus hyacinthinus* (Semper), 1879

Plates 26, figs. 6 (male), 6A (female), XX, common dusky blue.

DISTRIBUTION. Rockhampton and the southern Expedition Range, Queensland, to south-western Victoria and south-eastern South Australia; common; recently recorded in south-western Australia from Margaret River and Augusta to the Porongorup Range. In the south-east it occurs on the tablelands as well as in coastal areas, and in New South Wales as far west as Cowra, Coonabarabran (E. D. Edwards), and the Warrumbungles (J. V. Peters). Adults are on the wing from August to March, appearing in August even in the Blue Mountains and in the Australian Capital Territory.

The size and shade of the purple central areas in the female vary considerably. The males display a strong tendency to collect on hill-tops.

The larvae feed at night on the young shoots of *Cassytha* (Lauraceae). In coastal south Queensland they are found on the commonest and largest species *C. pubescens*. There the pupal duration varies from two weeks in summer to five weeks in early winter.

b. *Candalides hyacinthinus eugenia* Waterhouse and Lyell, 1914

DISTRIBUTION. North-eastern Queensland, from Kuranda to Byfield, 30 km north of Yeppoon. Inland it has been taken at the Burra Range, 136 km south-west of Charters Towers.

ADULT.   Male: similar to typical *hyacinthinus* but usually with less purplish suffusion above.

Female: above dull brown-black, central areas purplish bronze-brown. Beneath as in male.

Known specimens vary considerably, some being almost indistinguishable from typical *hyacinthinus*, while others are much duller above in both sexes. The adults fly near *Cassytha*, upon which the larvae doubtless feed.

c. *Candalides hyacinthinus simplex* (Tepper), 1882

Plate 26, fig. 7 (male).

DISTRIBUTION.   Western New South Wales, western Victoria, the southern parts of South Australia, including Kangaroo Island, and south-western Australia from Exmouth (S. Wallace, M. S. Moulds) to the Stirling Range, the Salmon Gums and Esperance areas (M. S. Upton, C. G. Miller) and Eucla (M. S. Moulds), but not known from the south-west corner of Western Australia south of Perth and west of the Porongorup Range; usually not common.

This subspecies has a very distinctive blue upperside. The Western Australian *cyanites* Meyr. does not appear to be separable from *simplex*.

In western New South Wales and South Australia the larvae feed on *Cassytha melantha* (Lauraceae).

84. *Candalides xanthospilos* (Hübner), 1817

Plate 26, figs. 9 (male), 9A (female), yellow-spot blue.

DISTRIBUTION.   Kuranda and the Atherton Tableland to eastern Victoria; common in coastal southern Queensland and New South Wales. In Queensland adults have been taken at Kuranda, Herberton, 16 km west of Paluma, Byfield, and commonly south of Mooloolah. They have been taken inland at the Carnarvon Range, Millmerran, Stanthorpe, Gunnedah, Coonabarabran, Mt Kaputar, and Murrurundi. Victorian specimens have been collected in the Mallacoota Inlet and Lake Tyers districts, and in the Wonnangatta Valley near Dargo. A specimen was recorded last century from Lord Howe Island.

The adults are easily recognized by the orange-yellow spot above and the white ground colour beneath the wings. They fly close to the ground amongst shrubs and undergrowth in wooded country.

The larvae feed at night on *Pimelea* (Thymelaeaceae), sheltering during the day under dead leaves on the ground. They are attended by numerous small black ants. In coastal New South Wales they have been reared from *Pimelea linifolia* and *P. ligustrina*. At Stanthorpe the food plants are *P. linifolia* and *P. colorans*. The pupa is fairly similar in shape to *Candalides hyacinthinus* (Plate XX, figs. 1D, 1E), but the abdomen is less flattened, with a rugose surface, and the anterior flange has only a very slight central indentation. The pupal duration is very variable. In December it ranges from eleven to fourteen days but, from pupae produced in southern Queensland in May, adults emerged over a protracted period of four to twelve months.

85. *Candalides heathi* (Cox), 1873

The species occurs from Queensland to Victoria, South Australia, Western Australia, and near Alice Springs, Northern Territory. Four subspecies are recognized. The species can be identified by its very broadly rounded wings and uniform greyish ground colour beneath, with only a series of subterminal dots. The forewing has a rather shorter cell than in the other species and in the male is without a sex-mark. The egg is very similar to that of *C. hyacinthinus* and *C. xanthospilos*, but the surface pits are not as coarse. The pupa most closely resembles *C. xanthospilos*, with a rugose surface, reduced anterior flange, and much less flattened abdomen than in *C. hyacinthinus*.

a. *Candalides heathi heathi* (Cox), 1873

Plate 26, fig. 4 (male), rayed blue.

DISTRIBUTION. Northern and central Queensland to central and western Victoria, South Australia, Western Australia, and the Alice Springs area, Northern Territory. It probably occurs over much of southern Australia south of the Tropic.

The larvae feed at night on native and introduced *Plantago* (plantain, Plantaginaceae), and are sometimes attended by two or three small black ants. They are also reported to feed on *Westringia fruticosa* (Lamiaceae), which grows close to the sea along the eastern coast, and the adults certainly fly around this plant. Near Howqua, Victoria, larvae feed on *Parahebe derwentiana* (Scrophulariaceae) and near Lake Hattah on *Westringia rigida*. In inland Queensland the larvae feed on *Myoporum deserti* (Ellangowan poison bush, Myoporaceae) (J.

Macqueen); they have also been found on *Pimelea* (Thymelaeaceae) at Millmerran, *Eremophila longifolia* (Myoporaceae) near Mitchell and *Parahebe perfoliata* at Canberra (A. F. Atkins). The pupal duration in coastal New South Wales during December and January is sixteen to nineteen days.

b. *Candalides heathi alpinus* **Waterhouse, 1928**

Plate 26, fig. 5 (male).

DISTRIBUTION.   Mt Kosciusko, New South Wales, and the Brindabella Range, Australian Capital Territory, above about 1,200 m; fairly common. Adults have been taken from November to January.

This subspecies is distinguished from typical *heathi* mainly by the darker colour beneath the wings and the paler veins above. However, adults are variable and some specimens differ little from typical *heathi*.

E. D. Edwards observed adults ovipositing and collected larvae in late December on *Parahebe derwentiana* (Scrophulariaceae), on the Brindabella Range. Larvae subsequently accepted *Plantago lanceolata* (Plantaginaceae). Plants on which larvae were present invariably carried populations of small black ants, *Iridomyrmex* sp. (*gracilis* group), but these did not appear to pay much attention to the larvae. At Mt Kosciusko the adults have been taken flying in areas where *Parahebe* grows, upon which the larvae no doubt feed.

c. *Candalides heathi doddi* **Burns, 1948**

DISTRIBUTION.   Ebor district, and Barrington Tops, New South Wales, at about 1,300 m.

ADULT.   Similar to typical *heathi*, but much larger, and forewing in male with broader dark brown termen, subterminal spots beneath wings more distinct.

The adults fly in very steep rocky areas where *Parahebe* (Scrophulariaceae) grows.

d. *Candalides heathi aeratus* **(Montague), 1914**

DISTRIBUTION.   Monte Bello Islands, Western Australia.

ADULT.   Similar to typical *heathi*, but males smaller and darker, females with less blue above, and both sexes with subterminal spots beneath more distinct.

Genus *ADALUMA* Tindale, 1922

This genus contains only one species from the Northern Territory. It is closely allied to *Nesolycaena* and should perhaps be merged with that genus. The form of the pupa shows that both are closely related to *Candalides*.

86. *Adaluma urumelia* Tindale, 1922

Plate 23, fig. 12 (male).

DISTRIBUTION. Northern Territory, including Groote Eylandt. On the mainland specimens have been taken at the Roper and King Rivers, Maningrida, and the Blyth, East and South Alligator and McArthur Rivers (E. D. Edwards). Adults have been taken in October, November, March, May and June.

Few specimens of this species are represented in collections. It may be identified easily by its short antennae, bluish white upperside in the male and grey in the female, and white underside with distinct black subterminal spots.

Larvae were found by E. D. Edwards resting on the lower surfaces of mature leaves of *Boronia lanceolata* (Rutaceae), and were attended by small black ants (*Monomorium* sp.). In May, at the end of the wet season, pupae were green in colour and were attached beneath the leaves of the food plant. However, pupae produced in captivity in semi-darkness from larvae collected in late May were mottled brown. The pupal duration in the wet season is probably quite short, but in two of the mottled brown pupae which developed in captivity it was 220 and 310 days, suggesting that in nature they would remain as pupae at least until the following spring or the wet season.

Genus *NESOLYCAENA* Waterhouse and Turner, 1905

This genus contains only one species confined to central and south-east Queensland.

87. *Nesolycaena albosericea* (Miskin), 1891

Plate 27, fig. 4 (male), satin blue.

DISTRIBUTION. Known from the Expedition Range, Isla Gorge, Eidsvold, the Carnarvon Range, Burrum Heads, and Fraser and Stradbroke Islands, Queensland; usually rare. At mainland localities it has been taken in October, November, March, at Fraser Island in

August, on Stradbroke Island in September. A specimen was observed but not caught at Goombungee, Q., in March by J. Macqueen.

The adults fly close to the ground near their food plant.

In the first two instars the larvae feed on the flowers of *Boronia*, but in later instars on the older leaves close to the ground, frequently consuming the whole leaf. Thus it is difficult to recognize the presence of larvae, which are not attended by ants. The pupa is rather similar to *Candalides xanthospilos*, being less flattened and more roughened than the other species of *Candalides*.

Near Eidsvold the larvae were discovered on *Boronia glabra* (Rutaceae) (D. P. Sands), at Expedition Range on *B. obovata* (A. F. Atkins), and at Isla Gorge on *Boronia* sp. aff. *rosmarinifolia* (A. F. Atkins). At Fraser Island J. Harslett observed adults flying in Wallum areas where *Boronia falcifolia* and *Banksia robur* grow, and noticed them feeding at the flowers of *Boronia* sp. aff. *rosmarinifolia*. This last *Boronia* also grows commonly at the Goombungee locality where J. Macqueen identified an adult flying.

Genus *ZETONA* Waterhouse, 1938

The genus contains only the one Australian species, which appears to be related to *Candalides* and allied genera.

88. *Zetona delospila* (Waterhouse), 1903

Plate 23, fig. 15 (male).

DISTRIBUTION. North-western Australia, Northern Territory and north-eastern Queensland. In W.A. it ranges from the Edgar Range, 145 to 186 km south-east of Broome, to the Ord River; in N.T. it has been taken 46 km south-south-west of Borroloola on the McArthur River (E. D. Edwards), and in northern Queensland 30 km west of Fairview.

South-east of Broome the adults flew close to the ground amongst grass and low shrubs growing in sandy areas above the sandstone escarpment of the Edgar Range. Similarly, at the locality west of Fairview, Q., adults were taken on the top of a sandstone escarpment. However, a specimen collected on the Ord River, W.A., was flying amongst grass growing on alluvial soils near the river. The early stages have not been discovered.

Genus *PETRELAEA* Toxopeus, 1929

This genus contains a single widely distributed species ranging from India and Burma, through south-east Asia to New Guinea, far northern Australia and the Solomons. The hindwing is without a tail.

89. *Petrelaea dana* (de Nicéville), 1883

Plate 14, fig. 15.

DISTRIBUTION.   Cobourg Peninsula, Northern Territory, Moa and Thursday Islands and Claudie River; very few specimens have been taken in Australia.

This is an inconspicuous dull species, resembling a small *Prosotas* or *Nacaduba*, but structurally distinct. At the Cobourg Peninsula, N.T., E. D. Edwards collected specimens along a track in rain forest, whereas a specimen collected near the Claudie River, Queensland, was flying in *Eucalyptus* forest about 30 m from the rain forest.

Genus *NACADUBA* Moore, 1881

This genus contains thirty-seven species, distributed from Sri Lanka, India, Taiwan and Japan, through south-east Asia to the Moluccas, New Guinea, Australia, the Solomons, Samoa, Fiji and Tonga.

90. *Nacaduba pactolus cela* Waterhouse and Lyell, 1914

DISTRIBUTION.   Darnley Island, Torres Strait; rare.

This subspecies is known from only a couple of specimens in the Australian Museum. The species is the largest in the genus and may be recognized by the very broad white edges to the markings beneath.

91. *Nacaduba berenice berenice* (Herrich-Schäffer), 1869

Plates 28, figs. 2 (male), 2A (female), XIX, six lineblue.

DISTRIBUTION.   Darnley and Murray Islands, and Cape York to Lake Macquarie, New South Wales; very common.

The females vary in the extent of the blue central areas above. There is also some seasonal variation, specimens caught in the summer tending to have more restricted blue areas of a deeper shade. Except for their whitish edges the markings beneath the wings are obscure.

The larvae usually feed on the flower-buds and young terminal leaves of *Cupaniopsis anacardioides*, *Alectryon coriaceus* and *Heterodendrum diversifolium* (all Sapindaceae). Their colour varies according to the colour of the flowers and shoots on which they feed. They are attended by small black ants (probably *Prolasius* sp.). In south-eastern Queensland they have been reared from the young foliage and flowers of *Macadamia tetraphylla* and *M. integrifolia* (Queensland nut, Proteaceae).

92. *Nacaduba kurava* (Moore), 1857

This species has a wide distribution from Sri Lanka, India, Japan and Taiwan, through south-east Asia to the Moluccas, the Kai and Aru Islands, New Guinea, northern and eastern Australia, the Bismarck Archipelago and the Solomons. Two subspecies are recognized from Australia, and twenty-four elsewhere.

a. *Nacaduba kurava parma* Waterhouse and Lyell, 1914

Plate 28, figs. 1 (male), 1A (female), white lineblue.

DISTRIBUTION.   Moa Island and Cape York, Queensland, to Grafton, New South Wales; sometimes very common.

This subspecies is variable in size, and in the female the white areas above are also variable. Dry season specimens are usually smaller, with the white areas reduced in size.

The known food plants all belong to the Myrsinaceae. In coastal southern Queensland one food plant is *Rapanea variabilis*, a shrub or small tree that grows at the edge of rain forest. Recently T. H. Guthrie has reared the species from eggs deposited on very young leaves of a mangrove, *Aegiceras corniculatum*, growing near Tweed Heads. The larvae remained fully exposed on the foliage of the food plant during the day. When young their dull pale yellow colour matched the leaf buds, and later their yellowish green colour matched that of the leaf petioles. The larvae reached maturity in about fourteen days. Near Cairns larvae feed on *Maesa haplobotrys*. They feed mainly on the underside of the leaves, eroding irregular patches, leaving only the upper epidermis. Pupation occurs in a curled dead leaf on the ground or, when the larvae were feeding on mangroves standing in water, in crevices in the trunk. The pupal duration in southern Queensland was found to be two to three weeks.

b. *Nacaduba kurava felsina* Waterhouse and Lyell, 1914

DISTRIBUTION.   Northern Territory, north from Katherine; rare.

ADULT.   Male: similar to *parma* but beneath with white central areas more restricted.

Female: above similar to *parma* but white central areas usually more restricted; beneath with white central areas similar to male of *parma*.

A female from Katherine in the Australian National Insect Collection has the white areas above more extensive than in *parma*, and is rather like the specimen figured as female *parma*. The white sub-terminal rings of the hindwing are much more prominent than in *parma*, and beneath the white areas are more restricted.

The Katherine female was reared from a larva feeding on the young foliage of a rain-forest tree growing on the bank of the Katherine River. The pupa was similar to typical *parma*.

93. *Nacaduba biocellata biocellata* (C. and R. Felder), 1865

Plate 28, figs. 7 (male), 7A (female), double-spotted lineblue.

DISTRIBUTION. Most and probably all of mainland Australia, inland as well as on the coast; usually common. It seems to be rarer in the far north but has been recorded from Darwin, Northern Territory (K. L. Dunn) and Yam Island, Torres Strait, but has not been taken on Cape York Peninsula, north of the Claudie River.

This is one of the smallest Lycaenidae and can be recognized by the absence of a tail to the hindwing and by the yellow-brown basal area of the forewing beneath nearly devoid of markings. Adults are sometimes extremely abundant, especially in inland areas, flying around *Acacia aneura* (mulga) and other *Acacia* trees. Males have been observed fluttering close to the ground beneath mulga trees, presumably searching for freshly emerged females.

The larvae are similar in shape to *Prosotas felderi* (Plate XX, figs. 5A, 5B). They are very variable in colour, matching the colour and form of the buds and flowers upon which they feed. The larvae have been found on several species of *Acacia* (Mimosaceae), including *A. penninervis*, *A. irrorata*, *A. deanei*, *A. betchei*, *A. salicina*, *A. osswaldii*, *A. aneura*, *A. brachybotrya*, *A. ligulata*, *A. rigens*, *A. sclerophylla*, *A. sowdenii* and *A. victoriae*. They can be collected in an upturned umbrella or on a sheet spread beneath a branch when they have been dislodged by tapping the branch sharply with a stick. Both larvae and pupae are attended by small black ants. Pupation occurs in curled leaves or under stones or pieces of bark on the ground beneath the food plant, or in ant galleries in the soil.

Genus *PROSOTAS* Druce, 1891

This genus contains eighteen species distributed from India to Taiwan, through south-east Asia to the Moluccas, New Guinea, the Bismarck Archipelago, the Solomons and northern and eastern Australia.

94. *Prosotas dubiosa dubiosa* (Semper), 1879

Plate 28, figs. 8 (male), 8A (female), small purple lineblue.

DISTRIBUTION. North-western Australia, north from Broome (M. S. Moulds), Northern Territory, including Groote Eylandt, Yam, Moa, Thursday and Prince of Wales Islands, and Cape York to Coffs Harbour; common.

This small species can be distinguished from the other two Australian species by its tailless hindwing.

Larvae have been recorded feeding on the flowers of litchi, *Litchi chinensis* (Sapindaceae) at Mareeba, Q., and the flowers of macadamia nut, *Macadamia integrifolia* (Proteaceae) at Gympie, Q. They have also been reported on the flower-heads of *Acacia leiocalyx* at Brisbane, when their colour closely matched that of the buds. Adults have been observed ovipositing on *Buckinghamia celsissima* (Proteaceae) (M. De Baar).

95. *Prosotas felderi* (Murray), 1874

Plates 28, figs. 9 (male), 9A (female), XX, Felder's lineblue.

DISTRIBUTION. Gympie, Queensland to Wollongong, New South Wales; sometimes very common on the coast north of the Richmond River, less common on the tablelands, but if good rains fall in January and February, may be taken commonly in late summer as far inland as Millmerran. Two specimens have been recorded from Berri, South Australia. The species is confined to Australia.

This species has rather more pointed wings than *P. dubiosa* in the male, and a shorter tail than *P. nora*.

The larvae have been found feeding on the buds and flowers of *Acacia leiocalyx* (Mimosaceae), and *Alectryon coriaceus* and *Cupaniopsis anacardioides* (both Sapindaceae). Those feeding on the last two plants in southern Queensland occurred with the larvae of *Nacaduba berenice* and *Erysichton lineata*. At Gympie, Q., the larvae have been reported feeding on the flowers of macadamia nut, *Macadamia integrifolia* (Proteaceae), and at Alstonville, N.S.W., on the flower buds of litchi, *Litchi chinensis* (Sapindaceae). The larvae match well the colour and form of the flower-buds upon which they feed. At Brisbane adults have been reported ovipositing on the flower-buds of *Buckinghamia celsissima* (Proteaceae) (M. De Baar).

96. *Prosotas nora auletes* (Waterhouse and Lyell), 1914
Plate 23, fig. 18 (female).

DISTRIBUTION. Islands of Torres Strait, and Cape York to Cairns, Kuranda, the Atherton Tableland and Mackay.

This species has a longer, thinner tail to the hindwing than *P. felderi*. The male is lilac-purple above with a narrowly brown-black termen.

Genus *IONOLYCE* Toxopeus, 1929

The genus contains two species, one of which is confined to the Solomons and the other ranges from Sri Lanka to New Guinea and northern Australia.

97. *Ionolyce helicon* (Felder), 1860
This species has a wide distribution and includes eight subspecies ranging from Sri Lanka, the Andaman and Nicobar Islands to Indonesia, the Moluccas, New Guinea and Australia. Within Australian limits two subspecies are recognized.

a. *Ionolyce helicon hyllus* (Waterhouse and Lyell), 1914
Plate 23, fig. 16 (male).

DISTRIBUTION. Cape York to near Rockhampton (C. G. Miller); very rare.

The female probably has a white patch above the forewing.

b. *Ionolyce helicon caracalla* (Waterhouse and Lyell), 1914

DISTRIBUTION. Darnley Island, Torres Strait; rare. This subspecies also occurs in New Guinea and the Bismarck Archipelago.

ADULT. Male: similar to *hyllus*, but above dull dark purple, beneath grey-brown, spots near tornus more prominent.

Female: forewing with a small central white patch. Beneath as in male.

The white patch on the forewing of the female is somewhat similar to that in *Erysichton lineata*.

Genus *CATOPYROPS* Toxopeus, 1929

The genus is distributed from India to the Solomons and Fiji, and contains four species, two of which are represented in Australia.

98. *Catopyrops ancyra mysia* (Waterhouse and Lyell), 1914

Plate 23, fig. 8 (male).

DISTRIBUTION. Darnley, Murray, Yam, Moa and Prince of Wales Islands, Torres Strait; common.

The female is brown-black above with a large pale blue central area in both wings.

99. *Catopyrops florinda* (Butler), 1877

This species occurs in Timor and nearby islands, northern and eastern Australia and Fiji.

a. *Catopyrops florinda halys* (Waterhouse), 1934

Plate 28, figs. 3 (male), 3A (female), speckled lineblue.

DISTRIBUTION. Townsville, Queensland to the Illawarra district, New South Wales; common north of the Manning River, N.S.W.

Specimens from Townsville are said to differ a little from those collected south of Mackay. Females taken in winter have larger blue areas above.

The larvae feed commonly on *Trema cannabina* (peach-leaf poison bush, Ulmaceae), but also on *Caesalpinia bonduc* (Caesalpiniaceae) and are usually attended by one or two small black ants. E. D. Edwards found that larvae accepted the leaves of *Celtis sinensis* (Ulmaceae) after their food supply was exhausted. The colour of the pupa matches the situation in which it is found, those attached beneath the leaves of the food plant being green, whereas those under dead leaves on the ground are brown.

b. *Catopyrops florinda estrella* (Waterhouse and Lyell), 1914

DISTRIBUTION. North-western Australia, north of the Edgar Range, 130 km south-east of Broome, Northern Territory, including Groote Eylandt, and Claudie River to Ingham, Queensland; fairly common.

ADULT.   Similar to *halys* but duller above, beneath pale grey-brown, usually not whitish towards termen in either wing, markings as in *halys*, but usually paler and less contrasted to ground colour.

Genus *ERYSICHTON* Fruhstorfer, 1916

The genus includes three species distributed through the Moluccas, New Guinea and its associated islands, eastern Australia and the Solomons.

100. *Erysichton lineata lineata* (Murray), 1874

Plate 28, figs. 5 (male), 5A (female), hairy lineblue.

DISTRIBUTION.   Cape York and the Claudie River to Kiama; rare south of Port Stephens, N.S.W., often common farther north; on the tablelands occurs as far south as Armidale.

The long hairs on the upperside of the male, and the underside in both sexes, readily distinguish this species from *E. palmyra*. The female above is somewhat similar to *E. palmyra*, but in that species both sexes have the terminal fringes chequered and a whitish subterminal area beneath both wings. The species occurs most commonly in or at the margins of rain forest, or amongst rain-forest trees which often grow behind sandy beaches.

The eggs are laid on the young flower-buds of the food plant. In shape the larva resembles that of *Prosotas felderi* (Plate XX, figs. 5A, 5B) but is not as humped dorsally. The larvae feed on the flowers of *Cupaniopsis anacardioides* and *Alectryon coriaceus* (both Sapindaceae), *Ehretia acuminata* (Boraginaceae), and *Macadamia integrifolia* (Queensland nut, Proteaceae), and are very difficult to detect. They are sometimes found together with larvae of *Nacaduba berenice*, *Prosotas felderi*, *Theclinesthes scintillata* and *Candalides absimilis*. Larvae collected at Port Macquarie in April pupated towards the end of that month and the adults emerged about a month later.

101. *Erysichton palmyra tasmanicus* (Miskin), 1890

Plate 28, figs. 6 (male), 6A (female), marbled blue.

DISTRIBUTION.   Cape York to Lake Macquarie; rare in New South Wales, more common in Queensland.

This species is found both in rain forest and open eucalypt forest.

The mature larvae are said to be reddish brown in colour and smoother than related species. In coastal south Queensland they feed on the flower-buds of the mistletoe *Dendrophthoe vitellina* (Loranthaceae), which they resemble closely in colour. Larvae have been found only on this species of mistletoe, being present from April to July, during its main flowering period. The pupal duration is about three weeks.

Genus *NEOLUCIA* Waterhouse and Turner, 1905

The genus contains three small species confined to southern mainland Australia and Tasmania.

102. *Neolucia agricola* (Westwood), 1851

This small brown species with chequered wing margins is distributed from central Queensland to South Australia, south-western Australia and Tasmania, occurring near the coast and at elevations up to about 1,500 m. Three subspecies are recognized.

a. *Neolucia agricola agricola* (Westwood), 1851

Plate 27, figs. 8 (male), 8A (female), fringed blue.

DISTRIBUTION. Expedition Range (A. F. Atkins), central Queensland, south-eastern Queensland up to 900 m, eastern New South Wales and the Australian Capital Territory up to 1,500 m, Victoria to about 1,370 m, southern parts of South Australia including Kangaroo Island and Eyre Peninsula. In coastal areas of southern Queensland, where it is one of the commonest early spring butterflies, and in central New South Wales adults are on the wing from September to November. Farther south and on the tablelands and mountains they appear as late as January and February.

At the higher altitudes this species is sometimes found in the same localities as *N. hobartensis*, from which it may be distinguished by the more pointed forewings in the male, and the pair of distinct black V-shaped spots beneath the hindwing. It is usually larger than *N. hobartensis*.

The larvae are attended by numerous small black ants and feed on the pea-like flower-buds of several Fabaceae, including *Dillwynia*, *Pultenaea*, *Daviesia*, *Bossiaea*, *Aotus* and *Eutaxia*. At the Expedition Range, Q., they have been found on *Bossiaea carinalis*, and in South Australia on *Pultenaea largiflorens* var. *latifolia* and *Eutaxia microphylla*.

They eat the inside of the buds through a hole they make in the flower calyx, and sometimes are found within the larger flowers. Their colours match the food plant so closely that the larvae are very difficult to detect, but they can be collected by beating the flowering shrub over a tray or upturned umbrella.

b. *Neolucia agricola insulana* **Waterhouse and Lyell, 1914**

DISTRIBUTION.   Tasmania, at altitudes up to at least 1,065 m. Adults are on the wing from mid-November to early February.

ADULT.   Very similar to typical *agricola*, but smaller, a little duller beneath, with obscure V-shaped spots on hindwing.

c. *Neolucia agricola occidens* **Waterhouse and Lyell, 1914**

DISTRIBUTION.   South-western Australia, from Geraldton to Twilight Cove (M. S. Upton) including the Stirling Range; common. Adults fly from September to November.

ADULT.   Above similar to typical *agricola*. Beneath forewing pale yellowish brown, markings usually obscure, pale brown edged whitish; hindwing pale greyish brown, markings usually obscure yellow-brown edged brown, a slight central greyish suffusion, subterminal V-shaped spots very obscure.

This subspecies is somewhat similar to *N. mathewi* beneath, but that species is much paler brown above, with only slightly chequered margins, and has not been recorded from Western Australia. In some specimens the markings beneath are just as distinct as in typical *agricola*.

103. *Neolucia hobartensis* **(Miskin), 1890**

This is an alpine or subalpine species found in south-eastern mainland Australia above about 1,370 m and in Tasmania above 300 m. Two subspecies are recognized.

a. *Neolucia hobartensis hobartensis* **(Miskin), 1890**

Plate 27, figs. 9 (male), 9A (female), mountain blue.

DISTRIBUTION.   The higher mountains of southern New South Wales, the Australian Capital Territory and Victoria above about 1,200 m, and Tasmania from about 300 m to 1,370 m. Adults are on the wing from January to early March. On the mainland they have

322

been taken at Boyd River, Kosciusko Plateau, Brindabella Range, and at Mt Hotham, Mt Erica, Mt Baw Baw and Lake Mountain.

This species may be separated from *N. agricola* taken in the same localities by the less pointed forewing in the male, and the absence of the distinct V-shaped subterminal spots beneath the hindwing.

At Mt Gingera, A.C.T. (1,650 m) the larvae feed on *Epacris petrophila* (Epacridaceae) growing along swampy creeks, and can be collected by beating the plants (E. D. Edwards). Young larvae feed on flowers and flower-buds, if present, in preference to foliage, but older larvae feed openly during the day on the foliage. They are not attended by ants. W. N. B. Quick found that *E. petrophila* is the major food plant at Lake Mountain, Victoria, whereas at Mt Baw Baw (1,550 m) and in the Mt Hotham area (about 1,500 m) *E. paludosa* seems to be favoured.

b. *Neolucia hobartensis monticola* **Waterhouse and Lyell, 1914**

Plate 27, fig. 10 (male).

DISTRIBUTION. Dorrigo Plateau and Barrington Tops, New South Wales, above about 1,200 m; common. Adults are on the wing in January and February.

This subspecies is much larger than typical *hobartensis* and about the size of *N. agricola*, but has a more rounded apex to the forewing than in that species. The long hairs covering most of the upperside of both wings are also present in *N. mathewi*, but are restricted to the basal area of the forewing in *N. agricola*. These hairs are more variable in typical *hobartensis*, but are seldom as extensive as in *monticola*. The latter is also darker above and beneath than in typical *hobartensis*. Near Ebor adults fly commonly near the edges of swamps.

The pupa is much less hairy than in *N. agricola* or *N. mathewi*. At Barrington Tops the larvae feed during January on the flower-buds of a white-flowered *Epacris* (Epacridaceae) growing at the edges of swamps. They may be collected by beating the food plant over a tray or sheet. Larvae collected in January and taken to Sydney pupated at the beginning of February and the adults emerged about two weeks later.

104. *Neolucia mathewi* **(Miskin), 1890**

Plate 27, figs. 11 (male), 11A (female), Mathew's blue.

DISTRIBUTION. The Blue Mountains, Mt Kosciusko (C. E.

Chadwick), and coastal New South Wales south from Port Stephens, the Mallacoota district and the Mt Baw Baw plateau (about 1,430 m) in Victoria, Flinders Island, King Island, and north-eastern Tasmania. The adults are common in September and October along the New South Wales coast, but have been taken during November at Blackheath, Mallacoota, and Flinders Island and in February and March at Mt Baw Baw.

This species is similar in size to *N. agricola*, but has much more obscure markings beneath, without the V-shaped spots on the hindwing, and the wings above are clothed with long hairs. These hairs are scarcely more extensive than in *N. hobartensis monticola*, from which it may be separated by the very obscure markings beneath and the greatly reduced chequered margins to the wings. The latter may not be noticed in slightly worn specimens. Specimens from near Mt Baw Baw are rather smaller and darker than those from coastal areas.

In early spring adults are often common where their food plant grows near sea beaches. Near Mt Baw Baw adults are very local in their occurrence near the food plants which grow on or near granite boulders. There it sometimes flies with *N. hobartensis*.

In coastal areas the eggs are laid in the spring on various parts of the food plant, *Monotoca elliptica* (red pigeon berry, Epacridaceae). They apparently remain dormant until early the following spring when this tough-leaved plant produces its new shoots and flower-buds, upon which the young larvae feed and develop rapidly. The larvae have also been reported feeding on the flowers of *Pultenaea daphnoides* (Fabaceae). At Sydney the pupal duration is sixteen to twenty-one days. Near Mt Baw Baw the food plant is *Monotoca scoparia*.

Genus *THECLINESTHES* Röber, 1891

The genus contains seven Australian species, one of which also occurs widely from Indonesia through New Guinea to the Bismarck Archipelago.

105. *Theclinesthes onycha* (Hewitson), 1865

This species occurs in eastern Australia, although a single specimen from Darwin is thought to belong here. Two subspecies are recognized by Sibatani and Grund (1978), who point out that the species has been known previously as *T. miskini*. The larvae feed on cycads (Gymnospermae).

a. *Theclinesthes onycha onycha* (Hewitson), 1865

Plate 27, figs. 16 (male), 16A (female).

DISTRIBUTION.   Eastern New South Wales, in coastal areas from Casino to Central Tilba, at times common south of Sydney; much less common on the tablelands and north-west slopes as far west as Coonabarabran and Mt Kaputar (G. Daniels); two specimens are known from the Australian Capital Territory. Specimens from Stanthorpe, Queensland, probably belong to this subspecies.

Adults may be found flying around the *Macrozamia* (Zamiaceae) food plant, or resting on the ground near by, in *Eucalyptus* forest. The males are often collected on hill-tops.

In southern coastal New South Wales the larvae are found more commonly on the numerous young *Macrozamia* (Zamiaceae) seedlings which appear after a bushfire. As many as two or three larvae may occur on any one frond, and during the day they rest on the undersides of the pinnae, probably feeding intermittently by eroding the lower surface. Affected fronds soon acquire a scorched appearance. There the food plant is *M. communis* (burrawang). Near Stanthorpe the larvae probably feed on *M. pauliguilielmi*. On *Macrozamia* in southern coastal New South Wales the larvae were attended only occasionally by one or two small black ants belonging to three species *Notoncus ectatommoides*, *Paratrechina ?bourbonica* and *Iridomyrmex glaber*.

b. *Theclinesthes onycha capricornia* Sibatani and Grund, 1978

DISTRIBUTION.   Eastern Queensland, from Cape York to Brisbane; mainly coastal but occurring as far inland as Springsure and the Carnarvon Range. A specimen from Darwin has been tentatively referred to this subspecies.

ADULT.   Similar to typical *onycha*, but androconia usually resembling those of *T. miskini*, longer than broad, with eleven or twelve ribs. Beneath with paler and more uniform brownish ground colour and paler markings more distinctly edged white; hindwing with more prominent black spots near tornus.

Both summer and winter forms are rather variable, but the winter form is often darker beneath with whitish areas beyond the post-median band. Adults are usually found flying near the *Macrozamia* (Zamiaceae) and *Cycas* (Cycadaceae) food plants, and males are frequently taken on hill-tops.

The life history and early stages are similar to typical *onycha*. At Cairns the larvae have been recorded on *Macrozamia lucida* and elsewhere no doubt other species of *Macrozamia* are attacked. Just north of Rockhampton larvae have been taken on *Cycas* sp. (Cycadaceae), and the adults have also been taken flying around *M. miquelii*. On *Cycas* the larvae were attended by large black ants, *Polyrhachis* sp. (*ammon* group).

106. *Theclinesthes miskini* (T. P. Lucas), 1889

This species ranges widely from Sumba, Flores and Timor, the Tanimbar, Kai and Aru Islands, New Guinea, the Admiralty Islands, New Britain, New Hanover, and the D'Entrecasteaux and Goodenough Islands. Six subspecies are recognized, three of which occur within Australian limits. The species has until recently been known as *T. onycha* (Sibatani and Grund, 1978). The larvae feed on *Acacia*, *Eucalyptus* and other plants.

a. *Theclinesthes miskini miskini* (T. P. Lucas), 1889

Plate 28, fig. 4 (male).

DISTRIBUTION. Throughout most of mainland Australia, including the inland, except south and south-eastern Victoria and north-eastern Queensland, approximately north from Townsville.

Adults of this subspecies are extremely variable and several geographic forms have been noticed, the differences apparently being clinal (Sibatani and Grund, 1978). Specimens vary considerably in size, in the lilac or blue coloration and the width of the dark terminal areas above, and in the shade of the grey or grey-brown ground colour beneath. Sometimes a section of the postmedian band beneath the hindwings is displaced outwards. In addition both summer and winter forms occur, as well as individuals intermediate between these two forms.

The adults fly around *Acacia*, upon which the larvae usually feed, and males are taken commonly on hill-tops.

The larvae have been found feeding on several species of *Acacia*

(Mimosaceae), but sometimes on Fabaceae. At Mitchell and St George, south Queensland, they fed on the phyllodes of *Acacia salicina*. In western Victoria and near Adelaide, S.A., they have been recorded on *A. pycnantha*, and in the Flinders Ranges, S.A., on *A. victoriae*. At Cobourg Peninsula, Northern Territory, E. D. Edwards found larvae on *A. auriculiformis*. In Western Australia the larvae have been reported on *A. tetragonophylla* near Geraldton, but at Dampier and Millstream adults have been reared from larvae feeding on *Sesbania cannabina* var. *sericea* (Fabaceae). J. F. R. Kerr found the larvae at Biloela and Bowen feeding on an unidentified legume growing on the roadside. The larvae are usually attended by small black ants.

b. *Theclinesthes miskini eucalypti* Sibatani and Grund, 1978

DISTRIBUTION.　North-eastern Queensland, from Cape York to Townsville, and westwards to Weipa in the north and at least as far as Mt Garnet farther south.

ADULT.　Similar to typical *miskini*, but male usually blue above or less frequently lilac, with more numerous and squat androconia; both sexes with postmedian band of hindwing beneath usually straight.

Along the western known limits of its distribution, this subspecies may show a gradual transition to typical *miskini* (Sibatani and Grund, 1978). The identity of females and lilac males can best be determined by the usually straight postmedian band beneath the hindwing. Specimens are usually smaller than *T. onycha capricornia* from which it may be distinguished by the blue colour above, the greyish rather than brown ground colour beneath, and bands prominently edged with dark grey as well as white.

The males are commonly taken on hill-tops.

The life history and early stages are similar to those of typical *miskini*.

At Tinaroo, near Atherton, the larvae were found feeding on *Eucalyptus polycarpa* (Myrtaceae) and between May and July they reached maturity in four weeks and the pupal duration was two weeks (N. McFarland, N. B. Tindale). Larvae have also been found on *Eucalyptus* sp. at Weipa (J. W. C. d'Apice). The males reared in both instances were blue above. Females only have been reared from larvae on *Atalaya variifolia* (Sapindaceae) collected between Weipa and the Claudie River (M. S. Moulds) and on *Acacia* sp. (Mimosaceae) at

Cooktown (A. F. Atkins) and near Paluma (D. P. Sands), but whether males from these two food plants would have been blue or lilac is not known. The larva on *Acacia* near Paluma was attended by green tree-ants, *Oecophylla smaragdina*.

c. *Theclinesthes miskini arnoldi* (Fruhstorfer), 1916

DISTRIBUTION.   Moa, Badu, Thursday, Horn and Prince of Wales Islands, Torres Strait. This subspecies also occurs in the Tanimbar, Kai and Aru Islands, Papua New Guinea, the Louisiade Archipelago, the D'Entrecasteaux Islands, New Britain and New Hanover.

ADULT.   Similar above in colour to the blue male form and female of *eucalypti*, but in male with a series of whitish subterminal rings on hindwing enclosing prominent dark spots, especially near tornus, androconia of male larger, longer than broad, with thirteen to sixteen ribs.

The males fly on hill-tops in *Eucalyptus* woodland.

The life history and early stages are not known from the Torres Strait islands, but near Port Moresby, Papua New Guinea, adults have been reared from larvae found by D. P. Sands on *Eucalyptus confertiflora* (Myrtaceae).

107. *Theclinesthes albocincta* (Waterhouse), 1903

Plate 14, fig. 2.

DISTRIBUTION.   North-western Victoria and coastal areas of South Australia, from Robe to Yorke Peninsula and Coffin Bay, Eyre Peninsula. The original specimens came from Peak Downs, near Emerald, central Queensland, and specimens are also known from the Simpson Desert, S.A., from 30 to 60 km south to south-west of Alice Springs, N.T. (E. D. Edwards), and from Dongara, from near Exmouth (M. S. Moulds), and from Dampier, W.A.

The distribution of this species is not yet fully understood, but it must have a wide distribution in inland Australia where the food plant grows.

Seasonal forms have been recognized in populations of this species from coastal South Australia. The winter form has more extensive blue areas above in both sexes, the pale subterminal rings of the hindwing are more distinct and bluish in the female, and present in the male, and the ground colour beneath is darker with a whitish suffusion beyond the postmedian band. Inland forms also differ from those in coastal South Australia. They are usually darker above, with more restricted dull greyish blue basal areas and pale subterminal rings on the hindwing of both sexes. They are variable beneath, but the postmedian band is usually darker, with more prominent pale suffusion beyond it. A male from near Dampier, W.A., and specimens from Alice Springs are brown above and have the pale subterminal rings of the hindwing only faintly indicated. Beneath the wings are pale yellowish brown with faint markings. One of the two original males from Peak Downs, Q., is unusual because it is extensively pale blue above, with more distinct pale subterminal rings on the hindwing than in the other forms.

In coastal South Australia the larvae feed on the flowers and flower-buds, or occasionally on the young fruit or leaves, of *Adriana klotzschii* (Euphorbiaceae) growing near the sea shore. They feed openly during the day, their colours matching the pink flowers and green stems. In captivity the larvae readily accepted young or old leaves, eating only the upper surface. Larvae are usually attended by one or two ants. At Port Gawler they were attended by *Polyrhachis (Campomyrma)* sp. and *Iridomyrmex* sp., and at Warooka by *Camponotus* sp. and *Notoncus ?gilberti*; at both localities they were also attended by *Rhytidoponera metallica*. Pupae are attached to dead leaves, stems or other debris on the ground beneath the food plant. The pupal duration during the summer ranged from seven to twelve days, and in autumn sixteen to twenty days. Larvae in all stages of development were collected in January and breeding may be continuous during the warmer months. At the Pink Lakes in western Victoria larvae have been found by A. D. Bishop on *Adriana hookeri*.

108. *Theclinesthes hesperia* Sibatani and Grund, 1978

This species occurs in southern Western Australia, where it apparently replaces *T. albocincta*. The larvae of both species feed on the one genus of plants. Two subspecies have been recognized.

a. *Theclinesthes hesperia hesperia* Sibatani and Grund, 1978

Plate 29, fig. 2 (male).

DISTRIBUTION. South-western Western Australia, from near

Perth to Bunbury; on coastal sand dunes.

ADULT.   Male: above dark blue-lilac, termen narrowly brown-black, terminal fringes brown, tipped white, androconia fairly numerous, elongate, with twelve ribs; forewing with veins and bar at end of cell narrowly brown-black; hindwing with obscure brown-black subterminal spots fused with blackish terminal line, spot near tail larger, black, tail short, thick, brown. Beneath grey-brown, bands slightly darker edged dark brown, bands on forewing also faintly edged white, an extensive grey postmedian suffusion, especially on hindwing, subterminal spots obscure.

Female: above dark blue, costa of forewing, termen and bar at end of cell brown-black; hindwing with distinct pale rings enclosing prominent black spots, especially near short, thick, brown tail. Beneath as in male.

Both sexes can be distinguished from *T. miskini* and *T. onycha* by the short thick tail, and males by the smaller elongate androconia. The species differs from *T. albocincta*, which has a similar tail, by the presence of androconia and less pointed forewings in the male, the unchequered terminal fringes, and the reduced white edging to the bands beneath.

The species appears to be common near Bunbury, where the males can be found hill-topping on the sand dunes.

The life history and early stages have not been described, but adults have been reared from larvae feeding on *Adriana quadripartita* (Euphorbiaceae).

b. *Theclinesthes hesperia littoralis* Sibatani and Grund, 1978.

DISTRIBUTION.   Known only from the coastal area near Esperance, Western Australia.

Adults differ from typical *hesperia* by the absence of lilac above, the absence of androconia in the male, and the darker undersides in both sexes. The early stages are unknown, but the larvae probably feed on *Adriana quadripartita* which grows commonly in the areas near the shore where the adults are found.

109. *Theclinesthes serpentata* (Herrich-Schäffer), 1869

This small species occurs throughout the southern half of Australia, and in south-eastern Tasmania. Two subspecies have been recognized.

a. *Theclinesthes serpentata serpentata* (**Herrich-Schäffer**), 1869

Plate 27, figs. 12 (male), 12A (female), chequered blue.

DISTRIBUTION.   Throughout mainland Australia south of a line drawn from Mackay, Queensland, to Roebourne, Western Australia, in the interior as well as in coastal areas; much commoner inland than near the coast. Adults have also been taken on Flinders Island.

The adults fly commonly around saltbushes (Chenopodiaceae) upon which the larvae usually feed. The species can be recognized at once by the blue central areas of the wings and the short thick tail to the hindwing.

The larvae and pupae are rather like those of *T. sulpitius* (Plate XX, figs. 4A–4C). The larvae feed on various saltbushes (Chenopodiaceae), including *Atriplex semibaccata*, *A. nummularia*, *A. leptocarpa*, *A. vesicaria*, *Rhagodia hastata*, *R. spinescens*, *R. nutans*, *R. parabolica*, and *Chenopodium album*. The dense spinules give the skin of the larva a dull mealy appearance resembling that of the saltbush leaves. In inland central Queensland adults have also been reared from larvae feeding on *Atalaya hemiglauca* (whitewood, Sapindaceae).

b. *Theclinesthes serpentata lavara* (**Couchman**), 1954

DISTRIBUTION.   South-eastern Tasmania. Adults have been collected at Cambridge, near Hobart, in March and early April.

ADULT.   Male: similar to typical *serpentata* but basal and central areas of wings above larger, blue without a purplish tint. Beneath forewing with markings more prominent; hindwing with markings more distinct, median band of spots black.

This is a smaller and more distinctly marked subspecies. Adults have been taken flying on the coastal mud-flats and feeding at the flowers of *Lycium ferocissimum*, the introduced African box-thorn.

110. *Theclinesthes sulpitius* (**Miskin**), 1890

This small dull species is confined to the east coast, occurring from Cooktown to Cairns, and from Yeppoon to eastern Victoria. Two subspecies are recognized.

a. *Theclinesthes sulpitius sulpitius* (Miskin), 1890

Plates 27, figs. 13 (male), 13A (female), XX, saltpan blue.

DISTRIBUTION.   Yeppoon to eastern Victoria; very common on the mud-flats near tidal estuaries. Near Brisbane adults have been taken in most months

The larvae feed on *Rhagodia, Salicornia quinqueflora, Tecticornia austra-lasica* (samphire or glasswort), and other saltbushes (Chenopodiaceae) which grow just above high-tide mark. Their coloration closely matches that of the food plant.

b. *Theclinesthes sulpitius obscura* (Waterhouse and Lyell), 1914

DISTRIBUTION.   Cooktown to Cairns; common on the salt mud-flats.

ADULT.   Similar to typical *sulpitius* but smaller and paler, with the apex of the forewing more rounded and the termen more convex.

The larvae feed on saltbushes growing on the salt flats near high-tide mark.

111. *Theclinesthes scintillata* (T. P. Lucas), 1889

Plate 27, figs. 17 (male), 17A (female), glistening blue.

DISTRIBUTION.   Northern Territory, Yam, Thursday and Prince of Wales Islands, and Cape York, Queensland, to Grafton, New South Wales; usually uncommon, especially towards the northern and southern limits of its range, but occasionally common elsewhere. A female in the Australian Museum was taken at Ebor, N.S.W., and another at Menangle Park, near Sydney, by E. D. Edwards.

Superficially the female resembles *Erysichton palmyra* or *E. lineata*, but has a smaller white patch on the forewing and is structurally distinct.

The mature larva is very similar in shape to *Prosotas felderi* (Plate XX, figs. 5A, 5B), dark green, blotched with cream, with a dorsal line of black spots (M. De Baar). Pupa somewhat like *P. felderi* (Plate XX, fig. 5C), but broader anteriorly and more hairy. The larvae are sometimes found feeding on the same flowers as *Nacaduba berenice, Erysichton lineata, Prosotas felderi* and *P. dubiosa*. In southern Queensland they have been found during the summer on flowers of *Alectryon coriaceus*, and in winter on *Cupaniopsis anacardioides* (both Sapindaceae).

They have also been reported on the flower-buds of *Acacia* (Mimosaceae), including *A. leiocalyx* at Brisbane (M. De Baar). The pupal duration varies from about eight days in summer to about twenty-two days in winter.

Genus *DANIS* Fabricius, 1807

This genus is distributed from the Philippines to the Moluccas, New Guinea, the Bismarck Archipelago, the Kai and Aru Islands, eastern Australia and the Solomons, but has its greatest development in New Guinea. Three species occur in Australia.

112. *Danis danis* (Cramer), 1775

This species ranges from the Moluccas to New Guinea and north-eastern Australia, where two distinct subspecies are found.

a. *Danis danis serapis* Miskin, 1891

Plate 26, figs. 12 (male), 12A (female), large green-banded blue.

DISTRIBUTION.   Cooktown to Ingham, and the Atherton Tableland to the Paluma Range; common near Cairns nearly throughout the year.

The larvae feed on the leaves of *Alphitonia excelsa* (red ash, Rhamnaceae).

b. *Danis danis syrius* Miskin, 1890

DISTRIBUTION.   Cape York Peninsula, from Cape York to the Claudie River; at Cape York adults have been taken throughout the year.

ADULT.   Male: above similar to *serapis* but a deeper blue, white central area of forewing smaller and base only slightly metallic green. Beneath as in *serapis*, but metallic bluish green rather than metallic green markings.

Female: above similar to *serapis* but white band narrower and without metallic green. Beneath as in male.

The adults are usually found flying in rain forest, where they are difficult to follow in the alternating patches of sunlight and shade.

113. *Danis hymetus* (C. and R. Felder), 1865

This species ranges from the Moluccas and the Kai and Aru Islands to New Guinea, the Bismarck Archipelago and eastern Australia. Several subspecies are known, including three from Australia.

a. *Danis hymetus taygetus* (C. and R. Felder), 1865

Plate 26, figs. 14, 14B (males), 14A (female), small green-banded blue.

DISTRIBUTION.   Mackay to Sydney; usually rare south of Port Stephens. It is mainly coastal, but in wetter seasons is sometimes found well inland at the Expedition Range, central Queensland, and on the tablelands at Glen Aplin and Millmerran, southern Queensland, and at Armidale, New South Wales.

The larvae feed on the leaves of *Alphitonia excelsa* (red ash, Rhamnaceae) and their hairs and pale green coloration harmonize well with the whitish and somewhat hairy underside of the leaves. The feeding of the larger larvae produces irregular holes in the foliage. The pupae are attached beneath the leaves, and the pupal duration in the autumn is about three weeks.

b. *Danis hymetus taletum* (Waterhouse and Lyell), 1914

DISTRIBUTION.   Mossman to Ingham, and the Atherton Tableland to the Paluma Range; common. At Kuranda adults have been taken in most months.

ADULT.   Male: similar to *taygetus* but with white central band of hindwing above usually obscure or absent.

Female: similar to *taygetus* but above with narrower white band, base of wings metallic blue, and hindwing without metallic green subtornal patch; occasionally a blue subterminal band in forewing.

The larvae feed on the leaves of *Alphitonia excelsa* (red ash, Rhamnaceae) and probably resemble those of *taygetus*.

c. *Danis hymetus salamandri* W. J. Macleay, 1866

DISTRIBUTION.   Islands of Torres Strait, and Cape York to the McIvor River, 40 km north of Cooktown, Queensland; common.

ADULT.   Male: similar to *taygetus*, usually with a white central band on hindwing above, terminal scale fringes chequered black and white; forewing beneath with white central area occasionally not extending to costa near apex (as in Plate 26, fig. 14B).

Female: above similar to *taletum*, terminal scale fringe chequered black and white. Beneath as in male.

114. *Danis cyanea arinia* (Oberthür), 1878

Plate 26, figs. 13 (male), 13A (female), tailed green-banded blue.

DISTRIBUTION. Darnley, Banks and Thursday Islands, and Cape York, Claudie River and Cooktown to Mackay, Queensland; common at times north of Ingham.

Mature larva bright apple-green, head pale brown. Pupa smooth, pale brown with darker brown markings; attached by anal hooks and central girdle. The larva feeds on *Entada phaseoloides* (matchbox bean, Mimosaceae) and is attended by small black ants.

Genus *JAMIDES* Hübner, 1819

The genus ranges from Sri Lanka, India and Taiwan, through south-east Asia to New Guinea, Australia, the New Hebrides, the Solomons, Fiji and Tonga. Five species occur within Australian limits.

115. *Jamides aleuas coelestis* (Miskin), 1891

Plate 22, figs. 12 (male), 12A (female), bright cerulean.

DISTRIBUTION. Claudie River to 30 km south-west of Ingham (P. S. Valentine) north-eastern Queensland; sometimes fairly common. At Kuranda adults have been taken in most months.

Both upper- and underside of this species bear a superficial resemblance to *Danis*. It is a rain-forest species, and the adults fly along the edges of tracks through the rain forest, settling on the flowers of vines. Numbers have been observed at night-time settled on dead twigs.

The life history and early stages have not been described, but at Cairns the species has been reared from larvae feeding on *Sarcopteryx stipitata* (Sapindaceae).

116. *Jamides nemophilus nemophilus* (Butler), 1876

DISTRIBUTION. Within Australian limits known from only a few specimens taken at Darnley Island.

It is rather similar to *J. aleuas*, with broad white central areas, but it is brown beneath with obscure subterminal spots.

117. *Jamides cytus claudia* (Waterhouse and Lyell), 1914

Plate 22, fig. 11 (male), pale cerulean.

DISTRIBUTION. Known only from the Claudie River, Cape York

Peninsula. Adults have been collected from August to January and in May.

The adults have been found flying amongst the trees at the edge of rain forest. Although their flight is weak, they are difficult to catch as they flit between the trees.

118. *Jamides phaseli* (Mathew), 1889

Plate 27, figs. 1 (male), 1A (female), dark cerulean.

DISTRIBUTION.   North-western Australia, Northern Territory, including Groote Eylandt, Moa, Horn and Prince of Wales Islands, and Cape York, Queensland, to Grafton, New South Wales; common. In north Queensland it occurs on the coast and up to 160 km inland, south of Mt Garnet (W. R. Hindson).

This species is found in rain forest and in open eucalypt forest.

The larvae feed within the flower-buds of several legumes and are sometimes attended by one or two small black ants. In coastal southern Queensland this species has a decided preference for *Canavalia maritima* (wild jack bean, Fabaceae) as the larval food plant. It is also a minor pest of garden beans, *Phaseolus vulgaris* (Fabaceae), attacking the flowers.

119. *Jamides amarauge* Druce, 1891

DISTRIBUTION.   Known from only a few specimens taken at Darnley Island, Torres Strait, in April and May. It also occurs in New Guinea and the Solomons.

ADULT.   Male: above shining pale blue; forewing with apex and termen broadly brown-black; hindwing with termen dull black, series of whitish subterminal rings, one near tail with black centre. Beneath grey-brown, markings indicated by whitish edges; forewing without central spot in cell; hindwing with a series of whitish subterminal rings, two near tornus enclosing black spots, dusted metallic green and crowned bright orange.

Female: above dull black, large blue central areas in both wings; hindwing with series of obscure whitish subterminal rings, one near tail with black centre. Beneath as in male.

Genus *CATOCHRYSOPS* Boisduval, 1832

The genus contains only a few species distributed from India and Sri

Lanka through south-east Asia to Australia, New Caledonia and the Society Islands. Two species occur in Australia.

120. *Catochrysops panormus platissa* (Herrich-Schäffer), 1869

Plate 28; figs. 13 (male), 13A (female), forget-me-not.

DISTRIBUTION. North-western Australia, Northern Territory, including Groote Eylandt, the islands of Torres Strait, and Cape York, Queensland, to Grafton, New South Wales; fairly common north of Brisbane.

The extent of the blue areas in the female above varies considerably, being larger and paler in winter specimens and smaller in summer specimens. The species may be recognized by the long slender tail, the very pale markings beneath, and the two small black subcostal spots of the hindwing beneath.

The larvae feed on the young shoots and pea-like flower-buds of Fabaceae; at Brisbane the food plant is *Crotalaria alata*. They are attended by one or two small black ants. From larvae collected at Kuranda in July and taken to Sydney, adults emerged in August.

121. *Catochrysops amasea amasea* Waterhouse and Lyell, 1914

Plate 23, fig. 17 (male).

DISTRIBUTION. Darnley, Murray and Moa Islands, and Cape York; known from very few specimens. The type specimens came from Cape York.

This species is smaller than *C. panormus*, with the apex of the forewing more rounded and the termen more convex.

Genus *LAMPIDES* Hübner, 1819

The genus contains only one species with a very wide distribution from western Europe and Africa, through the Indo-Malayan area to Australia, the islands of the west Pacific and as far as Hawaii. It is not known from America.

122. *Lampides boeticus* (Linnaeus), 1767

Plate 27, figs. 15 (male), 15A (female), pea blue.

DISTRIBUTION. Islands of Torres Strait, and mainland Australia, including central Australia; also at Lord Howe Island, very common; Tasmania, rare.

The adults fly rapidly but feed freely at flowers and frequent the leguminous larval food plants. In Europe the species is an acknowledged migrant.

The larvae feed on many plants with pea-like flowers (Fabaceae). The young larva tunnels into the flower and later into the pod, feeding on the developing seeds. They are sometimes attended by one or two small ants. *Crotalaria*, when available, including *C. pallida*, *C. alata* and *C. cunninghamii*, appears to be preferred to other native species, but in addition larvae have been reared from many other native and introduced Fabaceae, including *Sesbania cannabina*, *Clianthus formosus* (Sturt's desert pea), *Swainsona* spp. (Darling pea), *Kennedia prostrata*, *Lotus australis*, *Psoralea patens*, *Vicia faba* (broad bean), *V. sativa*, *Phaseolus vulgaris* (garden bean), *Virgilia oroboides*, *Dolichos*, *Lupinus* (lupin), *Chamaecytisus prolifer* (tree lucerne), *Lathyrus odoratus* (sweet pea), and *Pisum sativum* (garden pea). The larvae usually pupate unattached on the ground, under a leaf or stone, or if the food plant is growing in very sandy soil they may pupate a centimetre or two below the surface. The pupae can then be collected by sieving the sand, and it has been suggested that deformed adults will emerge from them if they are not again covered with a light layer of sand.

Genus *SYNTARUCUS* Butler, 1901

This genus is widely distributed from southern Europe and Africa to the Indo-Malayan and Australian regions, but is best represented in Africa. Only one species occurs in Australia.

123. *Syntarucus plinius pseudocassius* (Murray), 1873

Plates 27, figs. 2 (male), 2A (female), XX, zebra blue, plumbago blue.

DISTRIBUTION. From Port Stewart, Queensland, to Sydney, New South Wales, extending as far inland as Cunnamulla, Q., and Cowra, N.S.W. Specimens have also been taken near Alice Springs, N.T. (G. Griffin). At Sydney adults are seldom taken before autumn, when the food plant is flowering, whereas in Queensland they are on the wing from September to May.

The females vary in the extent of the white markings above, and in Queensland this may be seasonal in nature. The adults fly commonly around *Plumbago*, the larval food plant.

Eggs laid at Sydney in February and March may remain dormant until the *Plumbago* begins to bud towards the end of the year. The larvae feed on the flower-buds of *P. auriculata* and *P. zeylanica* (Plumbaginaceae), blue flowering shrubs or hedge plants grown commonly as ornamentals. The latter may have a natural occurrence in northern Australia. The larvae are occasionally attended by ants.

Genus *ZIZEERIA* Chapman, 1910

This is another very small genus with a wide distribution from southern Europe, Africa, southern and south-east Asia to northern and eastern Australia. There is one Australian species.

124. *Zizeeria karsandra* (Moore), 1865

Plate 28, figs. 15 (male), 15a (female), dark grass-blue.

DISTRIBUTION. North-western Australia, north from Fitzroy Crossing (M. S. Moulds), Northern Territory, including Groote Eylandt, and eastern Australia from Weipa to as far south as Bundanoon near the coast, and south to the Murray Valley and Lake Hattah west of the Divide, just reaching north-western Victoria and South Australia; at times common, especially in the north.

The spots beneath the wings are much more distinct than those of *Zizina labradus*, which it resembles superficially. The adults fly close to the ground in open areas, amongst low-growing herbage, especially near the larval food plants.

The larvae usually feed on the flowers, young fruit, and young leaves of *Tribulus* spp., including *T. terrestris* (bullshead burr or caltrop), and *T. cistoides* (Zygophyllaceae). They are said to feed also on the pea-like flowers of legumes (Fabaceae) and in southern inland New South Wales have also been found on *Glinus lotoides* (Molluginaceae) (E. D. Edwards). They are attended by one or two small black ants. In Okinawa this subspecies feeds on *Amaranthus* (Amaranthaceae).

Genus *ZIZINA* Chapman, 1910

The genus contains few species with a wide distribution in Africa, southern Asia, Japan, Australia, New Guinea, New Zealand and the islands of the western Pacific. Only one species occurs in Australia.

125. *Zizina labradus* (Godart), 1824

This species is widely distributed in the Australian region.

a. *Zizina labradus labradus* (Godart), 1824

Plates 28, figs. 14 (male), 14A (female), XX, common grass-blue.

DISTRIBUTION. Throughout continental Australia and Tasmania, except the northern half of the Cape York Peninsula; also at Lord Howe and Norfolk Islands and New Zealand; very common.

This is the commonest butterfly in Australia, being found in most open spaces and in gardens, wherever legumes are growing. Adults are on the wing in most areas throughout the year. The species is a minor pest of lucerne, garden beans and other leguminous plants.

The larvae feed on the young leaves, flower-buds, and seed-pods of Fabaceae including *Trifolium* (clover), *Medicago* (lucerne, medics), *Desmodium*, *Glycine*, including the cultivated *G. max* (soybean) and *G. wightii*, *Swainsona*, *Phaseolus vulgaris* (garden beans), *Macroptilium lathyroides*, *Pisum sativum* (garden peas), *Indigofera*, *Trigonella*, *Psoralea adscendens*, *P. patens*, *Lotus australis*, *Vicia faba* (broad bean), *Virgilia oroboides* and other low-growing legumes. At Queanbeyan, N.S.W., a larva was found on a seedling *Acacia baileyana* (Mimosaceae). They are often associated with one or two small black ants.

b. *Zizina labradus labdalon* Waterhouse and Lyell, 1914

DISTRIBUTION.   Islands of Torres Strait, and Cape York to the Claudie River; apparently uncommon.

ADULT.   Male: similar to *labradus* but above with broader brown margins, and beneath with markings more distinct.

Female: similar to *labradus* but seldom any blue. Beneath as in male.

Genus *FAMEGANA* Eliot, 1973

This genus contains only one small species referred until recently to *Zizeeria*.

126. *Famegana alsulus alsulus* (Herrich-Schäffer), 1869

Plate 28, fig. 17 (male), black-spotted grass-blue.

DISTRIBUTION.   North-western Australia, Northern Territory, including Devils Marbles (K. L. Dunn) and Groote Eylandt, the islands of Torres Strait and Cape York, Queensland, to Gloucester, New South Wales, and near Hermannsburg, central Australia; in the west it is now known from as far south as Karratha (M. S. Moulds), Dampier and the Hamersley Range (M. Hutchison).

The species may be identified at once by the absence of the normal pattern of spots beneath the wings, and the presence of a single round black spot beneath the hindwing. The adults fly close to the ground.

The larvae feed on the pea-like buds and flowers of legumes (Fabaceae) and are attended by small ants. The eversible whitish glands of the eighth abdominal segment are often protruded as the larva feeds with its head deep within a flower-bud.

Genus *ZIZULA* Chapman, 1910

The genus contains only a couple of small species and is distributed almost throughout the equatorial regions of the world.

127. *Zizula hylax attenuata* (T. P. Lucas), 1890

Plate 28, fig. 16 (male), tiny grass-blue.

DISTRIBUTION. Northern Territory, Prince of Wales Island, Cape York to Sydney, and at Hermannsburg and Palm Valley, central Australia; common north of Brisbane.

This is one of the smallest Australian butterflies and can be easily identified by the pattern of spots beneath the wings. The adults fly close to the ground in open areas amongst grass or other herbage.

The life history and early stages have not been recorded in Australia. The larva of the African subspecies is said to feed on the flowers of *Oxalis* (Oxalidaceae).

Genus *EVERES* Hübner, 1819

The genus contains few species scattered through Europe, Asia, Australia and America. One is represented in Australia.

128. *Everes lacturnus australis* Couchman, 1962

Plate 27, figs. 3 (male), 3A (female), tailed cupid.

DISTRIBUTION. Northern Territory, including Groote Eylandt, Darnley Island, and Cape York, Queensland, to Port Macquarie, Barrington Tops, Cessnock, Mt Kaputar, New South Wales; common north of Brisbane.

This species may be distinguished by its small size and fragile appearance, its long slender tail, smooth eyes, the prominent orange patch above and beneath the hindwing, and the presence and position of the small black spots in the basal half of the hindwing beneath. The adults fly close to the ground in grassy places.

The life history and early stages have not yet been described from Australia. The larva of the Indo-Malayan subspecies *rileyi* Godfrey is described as green with darker dorsal and subdorsal lines and paler lateral line. It feeds on Fabaceae, including *Lotus* (trefoil). The pupa is sparsely covered with long hairs and may be either green with darker green markings or dark green with black spots.

Genus *PITHECOPS* Horsfield, 1828

The genus contains four species distributed through the Oriental region, with one species occurring in the Australian region.

129. *Pithecops dionisius dionisius* (Boisduval), 1832

Plate 14, fig. 16 (male).

DISTRIBUTION.   Darnley Island, Torres Strait, Cape York (M. Walford-Huggins) and the Claudie River (J. W. C. d'Apice, A. Hiller), Queensland.

This species is somewhat similar to *Neopithecops lucifer*, but is much larger. The adults fly close to the ground in rain forest. The early stages are not known.

Genus *CELASTRINA* Tutt, 1906

This genus is widely distributed from India and Sri Lanka, through south-east Asia to New Guinea and the Solomons. One species occurs in Queensland.

130. *Celastrina tenella tenella* (Miskin), 1891

Plate 22, figs. 10 (male), 10A (female), Australian hedge blue.

DISTRIBUTION.   Known only from Kuranda, north Queensland. Adults have been taken fairly commonly at times from February to July, but mainly in June and July.

This small species is easily identified in both sexes by the spotted underside, and by the whitish subterminal rings of the hindwing above not being duplicated beneath. The adults are usually taken feeding at flowers, especially on hill-tops, or along pathways through the rain forest. The early stages are not known.

Genus *NEOPITHECOPS* Distant, 1884

The genus is distributed from India and Sri Lanka to southern China, Taiwan and the Philippines, and through south-east Asia to the Kai and Aru Islands, New Guinea and Cape York.

131. *Neopithecops lucifer heria* (Fruhstorfer), 1919

Plate 26, fig. 15 (female), quaker.

DISTRIBUTION.   Islands of Torres Strait and Cape York; rare.

The single prominent black apical spot beneath the hindwing at once places this small species.

The life history and early stages have not been recorded from Australia. The Indian subspecies feeds on *Glycosmis pentaphylla* (Rutaceae) and is sometimes attended by ants. This shrub is recorded from north Queensland and is probably the food plant of the Australian subspecies.

Genus *MEGISBA* Moore, 1881

132. *Megisba strongyle nigra* (Miskin), 1890

Plate 22, fig. 13 (male), Malayan.

DISTRIBUTION.   Cape York to Bingil Bay, near Tully (W. R. Hindson), Paluma (J. R. Turner) and the Atherton Tableland; common.

This small species is very easily recognized. The size of the white patches above varies to some extent.

The life history and early stages have not been recorded from Australia. The larvae of the subspecies from Sri Lanka are reported to feed on plants of the family Sapindaceae, and those of the Okinawa subspecies on *Mallotus japonicus* (Euphorbiaceae).

Genus *EUCHRYSOPS* Butler, 1900

The genus occurs in Africa and from India and Sri Lanka through south-east Asia to Australia and the south-west Pacific.

133. *Euchrysops cnejus cnidus* Waterhouse and Lyell, 1914

Plate 28, fig. 10 (male), cupid.

DISTRIBUTION.   North-western Australia, Northern Territory, including Groote Eylandt, the islands of Torres Strait, and Cape York to Coffs Harbour; common.

From the somewhat similar *Catochrysops panormus* this species may be distinguished by the much shorter tail, the smooth eyes, and the small black spots in the basal half of the hindwing beneath. The colour of the male above is very different in the two species.

The larvae feed on the trailing legumes *Vigna vexillata*, *V. luteola* and *V. unguiculata* (cowpea), and on *Phaseolus vulgaris* (garden bean), *Macroptilium lathyroides* and *Sesbania* sp. (all Fabaceae). At first they eat the flower-buds and flowers, but later are found within the seed-pods. They are attended by various species of ants, including *Polyrhachis ammon*, *Camponotus*, *Iridomyrmex*, and *Crematogaster*. In coastal southern Queensland larvae were most abundant in February and March, and at that time of year the pupal duration was nine to twelve days.

Genus *FREYERIA* Courvoisier, 1920

The genus ranges from Europe and Africa, to the Oriental and Australian regions.

134. *Freyeria trochylus putli* (Kollar), 1844

Plate 27, fig. 14 (male), grass jewel.

DISTRIBUTION.   Northern Territory, including Groote Eylandt, and the Claudie River to Burleigh Heads, Queensland; sometimes very common north of Rockhampton.

The row of black spots beneath the hindwing readily identifies this very small species. The adults fly close to the ground in open places.

The eggs are laid singly on the flower-buds or flowers of the food plant, or beneath a leaf or on a leaf axil. In coastal southern Queensland the larvae feed within the small pink flowers of a herbaceous legume, *Indigofera hirsuta* (hairy indigo, Fabaceae), but probably also on other legumes. They are attended by small black ants, *Iridomyrmex* sp. The pupal duration in July and August was five weeks.

Subfamily RIODININAE

(Plate 20)

The subfamily reaches its greatest development in North and South

America, but a few small genera occur in the Oriental and Papuan areas. Only one species extends to the far north-east of Australia. As in most of the Lycaenidae the fore tarsi in the male are reduced to a single elongate segment, but in the female have five segments and paired claws. The male genitalia resemble those of the Lycaeninae, with a pair of curved gnathos hooks. The larva is onisciform and is clothed with short hairs. In some species the larvae are attended by ants. The pupa is sometimes clothed with short hairs, and is usually attached by anal hooks and a central girdle.

Genus *PRAETAXILA* Fruhstorfer, 1914

The genus occurs in New Guinea, the Kai and Aru Islands, and in the Cape York Peninsula.

135. *Praetaxila segecia punctaria* (Fruhstorfer), 1904

Plate 20, figs. 13 (male), 13A (female), Australian harlequin.

DISTRIBUTION.   Cape York to Coen; not common. Other subspecies are known from New Guinea and the Aru Islands.

The adults fly close to the ground in rain forest and settle on the leaf litter (M. S. Moulds). The early stages are unknown.

Collection and Study

BUTTERFLY collecting is an interesting, instructive and relatively inexpensive hobby which in recent years has steadily gained in popularity. It provides healthy outdoor exercise in pleasant surroundings and can lay the foundation for a lasting appreciation of our rich natural environment. A butterfly collection has both aesthetic appeal and scientific value. It can provoke in young and old alike a real curiosity in natural phenomena, and develop in the more persistent student a life-time interest in natural science. Many a professional scientist owes his initial interest in biology to butterfly collecting.

The basic equipment needed by any collector includes a collecting net, a killing bottle, pins, entomological forceps, setting boards and store boxes.

COLLECTING NET

Adult butterflies are usually collected in a net, the size and design of which varies considerably. Most collectors have their favourite net.

The simplest and one of the most effective nets has a circular frame of steel wire, about 45 cm in diameter, attached in some way to a wooden, bamboo, or aluminium handle. If the frame is detachable it is much easier to pack when travelling and the fine net bag can be replaced more readily if it should be torn. Three simple methods for attachment are often used: (*a*) the two ends of the steel wire can each be bent to form small rings which fit over a screw thread of appropriate diameter

346

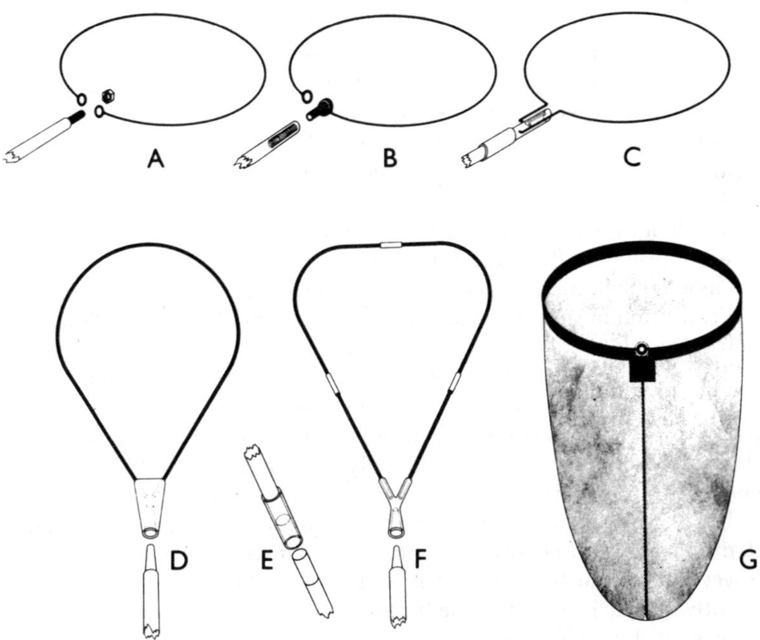

Fig. 20. Collecting nets: A-C, Wire frames; D, Fixed cane frame; E, F, Collapsible kite net, with detail of brass ferrule; G, Net bag.

protruding from the end of the handle, and held firmly in place by a nut (Fig. 20A). (*b*) One end of the steel wire can be bent into a small ring, while the other is welded to a bolt which, when screwed into a thread in the end of the handle holds the ring in place (Fig. 20B). (*c*) The ends of the wire can be bent as shown (Fig. 20C), fitted into similar grooves in the end of the wooden handle and held in place by sliding a metal sleeve over them.

A light and strong net frame may also be made from cane. The cane should be steamed and bent around to form a large loop, with its ends firmly glued into two slanting holes bored in a shaped block of wood (Fig. 20D). A larger hole is bored in the other end of the block to admit the handle. The jointed kite net (Fig. 20F) is also made from cane and can be taken to pieces for packing. Its frame is made of four pieces of cane with brass ferrules at each end (Fig. 20E). The ends of the longer straight sections fit into a brass Y-piece, into which the handle can also be fitted.

The length of the handle will depend upon the height at which the desired specimens fly. For most purposes it need not exceed 120 or 150 cm in length. Jointed extension handles can be a considerable help for the collection of specimens that settle on flowers or foliage well out of normal reach.

The net bag should be made from a strong soft fabric, preferably light in weight. The best net bags nowadays are made of fine mesh nylon or other synthetic material, which has a low air resistance and dries rapidly if it should be accidentally wetted. A fine cotton netting is effective, but is more difficult to dry and is more easily torn.

For a frame 45 cm in diameter, the net bag should be about 75 cm in depth, to allow ample room for the enclosed insect to be trapped in the lower half of the bag when it is folded over. The bag should taper slightly and be rounded at the bottom (Fig. 20G). The mouth should have a machined hem about 4 cm in width through which the wire or cane frame can be threaded. If the frame is a fixture the hem should be broader and be folded over the frame and coarsely stitched in place or be held by press studs. In either case a strong gusset should be provided to prevent tearing the mouth of the net nearest the handle.

KILLING THE SPECIMEN

Many collectors prefer to kill or immobilize larger specimens while they are still in the net, by giving the thorax a sharp pinch through the net, while the wings are held back to back above the body. When done quickly and deftly this method has the advantage of preventing undue damage to the specimen by its fluttering in the net or the killing bottle, and also makes its escape less likely. With practice the correct amount of pressure can be applied to the thorax with the thumb and index finger to stun it, without causing noticeable external damage.

After a specimen is immobilized in this way it should either be placed in a killing bottle to ensure that it is dead, or be pinned after the point of the pin has first been dipped into nicotine sulphate, obtainable from a horticultural supply shop. If treated in this way, the dead specimen becomes fully relaxed and ready for spreading within about fifteen minutes. It can be kept relaxed prior to spreading by pinning it in a slightly moist relaxing box. If killed in a killing bottle *rigor mortis* sets in very rapidly and a period of about six to twelve hours must then elapse before it again becomes fully relaxed and suitable for pinning and spreading.

The best killing bottle can be prepared by placing some crystalline potassium cyanide in the bottom of a wide-mouthed jar or a similar plastic container, mixing it with about 12 mm of granular vermiculite, and then pouring over the mixture about 12 mm of freshly mixed plaster of Paris. The jar should have a close-fitting stopper of rubber or waxed cork. When the plaster has set firmly, a pad of cellulose tissue should be cut to fit firmly in the bottom of the jar over the plaster to absorb surplus moisture.

As cyanide is a dangerous poison, the killing bottle should be clearly labelled and kept well out of reach of small children. If used sensibly, removing the stopper only momentarily when specimens are to be inserted or removed, and avoiding inhalation of the gas when the bottle is open, this killing bottle can be used with complete safety. With care such a bottle should be effective for a year or more with average use. When it ceases to kill insects with reasonable speed, it should be discarded by burying it without its stopper in moist soil in a disused corner of the garden.

If cyanide is not available, an effective killing bottle can be made by pouring 12 mm of plaster of Paris into the bottom of the jar and, when hard, adding a few drops of ethyl acetate to the plaster. It is as well to cover the plaster with a pad of cellulose tissue to prevent the specimens coming into contact with any of the liquid. This killing bottle needs to be recharged before each day's collecting. Ethyl acetate is relatively harmless, but it is as well to avoid inhaling its vapour any more than necessary.

RELAXING OR FIELD BOXES

Traditional pocket boxes are oval in shape and made of zinc or galvanized iron, hinged in the middle, with each half lined with cork or 6 mm balsa (Fig. 21A). Plastic containers (Fig. 21B) of suitable shape, such as those designed to take a 500 gm block of butter, or the deeper types used for ice-cream, are effective if they have an air-tight lid. The bottom is covered by a piece of plastic sponge cut to shape and a piece of 3 mm balsa of the same dimensions. The latter can be fixed in with small metal brackets fused to the inside of the plastic with a hot soldering iron. Before collecting is begun, the cork or balsa is slightly moistened with water, a crystal or two of chlorocresol being added to prevent mould growth. The specimens placed in the box will then become fully relaxed ready for pinning and setting.

PINS AND PINNING

Butterflies should be pinned on stainless steel entomological pins. These are usually 38 mm in length, and are manufactured especially

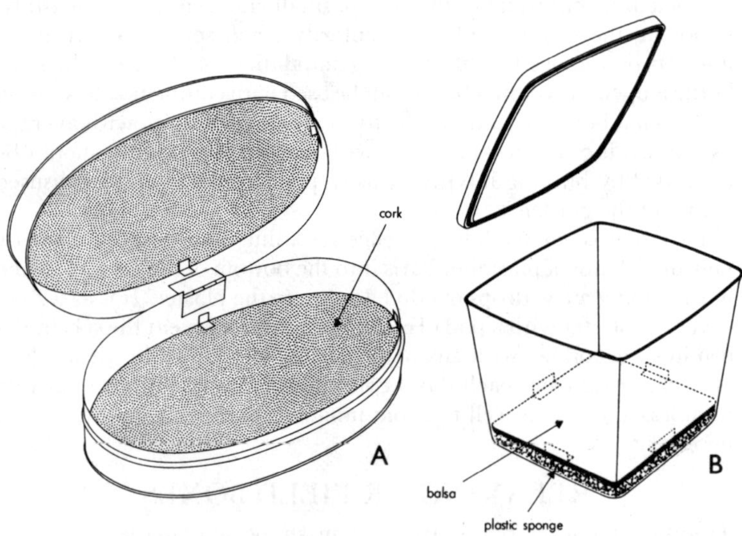

Fig. 21. Relaxing or field boxes: A, Hinged metal box with cork lining; B, Plastic butter box with plastic sponge and balsa lining.

for this purpose. Asta No. 3 pins will be found suitable for most species, but the largest species, such as birdwings, require No. 5.

To pin the specimen, when the wings are held back-to-back above the body, it is gently gripped between the thumb and forefinger of the left hand and the point of the pin is carefully inserted into the middle of the thorax above and passed *vertically* through the thorax.

SETTING OR SPREADING

(a) Setting Boards

A setting board is used to spread the wings of the specimen out flat in a standard way, and to hold them in this position until the body is thoroughly dry. It provides a central groove for the pin and the body of the insect, and a pair of flat surfaces either side on which the wings are to be extended flat. Boards should be made in various sizes according to the size of the insects to be set.

Cheap but effective setting boards can be made from 12 mm fibrous building board (Caneite). Two identical strips of board are cut and glued, parallel to one another, to a broader strip, leaving a central groove between their two inner edges (Fig. 22A). When the glue is dry the upper sides are carefully sandpapered and, preferably, covered with white, plain or squared, paper pasted to the surface.

More sophisticated boards, which last much longer, can be made by gluing a strip of polystyrene foam insulating board, 12 mm in thickness, to a piece of plywood of similar size, to which are glued two narrower strips of balsa, 9 mm in thickness, leaving a groove between them (Fig. 22B).

Some collectors prefer wooden boards (Fig. 22C).

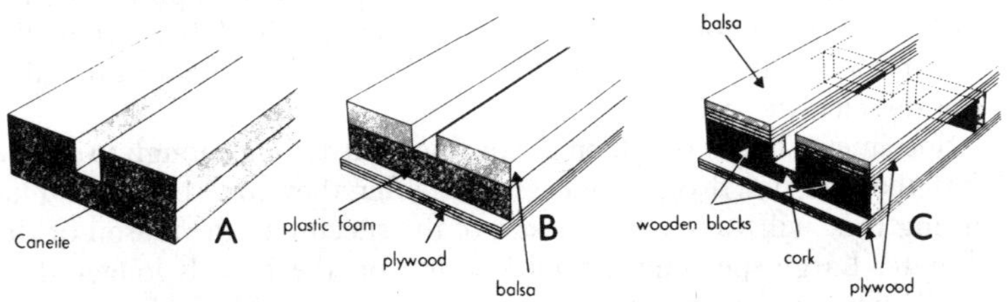

Fig. 22. Setting boards: A, simple Caneite board; B, Balsa and plastic foam board; C, Plywood board with balsa surface.

(b) Setting or Spreading Procedure

After selecting a board, wide enough to enable the wings to be fully extended without overhanging the sides and with a groove of suitable width, the specimen is carefully pinned in the groove. The pin should be *vertical* and the underside of the wing-bases should just clear the edges of the groove. Two strips of grease-proof paper, cut slightly narrower in width than each half of the board, are pinned through their leading edges well in advance of the body of the specimen, and carefully lowered on to the wings on each side (Fig. 23). The wings are then gently pressed down beneath the paper by holding the rear edge of the paper with the forefinger of the left hand. Tension on the paper can be varied by varying the pressure of the finger. With the point of a sharp needle or pin placed behind one of the main veins near the base of the forewing, the wing is slid forward beneath the paper until its inner margin or dorsum is at right angles to the groove. Care should be taken to avoid piercing the wing membrane. By applying tension to the paper, the wing may be held in this position until a pin is inserted in the paper just beyond the termen. The hindwing is then slid forward similarly making sure its costa passes beneath the forewing and it assumes a natural position in relation to the forewing. Further pins are then inserted around the margin of the two wings so that the paper holds them perfectly flat. This procedure is repeated with the other pair of wings. The antennae should then be arranged in position and, if necessary, pinned down with small squares of paper. Finally, cross-pins are inserted below, and if necessary above, the abdomen to hold it in a central horizontal position.

When each specimen is set, a label showing the locality, the date of capture and the collector's name should be printed and pinned beside it. Never trust your memory; always prepare the labels when the information is fresh in your mind. A specimen without a locality label is of no scientific value.

Specimens should be left on the setting boards long enough to ensure that they do not droop unnecessarily after they are removed. This drying time will vary with the size of the specimen, the season or the climate. Large specimens should be left on the boards longer than small ones, and drying always takes longer in humid conditions compared with dry conditions. Whereas two or three weeks may be necessary for large specimens, small ones may be thoroughly dry in less than a week during summer.

When drying is completed the paper cover strips and cross-pins are

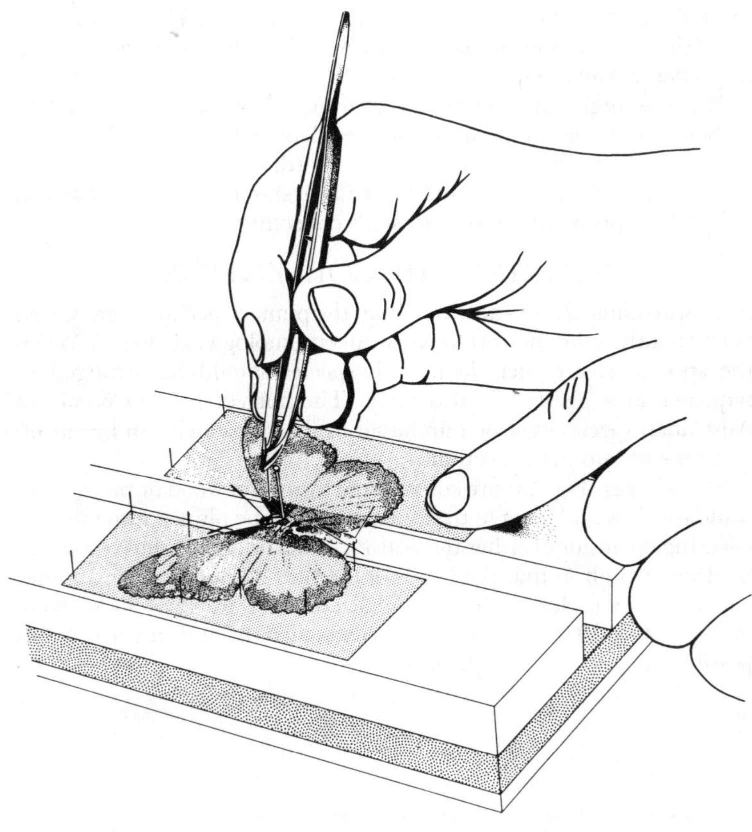

Fig. 23. Method of setting a butterfly.

removed and the central pin carrying the dried specimen with wings outspread is withdrawn. The point of the central pin is then passed through the centre of the prepared label so that the label is always attached to the specimen. The label should be neatly and symmetrically pinned, with printed side uppermost.

PAPERING SPECIMENS

During an extended collecting trip it may not be convenient to spread more than a few of the specimens that are captured. The surplus may be labelled and pinned in store boxes or, if the collector prefers, placed unpinned in small paper triangles.

Paper triangles are usually made from rectangles of firm but absorbent paper cut to a variety of sizes depending on the size of the specimens. Convenient sizes for most butterflies are 15 by 11 cm, 13 by 9 cm and 10 by 6 cm. A supply of these should be cut and folded (Fig. 24) in preparation for a season's collecting.

STORING THE COLLECTION

After spreading, drying and labelling, the pinned specimens are stored permanently either in store-boxes or in entomological cabinets. When the specimens are identified, each species should be arranged in sequence as suggested in this book. The correct printed names of Australian species may be purchased in sheets* which can be cut up and pinned into the collection.

Store-boxes (Fig. 25) are constructed from thin wood or plywood in standard sizes and open so that specimens can be pinned into the cork covering the inside of either the bottom or the lid. Boxes must therefore be deep enough so that the heads of the pins do not foul one another when the box is closed. Convenient sized boxes are 45 × 30 × 10 cm or 35 × 30 × 10 cm. They should be as airtight and insect proof as possible and, of course, light-proof.

*Sets of the printed names of Australian butterflies, as used in this book, should be available from Australian Entomological Press, 14 Chisholm St., Greenwich, N.S.W. 2065, by the time this book is published.

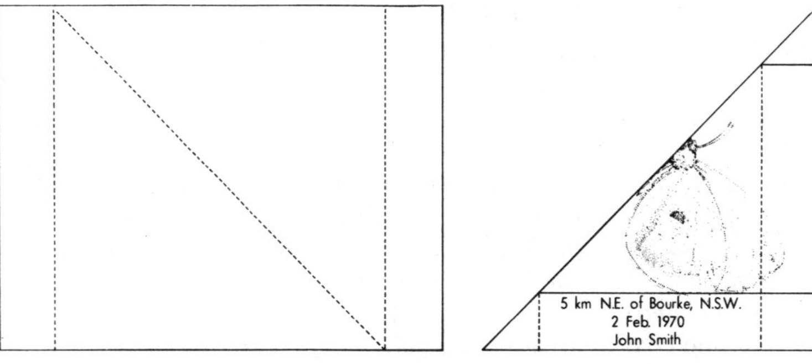

Fig. 24. Paper pattern and method of folding paper triangles.

Entomological cabinets are chests of shallow drawers, each with a more or less air-tight glass lid. The better cabinets have a cell surrounding each drawer, into which flake naphthalene is put to protect the collection from pests, and there may be a door to close each cabinet to provide extra protection. Cabinets are made from either wood or steel and are usually fairly expensive. However, they are well worth having, because they greatly reduce the problems of damage by insect pests and mould, and damage due to handling or dropping boxes. In addition, they permit easy examination of specimens through the glass lid and greatly enhance the appearance of the collection.

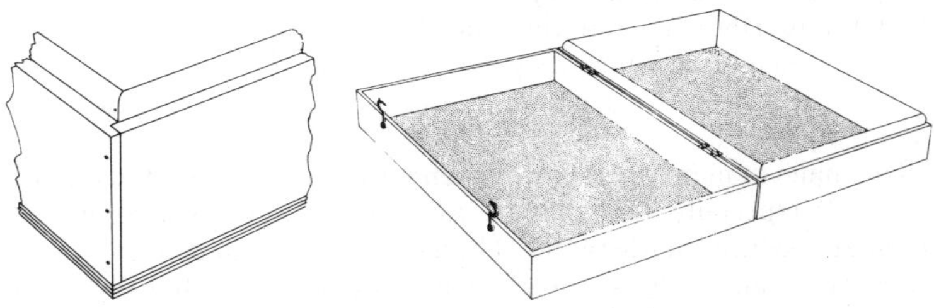

Fig. 25. An insect store-box, with detail of one corner.

CONTROLLING PESTS

Dried insect specimens are attacked by several small insect pests which soon invade the collection unless adequate measures have been taken to discourage them. They include psocids or book lice, and museum beetles (*Anthrenus* and related genera).

The collection should be stored constantly in the presence of naphthalene, either in flake form or in moth-balls. Flake naphthalene can be tied or sewn into small muslin bags pinned into the corners of store-boxes, or carefully melted and poured into one end of the box while it is slightly inclined.

In humid coastal areas of Australia mould can cause considerable damage to collections and eventually complete loss. A crystalline chemical, chlorocresol, has proved to be effective in preventing the establishment of mould in a collection but, like naphthalene, should be renewed periodically. This material will also prevent mould in relaxing boxes.

COLLECTING ADULTS

Contrary to popular belief, the butterfly collector seldom needs to run after his quarry. In fact, haste often results in the loss of a specimen which, by careful stalking and deft manipulation of the net, he would have secured. The specimen can be caught by a short rapid sweep of the net, followed by a twist of the handle to fold over the net bag and trap the butterfly within. If the specimen should be resting on the ground it is best, while holding the bag up with the left hand, to place the net suddenly over it and then to allow the insect to fly upwards into the net before folding over the bag.

The most successful collector is one who studies the flight behaviour of adult butterflies, noticing the time of day when they are active, the places and heights at which they usually fly, the flowers at which they feed and the larval food plants around which the adults are most commonly found.

The males of many species are taken most frequently on the tops of steep hills, especially those that are set apart a little from other hills or dominate extensive flatter areas. The females of these and other species are more usually collected near the larval food plants, but they also visit the hill-tops briefly for mating.

REARING AND THE STUDY OF LIFE HISTORIES

Much remains to be learnt about the life histories, food plants and habits of many Australian butterflies. The available information about each species is summarized in this book, and this should be used as a guide to enable collectors to study the biology more closely.

Most collectors have reared adults from larvae or pupae, but few have taken the trouble to record their observations, in the mistaken belief that the biology of our species is well understood. On the contrary, we need to know much more about most aspects of the biology of our butterflies. The collector can contribute valuable information about a great variety of subjects such as: the egg-laying behaviour of most species, the appearance of the eggs, the incubation period, the behaviour of larvae at various stages in their growth, their feeding behaviour, and food plant preferences, the number of larval instars and the duration of each, variation in larval colour and pattern, perhaps in relation to the food plant, the detailed structure and appearance of each species, protective devices in eggs, larvae and pupae, the parasites, predators and diseases of eggs, larvae and pupae, the relationship of larvae and ants, the choice of pupation sites, pupal duration, structure and coloration, especially in relation to background colour, sound production in pupae and its significance, the seasonal history of each species and survival mechanisms during winter and adverse weather. By thoughtful observation, accurately recorded in a note book, as well as by adequately preserving specimens, anyone interested in butterflies can make worthwhile contributions to our knowledge of these subjects, and at the same time derive a deeper and richer satisfaction from his hobby.

In collecting and studying butterflies the beginner will soon realize how important it is to have a little knowledge of botany. Unfortunately, we are not well supplied with botanical books that will enable easy identification of even our common Australian plants. Several useful books have been published recently, including Cochrane *et al.* (1968), Beadle *et al.* (1972), Burbidge (1963), Burbidge and Gray (1970), Erickson *et al.* (1973), Galbraith (1977), Rotherham *et al.* (1975), and Anderson (4th edit., 1968). It is as well to collect and press samples of the larval food plants, including where possible flowers and fruits, so that they can be identified by one of the herbaria located in the State capitals. It is only in recent years, for example, that collectors have taken the trouble to have the mistletoe food plants of Pieridae and Lycaenidae authoritatively identified, and there is still much to be done. This will probably show that there are marked, but still unrecognized, food preferences in the various species.

Once the food plant is recognized, larvae should be sought as indicated in the descriptions of each species. However, do not restrict your search to the known food plants, but watch out for related plants and examine those around which adult females are observed flying.

In rearing larvae it is essential that the food supply be kept in fresh condition and renewed frequently. It is usually not satisfactory to use cardboard containers, for in these the food plant dries out too quickly. Plastic bags are very useful for rearing many species if the neck is securely closed with a rubber band. However, they must be kept clean, the droppings being removed daily and preferably a clean bag being substituted each time the food is changed. Newly hatched larvae are best placed on some fresh young growth of their food plant in clean screw-top jars. The food should keep fresh for several days in this way, but care should be taken to prevent the very small larvae drowning in the water that tends to condense on the sides. Young larvae can be transferred without damage from one jar to another with a soft water-colour brush. Both plastic bags and glass jars should be kept out of direct sunlight.

A cage is usually desirable for pupal emergence, especially for the larger species, to ensure that the newly emerged adult can spread its wings fully before drying. Special care should be taken with the pupae of Hesperiidae. As the leaves of grasses and sedges tend to become tightly rolled on drying, pupae that are left in their shelters after the leaves have been cut will be crushed. They should therefore be carefully removed. However, if kept loose, the adults which emerge are usually deformed. These pupae should therefore be slipped head upwards into paper cylinders, formed by rolling a rectangle of stiff paper around a lead pencil and twisting one end. Several of these can be stood upright in a small tin in the rearing cage.

Glossary

Aberration—A relatively rare form that can be distinguished from the normal in any species.

Aedeagus—The sclerotized tube in male Lepidoptera through which the vesica or penis passes (Fig. 5).

Aestivate—To pass the summer in a dormant state.

Anal hooks—Minute, often clubbed, hooks beneath the anal segments of the pupa in most Lycaenidae, attaching it to a pad of silk spun by the larva (Fig. 2).

Anal plate—A sclerotized middorsal plate of the tenth abdominal segment in larvae (Fig. 1).

Anal veins—One to three longitudinal wing veins (1A to 3A) situated posterior to the cubitus or cubital vein (Fig. 4A).

Androconia—Special wing scales in male Lepidoptera which are thought to be important in the dissemination of pheromones.

Apex—Tip of the wing (Fig. 4B).

Apical—Area at or adjacent to the tip of a wing or other structure (Fig. 4B).

Apiculus—Tapering apical portion of the antennal club in Hesperiidae.

Apodeme—Infolded projection of the insect exoskeleton to which muscles are attached.

Apophyses (anteriores and posteriores)—Apodemes or attachments for the muscles operating the female ovipositor and genital apertures (Fig. 5B).

Aposematic—Warning; usually referring to the bright contrasting coloration of distasteful or harmful animals.

Autosomes—The chromosomes that are not involved in sex determination.

Basal—Area at or adjacent to the base of the wing or other structure.
Bifid—Forked.
Biordinal—Of two sizes; referring to the larval crochets.

Bursa copulatrix—The copulatory ducts and sac in adult female Lepidoptera, comprising the ostium bursae, ductus bursae and corpus bursae (Fig. 5B).

Carina (pl. carinae)—A longitudinal ridge.

Central girdle—The rope of silk that gives central support to some pupae in addition to the cremaster.

Chaetosema (pl. chaetosemata)—A hairy sense organ found on the head of butterflies and many moths.

Chemoreceptor—A sense organ responsive to chemical stimuli.

Chlorocresol (4-Chloro-*m*-cresol)—A chemical used for preventing infection of specimens by mould.

Chorion—The egg shell.

Chromosomes—Strands of genetic material in the cell nucleus carrying the genes.

Cline—A character gradient in which the structure or physiology of an organism changes over its geographical range.

Club—Thickened or swollen apical end of the antenna or other structure.

Cornutus (pl. cornuti)—A strongly sclerotized spine attached to the outside of the everted vesica of the male.

Corpus bursae—The membranous sac of the female genitalia in which is deposited the male spermatophore during copulation (Fig. 5B).

Costa—The anterior or leading margin of the wing, joining the base and the apex (Fig. 4B).

Costal—Area along or immediately adjacent to the costa; the first vein (C) of the wing, running along the costa.

Coxa—The first or basal segment of the leg (Fig. 3A).

Cremaster—Constricted anal segments of the pupa, bearing hooked hairs used for attaching the posterior end of the pupa to the silk pad spun by the larva (Fig. 2).

Cremastral hooks—See anal hooks.

Crenulate—With small regular indentations, scalloped.

Crochets—Series of terminal sclerotized hooks on larval proleg (Fig. 1).

Cubitus—The fifth main longitudinal vein (Cu) of the wing usually forming the posterior margin of the discal cell (Fig. 4A).

Cuticle—The non-cellular skin of insects consisting of chitin and protein; it is cast at moulting to allow increase in size.

Dentate—Toothed.

Diapause—A period of suspended growth or development, usually accompanied by a greatly decreased metabolism, permitting survival during an unfavourable period, such as winter or adverse weather.

Diaphragma—Membrane closing the posterior end of the male abdomen between the bases of the valvae.

Dimorphic—Occurring in two forms; differences in form or coloration between the sexes are called sexually dimorphic.

Discal—The central area of the wing.

Discal cell—The large central area of the wing partly or completely bounded by veins, anteriorly by R or Rs, posteriorly by CuA, and distally by the discocellular veins (Fig. 4A).

Discocellulars—The short and weak transverse veins which form the distal margin of the discal cell in the wing (Fig. 4A).

Distal—Farthest from the base or centre.

Diverticulum—A blind offshoot or sac from the alimentary canal or other tube.

Dorsal gland—An eversible glandular organ, situated middorsally on the seventh abdominal segment in the larvae of most Lycaenidae.

Dorsal vessel—The elongate, tubular organ or "heart" responsible for circulating the blood in insects.

Dorsolateral—Area between the dorsal and lateral surfaces.

Dorsolateral organs—A pair of eversible tubular organs situated dorsolaterally on the eighth abdominal segment in the larvae of most Lycaenidae.

Dorsum—Upper surface; sometimes applied to the inner margin of the wing, which in many moths is dorsal when the wings are at rest.

Ductus bursae—The duct joining the corpus bursae in the female genitalia to the copulatory aperture or ostium bursae (Fig. 5B).

Ductus seminalis—The duct leading from the ductus bursae to the vagina of the female (Fig. 5B).

Ecdysis—Moulting of the larval or pupal cuticle.

Epidermal glands—Secretory glands of the cuticle.

Epiphysis—A movable lobe of the fore tibia found in Hesperiidae, Papilionidae and most moths (Fig. 3A).

Exuviae—Moulted or cast cuticle of the larva or pupa.

Eye-spot—A coloured spot, ringed with a contrasting colour.

Family—An assemblage of related genera.

Femur—The third segment of the leg between the trochanter and the tibia (Fig. 3).

Flagellum—That portion of the antenna distal to the basal two segments, the scape and pedicel.

Frenulum—A strong bristle or bristles arising from the base of the hindwing costa, held beneath the forewing by the retinaculum, which serves to hold the wings together in flight.

Fringe—Projecting marginal scales of the wing, especially between the apex and tornus; sometimes called cilia.

Frons—Anterior sclerite of the head.

Frontoclypeus—Fused frons and clypeus, two anterior sclerites of the head, found in larvae.

Ganglion (pl. ganglia)—A swelling on a nerve due to the presence at one place of the nuclei of a number of nerve cells.

Gene—The hereditary unit that occupies a fixed position on the chromosome and is responsible for transmitting the inherited characters of the parents to the offspring.

Genitalia—Organs used for either mating or oviposition (Fig. 5).

Genus—A group of closely related species.

Gnathos—A single or paired structure of the male genitalia beneath the anus (Fig. 5A).

Granulose—Having a granular surface.

Ground colour—The dominant colour of the wing contrasted to that of the pattern.

Gynandromorph—An abnormal specimen made up of both male and female structures.

Haemolymph—Insect blood.

Hair-pencil—A dense tuft of long parallel hairs in the males of some species; the hairs are displayed during courtship and may carry pheromones.

Haustellum—Coiled proboscis in Lepidoptera.

Hill-topping—Behaviour exhibited by certain insects whereby they fly to hill-tops or elevated points.

Holotype—The specimen selected and labelled as the fixed reference specimen for each species.

Humeral vein—A short vein arising near the base of the hindwing and running towards the base of the costa (Fig. 4A).

Hyaline—Transparent; in Hesperiidae often applied to semitransparent or translucent spots.

Inner margin—The trailing edge of the wing, joining the base to the tornus (Fig. 4B).

Instar—The period between the hatching of the egg and the first larval moult or ecdysis, and the period between two successive larval moults or ecdyses.

Integument—The exoskeleton or wall of the insect body.

Juxta—A sclerotized plate beneath the aedeagus of the male genitalia; part of the fultura inferior.

Labial palpi—Paired appendages arising from the insect labium, three-segmented in butterflies.

Labium—Fused mouth-parts forming the posterior (ventral) wall of the mouth.

Larva—Caterpillar; the feeding, sexually immature, developmental stage of an insect.

Lunule—Crescent-shaped spot.

Mandible—Sclerotized dentate jaw of the larva (Fig. 1).

Maxillae—Paired mouth-parts situated between the mandibles and the labrum, serving as accessory jaws and assisting in the selection of food by touch and taste.

Meconium—Liquid containing the products of pupal metabolism, expelled from the anus of the newly emerged adult.

Media—The fourth main longitudinal vein (M) of the wing; the basal half is lost in butterflies and most moths (Fig. 4A).

Median—Applying to the media; the central area of the wing or other structure (Fig. 4).

Melanism—Hereditary darkening caused by an increased deposition of the black pigment melanin.

Mesoseries—The arrangement of the crochets of the larval proleg in a single longitudinal band.

Mesothorax—The middle segment of the thorax, bearing the midlegs and, in adults, the forewings.

Metamorphosis—The transformation from larval to adult form; referred to as "complete" in those insects, including Lepidoptera, in which there is a non-feeding pupal stage.

Metathorax—The third segment of the thorax, bearing the hindlegs and, in adults, the hindwings.

Micropyle—Microscopic aperture in the chorion of the egg through which the sperm enters.

Migration—A mass movement, often in one general direction at any one time.

Mimicry—The resemblance of one species to another occurring in the same area.

Multiordinal—Of many sizes, referring to the crochets.

Mutation—An inherited change in one or more genes resulting in a physical or behavioural modification in successive generations.

Myrmecophilous—Associated with ants.

Nominotypical—The first of two or more subspecies in any species to be named.

Notum—The upper surface of the thorax.

Ocellus (pl. ocelli)—A simple organ of sight found above the eye in some moths, or at the side of the larval head (see stemmata); also used by lepidopterists to refer to an eye-spot on the wing.

Ommatidium (pl. ommatidia)—One of the facets of the compound insect eye.

Onisciform—Shaped like a wood-louse or slater.

Osmeterium (pl. osmeteria)—A forked eversible scent-producing organ arising from the anterior margin of the prothorax in larvae of Papilionidae (Fig. 1).

Ostium bursae—The copulatory aperture at the entrance to the bursa copulatrix (Fig. 5B).

Oviposition—Egg-laying.

Ovipositor—A tubular organ used for oviposition by some insects; not developed in butterflies—see papillae anales.

Ovum (pl. ova)—An egg.

Palp, palpus (pl. palpi)—One of a pair of segmented organs of the maxilla or labium in insects, bearing sensilla.

Papillae anales—A pair of hairy lobes at the posterior end of the female abdomen, used in oviposition (Fig. 5B).

Parasite—An animal completely dependent on the tissues of another for its food; in insects usually internal and causing the ultimate death of its host, and then known as a "parasitoid".

Paratype—A specimen other than the holotype upon which the description of a species was originally based.

Patagia—A pair of anterior sclerites of the prothorax.

Pedicel—The second segment of the antenna.

Pheromone—A substance secreted by an individual that produces a response in other individuals of the same species.

Phytophagous—Plant-eating.

Planta—The membranous "sole" of the abdominal proleg.

Pleural—Referring to the lateral surfaces of the body.

Polymorphism—The existence of two or more genetically different forms in the same interbreeding population.

Postmedian—Just beyond the central area of the wing or other structure (Fig. 4B).

Predator—An animal that utilizes two or more other animals for food.

Proboscis—Tubular organ of insects used for sucking in liquid food; in Lepidoptera it is called the haustellum, and is formed from the paired galeae, the two members being hooked together to form a tube, which is coiled when not in use.

Process—A protruding portion of the exoskeleton.

Procryptic—Form, pattern or colour that assists concealment.

Prolegs—Fleshy leg-like organs of abdominal segments 3 to 6 and 10 in larvae (Fig. 1).

Prothoracic plate—A sclerotized middorsal plate on the larval prothorax (Fig. 1).

Prothorax—The first segment of the thorax, bearing the forelegs and a pair of spiracles.

Proximal—Nearest to the base or centre.

Pupa—The non-feeding post-larval immature stage in development (Fig. 2).

Pupal cap—The anterior portion of the pupal cuticle covering the head of the future adult.

Radius—The third main longitudinal wing vein (R), usually with three to five branches in the forewing.

Retinaculum—The hook beneath the forewing into which the frenulum fits.

Rigor mortis—A state of muscle rigidity which follows death in most organisms.

Rugose—Having a roughened surface.

Saccus—A usually hollow apodeme directed forwards midventrally from the vinculum into the body cavity of the adult male (Fig. 5A).

Scale—A flattened hair forming the characteristic clothing of Lepidoptera and a few other insects.

Scape—The basal segment of the antenna.

Sclerite—A sclerotized plate, forming part of the exoskeleton or body wall of an insect, and surrounded by membrane.

Sclerotin—The brown substance produced by tanning of protein, deposited in certain areas of the exoskeleton producing hard, often thickened plates or sclerites.

Sclerotized—Hardened by the formation of sclerotin.

Scoli—An outgrowth of the body wall in some larvae, bearing branches or setae.

Segment—A ring-like or tubular section of the body or its appendages.

Sensillum (pl. sensilla)—A sense organ.

Seta (pl. setae)—A hair, arising from a socket; the hairs present in first instar larvae are primary setae, those acquired in later instars are secondary setae.

Sex-brand, sex-mark—A patch of dense androconia or specialized scent scales on the male wing.

Shaft—That part of the flagellum of the antenna between the pedicel and the club.

Signum (pl. signa)—A sclerotized patch or structure on the internal surface of the corpus bursae (Fig. 5B).

Spermatheca—A storage sac for sperms, opening into the vagina of the female.

Spermatophore—The sac containing sperms, deposited by the male in the bursa copulatrix during mating.

Sphragis—A hardened male secretion deposited during mating around the ostium bursae of the female in certain butterflies (Fig. 6).

Spinneret—The tubular organ beneath the larval head through which silk is extruded from the silk glands (Fig. 1).

Spinules—Minute thorn-like projections of the larval skin.

Spinulose—Covered with spinules.

Spiracle—A lateral aperture of the tracheae, forming the respiratory system of insects (Figs. 1, 2).

Spur—A movable spine-like structure, usually clothed with scales, found on the mid and hind tibiae.

Stemmata—Primitive organs of sight situated on each side of the larval head (Fig. 1).

Sterigma—A sclerotized plate or plates surrounding the ostium bursae of the female.

Sternum—The lower surface of the body.

Striated—Covered with striae or minute lines.

Stridulation—Sound production by rubbing a series of hard projections against a ribbed or file-like surface.

Subapical—Before the apex (Fig. 4B).

Subbasal—Near the base (Fig. 4B).

Subcosta—The second main longitudinal vein (Sc) of the wing (Fig. 4A).

Subcostal—Near the costa of the wing (Fig. 4B).

Subdorsal—Near the dorsum; in larvae on either side of the middorsal longitudinal line.

Subfamily—A section of a family, containing genera more closely related to one another than to the other genera of the family.

Submedian—Before the middle or central area of the wing or other structure (Fig. 4B).

Subspecies—A geographical race; one or more taxonomically distinct populations within a species, occupying only part of the whole geographical range.

Subspiracular—Just beneath the spiracles.

Subterminal—Before the termen of the wing, or the tip of another structure (Fig. 4B).

Sulcus (pl. sulci)—A groove or furrow.

Superfamily—A group of related families.

Synonym—One of two or more names that have been applied to the one species.

Tail—An outgrowth from the margin of the hindwing.

Tarsus (pl. tarsi)—The foot or the most distal division of the leg, normally formed of five segments.

Tegumen—The upper basal section of the male genitalia, the modified ninth tergum of the abdomen.

Tergum—The upper surface of the body, joined by the membranous lateral pleuron to the sternum.

Termen—The outer margin of the wing, joining the apex and the tornus (Fig. 4B).

Terminal—At or adjacent to the termen of the wing, or tip of another structure (Fig. 4B).

Territoriality—Behaviour in which an animal patrols and defends an area from other animals of the same species.

Tibia (pl. tibiae)—The fourth segment of the leg (Fig. 3).

Tornus—The outer angle of the wing between the termen and the inner margin (Fig. 4B).

Trachea (pl. tracheae)—Respiratory tubes leading from the spiracles to supply air to the various internal organs.

Trochanter—The small second segment of the leg, between the coxa and the femur (Fig. 3A).

Tubercle—A small, usually rounded, outgrowth of the body wall.

Uncus—The upper hook-like structure of the male genitalia (Fig. 5A).

Uniordinal—Of one size, referring to the crochets.

Valva (pl. valvae)—The paired lateral claspers of the male genitalia (Fig. 5A).

Vein—A structural strut of the wing, usually tubular and containing a trachea and blood.

Venation—The system of wing veins (Fig. 4A).

Vesica—The penis, a membranous eversible organ partly or wholly contained in the tubular aedeagus.

Vinculum—The ventral U-shaped body of the male genitalia, the modified ninth sternum of the abdomen (Fig. 5A).

Selected Bibliography

ANDERSON, R. H. (1968).—*The Trees of New South Wales*, 4th edn., 510 pp., 3 pls., 152 figs. (Government Printer: Sydney).

BEADLE, N. C. W., EVANS, O. D., CAROLIN, R. C. and TINDALE, M. D. (1972).—*Flora of the Sydney Region*, 724 pp., 16 pls., 56 figs. (A. H. and A. W. Reed: Sydney).

BURBIDGE, N. T. (1963).—*Dictionary of Australian Plant Genera*, 345 pp., 2 figs. (Angus & Robertson: Sydney).

BURBIDGE, N. T. and GRAY, M. (1970).—*Flora of the Australian Capital Territory*, 447 pp., 407 figs. (Australian National University Press: Canberra).

COCHRANE, G. R., FUHRER, B. A., ROTHERHAM, E. R. and WILLIS, J. H. (1968).—*Flowers and Plants of Victoria*, 216 pp., 543 pls. (A. H. and A. W. Reed: Sydney).

COMMON, I. F. B. (1964).—*Australian Butterflies*, 131 pp., 4 pls., 509 figs. (Jacaranda Press: Brisbane).

COMMON, I. F. B. and WATERHOUSE, D. F. (1981).—*Butterflies of Australia*, 682 pp., 49 pls., 25 figs. (Angus & Robertson: Sydney).

CORBET, A. S. and PENDLEBURY, H. M. (1978).—*The Butterflies of the Malay Peninsula*, 3rd edn. revised by J. N. Eliot, 578 pp., 36 pls. (Malayan Nature Society: Kuala Lumpur).

COUCHMAN, L. E. and COUCHMAN, R. (1977).—The butterflies of Tasmania. Tasmanian Year Book, 1977, pp. 66–96, 6 pls.

C.S.I.R.O. (1970).—*The Insects of Australia*, 1029 pp., 9 pls., 704 figs. (Melbourne University Press: Melbourne).

D'ABRERA, B. (1977).—*Butterflies of the Australian Region*, 2nd edn., 415 pp. (Lansdowne: Melbourne).

ERICKSON, R., GEORGE, A. S., MARCHANT, N. G. and MORCOMBE, M. K. (1973).—*Flowers and Plants of Western Australia*, 216 pp., 538 pls. (A. H. and A. W. Reed: Sydney).

FISHER, R. H. (1978).—*Butterflies of South Australia*, 272 pp., 16 pls., 83 figs. (Government Printer: South Australia).

GALBRAITH, J. (1977).—*A Field Guide to the Wild Flowers of South-East Australia*, 450 pp., 105 pls., 57 figs. (Collins: Sydney, London).

McCUBBIN, C. (1971).—*Australian Butterflies*, 206 pp., 159 pls. (Nelson: Melbourne).

MOULDS, M. S. (1977).—*Bibliography of the Australian Butterflies*, 239 pp. (Australian Entomological Press: Sydney).

ROTHERHAM, E. R., BRIGGS, B. G., BLAXELL, D. F. and CAROLIN, R. C. (1975).—*Flowers and Plants of New South Wales and Southern Queensland*, 191 pp., 556 pls. (A. H. and A. W. Reed: Sydney).

SIBATANI, A. and GRUND, R. (1978).—A revision of the *Theclinesthes onycha* complex (Lepidoptera: Lycaenidae). *Trans. lep. Soc. Japan* **29** (1): 1–34.

WATERHOUSE, G. A. (1932).—*What Butterfly is That?*, 291 pp., 34 pls., 4 figs. (Angus & Robertson: Sydney).

WATERHOUSE, G. A. and LYELL, G. (1914).—*The Butterflies of Australia*, 239 pp., 888 figs., 1 map. (Angus & Robertson: Sydney).

Distribution Maps

BUTTERFLIES OF AUSTRALIA
Chaetocneme beata
Chaetocneme critomedia
Chaetocneme porphyropis
Netrocoryne repanda
Tagiades japetus
Tagiades nestus
Exometoeca nycteris
Rachelia extrusa
Trapezites symmomus
Trapezites heteromacula
Trapezites macqueeni
Trapezites eliena

Trapezites iacchus
Trapezites iacchoides
Trapezites maheta
Trapezites phigalioides
Trapezites phigalia
Trapezites sciron
Trapezites petalia
Trapezites luteus
Anisyntoides argenteoornatus
Anisynta sphenosema
Anisynta albovenata
Anisynta cynone

Anisynta tillyardi
Anisynta monticolae
Anisynta dominula
Dispar compacta
Pasma tasmanica
Signeta flammeata
Signeta tymbophora
Toxidia peron
Toxidia thyrrhus
Toxidia parvula
Toxidia doubledayi
Toxidia rietmanni
BUTTERFLIES OF AUSTRALIA

Toxidia melania
Toxidia inornata
Toxidia andersoni
Neohesperilla xiphiphora
Neohesperilla crocea
Neohesperilla senta
Neohesperilla xanthomera
Hesperilla sarnia
Hesperilla furva
Hesperilla crypsigramma
Hesperilla sexguttata
Hesperilla malindeva

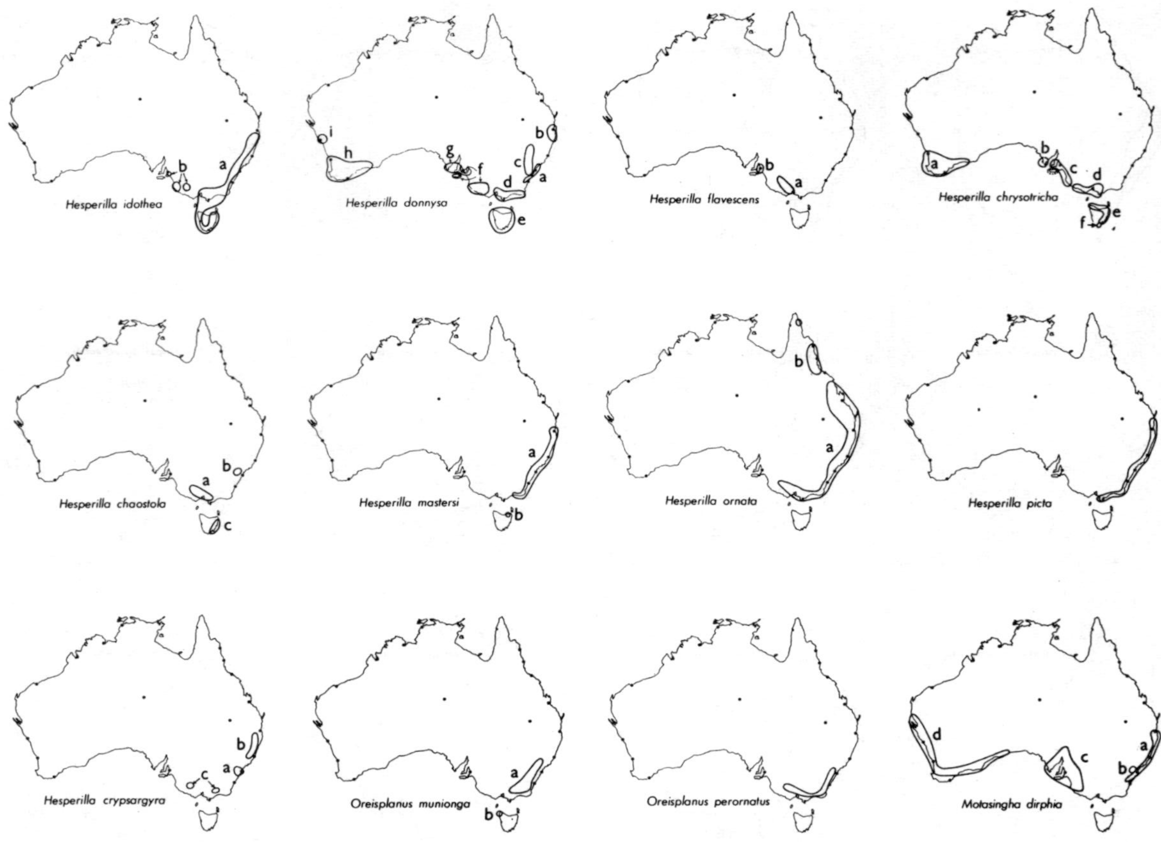

Hesperilla idothea
Hesperilla donnysa
Hesperilla flavescens
Hesperilla chrysotricha
Hesperilla chaostola
Hesperilla mastersi
Hesperilla ornata
Hesperilla picta
Hesperilla crypsargyra
Oreisplanus munionga
Oreisplanus perornatus
Motasingha dirphia

Motasingha atralba
Mesodina halyzia
Mesodina aeluropis
Proeidosa polysema
Croitana croites
Croitana arenaria
Croitana aestiva
Notocrypta waigensis
Taractrocera papyria
Taractrocera anisomorpha
Taractrocera ilia
Taractrocera dolon

BUTTERFLIES OF AUSTRALIA
Taractrocera ina
Ocybadistes flavovittatus
Ocybadistes walkeri
Ocybadistes hypomeloma
Ocybadistes ardea
Suniana lascivia
Suniana sunias
Oriens augustulus
Arrhenes dschilus
Arrhenes marnas
Telicota eurotas
Telicota colon

Telicota augias
Telicota anisodesma
Telicota ohara
Telicota ancilla
Telicota mesoptis
Telicota brachydesma
Cephrenes trichopepla
Cephrenes augiades
Sabera caesina
Sabera fuliginosa
Sabera dobboe
Mimene atropatene

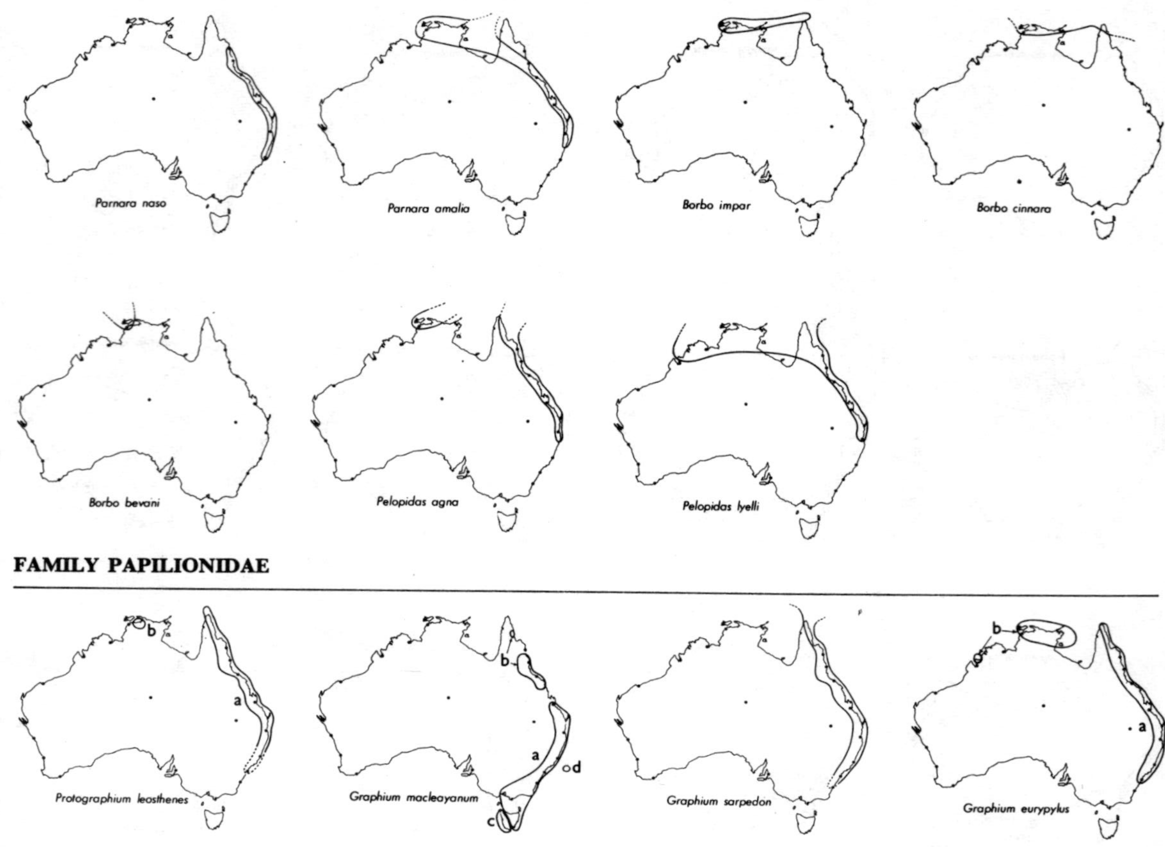

FAMILY PAPILIONIDAE

Graphium macfarlanei
Graphium agamemnon
Graphium aristeus
Papilio anactus
Papilio aegeus
Papilio fuscus
Papilio canopus
Papilio ambrax
Papilio demoleus
Papilio ulysses
Cressida cressida
Pachliopta polydorus

FAMILY PIERIDAE

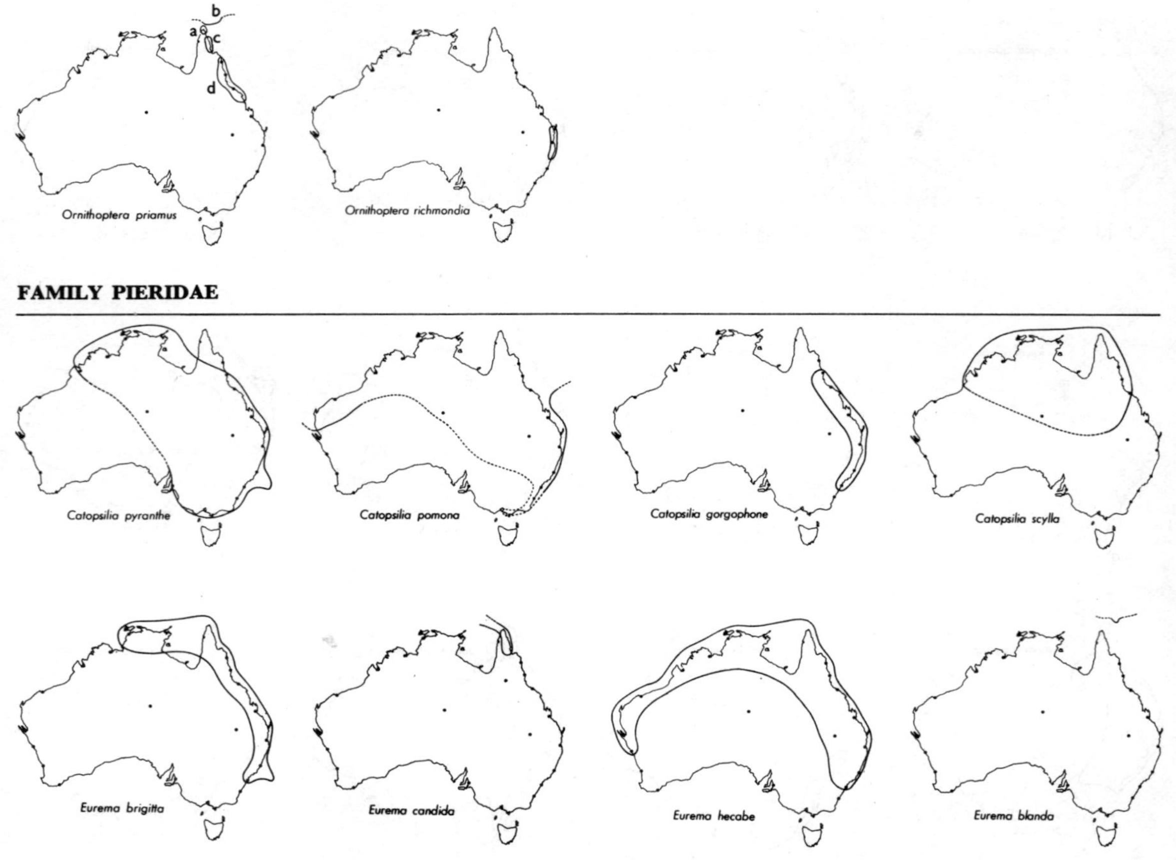

Eurema smilax
Eurema sana
Eurema laeta
Eurema herla
Elodina parthia
Elodina angulipennis
Elodina padusa
Elodina perdita
Delias argenthona
Delias mysis
Delias ennia
Delias aganippe

Delias aruna
Delias harpalyce
Delias nigrina
Delias nysa
Anaphaeis java
Cepora perimale
Appias paulina
Appias albina
Appias melania
Appias celestina
Appias ada
Pieris rapae

FAMILY NYMPHALIDAE

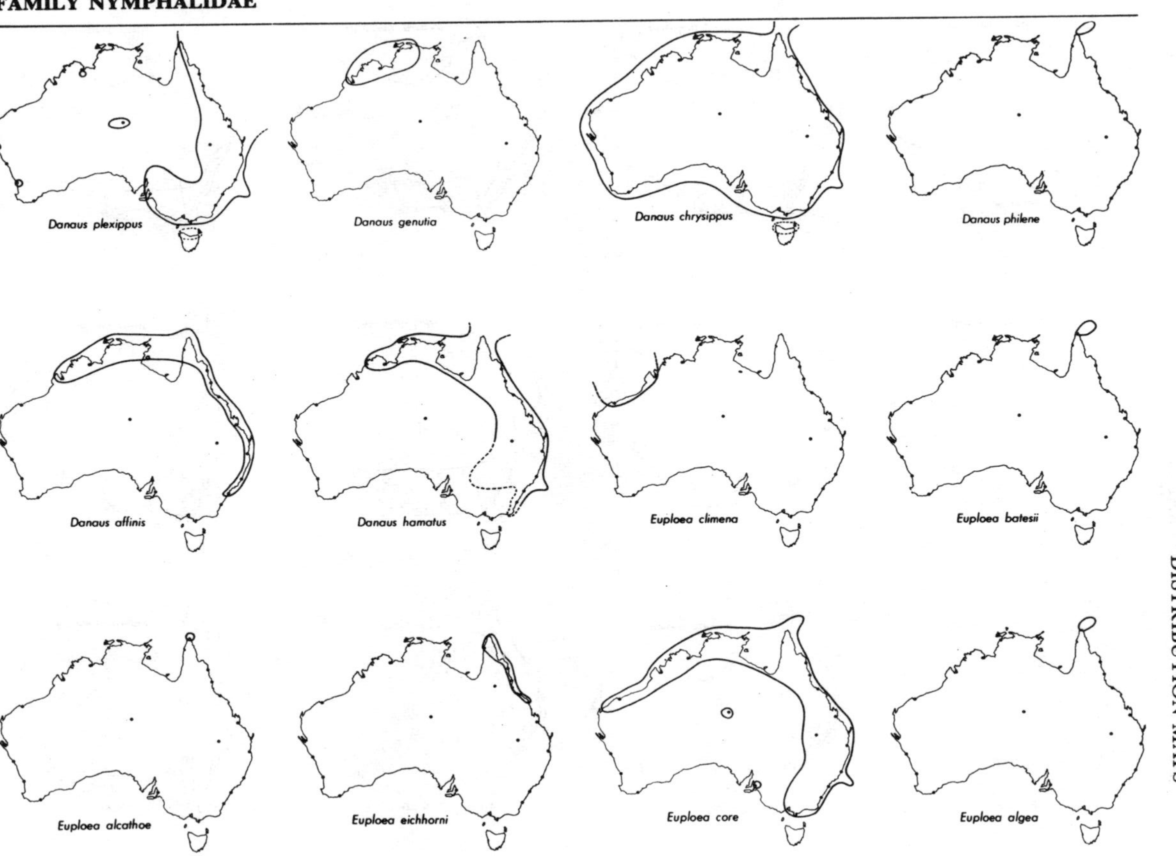

BUTTERFLIES OF AUSTRALIA
Euploea sylvester
Euploea tulliolus
Euploea darchia
Euploea usipetes
Tellervo zoilus
Melanitis leda
Melanitis amabilis
Elymnias agondas
Mycalesis sirius
Mycalesis terminus
Mycalesis perseus
Orsotriaena medus

Hypocysta euphemia
Hypocysta irius
Hypocysta metirius
Hypocysta pseudirius
Hypocysta adiante
b
a
Hypocysta angustata
Argynnina cyrila
Argynnina hobartia
a
b
c
Geitoneura acantha
a
b
Geitoneura minyas
b
a
Geitoneura klugii
c
b
a
Heteronympha merope
a
c
b
DISTRIBUTION MAPS

Heteronympha mirifica
Heteronympha paradelpha
Heteronympha penelope
Heteronympha banksii
Heteronympha solandri
Heteronympha cordace
Nesoxenica leprea
Oreixenica orichora
Oreixenica ptunarra
Oreixenica latialis
Oreixenica lathoniella
Oreixenica correae

Oreixenica kershawi
Tisiphone helena
Tisiphone abeona
Xois arctoa
Taenaris catops
Taenaris artemis
Polyura pyrrhus
Charaxes latona
Apaturina erminea
Lexias aeropa
Phaedyma shepherdi
Neptis praslini
DISTRIBUTION MAPS

BUTTERFLIES OF AUSTRALIA
Pantoporia venilia
Pantoporia consimilis
Argyreus hyperbius
Mynes geoffroyi
Doleschallia bisaltide
Hypolimnas bolina
Hypolimnas misippus
Hypolimnas alimena
Hypolimnas anomala
Yoma sabina
Vanessa kershawi
Vanessa cardui

Vanessa itea
Junonia erigone
Junonia hedonia
Junonia villida
Junonia orithya
Cethosia penthesilea
Cethosia cydippe
Vindula arsinoe
Vagrans egista
Phalanta phalantha
Cupha prosope
Acraea andromacha

FAMILY LIBYTHEIDAE

Libythea geoffroy

FAMILY LYCAENIDAE

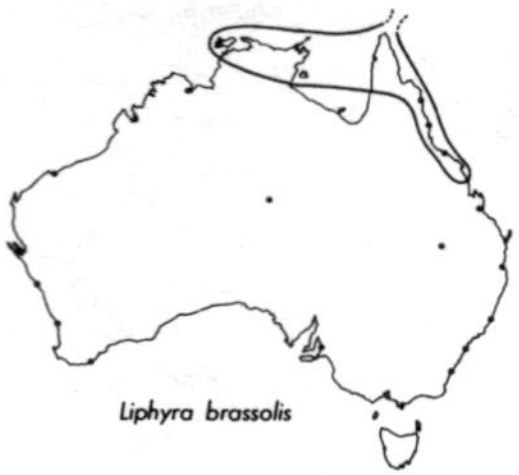

Liphyra brassolis

Lucia limbaria

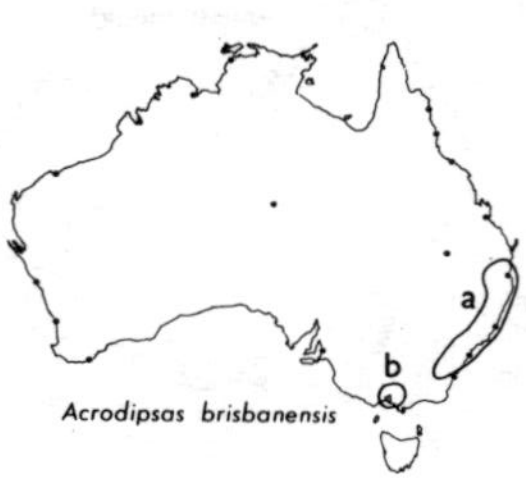

Acrodipsas brisbanensis

Acrodipsas cuprea

Acrodipsas myrmecophila

Acrodipsas illidgei

Acrodipsas arcana

Acrodipsas hirtipes

Acrodipsas melania
Paralucia pyrodiscus
Paralucia aurifera
Paralucia spinifera
Pseudodipsas eone
Pseudodipsas cephenes
Hypochrysops theon
Hypochrysops hippuris
Hypochrysops halyaetus
Hypochrysops delicia
Hypochrysops ignitus
Hypochrysops epicurus

BUTTERFLIES OF AUSTRALIA

Philiris diana
Philiris nitens
Philiris fulgens
Philiris ziska
Philiris innotata
Arhopala wildei
Arhopala centaurus
Arhopala madytus
Arhopala micale
Ogyris genoveva
Ogyris zosine
Ogyris idmo

BUTTERFLIES OF AUSTRALIA

Jalmenus pseudictinus
Jalmenus daemeli
Jalmenus lithochroa
Jalmenus inous
Jalmenus icilius
Jalmenus clementi
Pseudalmenus chlorinda
g
f
e o
a
d
b
c
Hypolycaena phorbas
b
a
Hypolycaena danis
Deudorix epijarbas
b
a
Deudorix epirus
Virachola democles
DISTRIBUTION MAPS

BUTTERFLIES OF AUSTRALIA
Virachola smilis
Rapala varuna
Bindahara phocides
Anthene seltuttus
Anthene lycaenoides
Candalides gilberti
Candalides margarita
Candalides helenita
Candalides absimilis
Candalides consimilis
Candalides cyprotus
Candalides erinus

Candalides geminus
Candalides acastus
Candalides hyacinthinus
Candalides xanthospilos
Candalides heathi
Adaluma urumelia
Nesolycaena albosericea
Zetona delospila
Petrelaea dana
Nacaduba pactolus
Nacaduba berenice
Nacaduba kurava

BUTTERFLIES OF AUSTRALIA

Theclinesthes onycha
Theclinesthes miskini
Theclinesthes albocincta
Theclinesthes hesperia
Theclinesthes serpentata
Theclinesthes sulpitius
Theclinesthes scintillata
Danis danis
Danis hymetus
Danis cyanea
Jamides aleuas
Jamides nemophilus

BUTTERFLIES OF AUSTRALIA

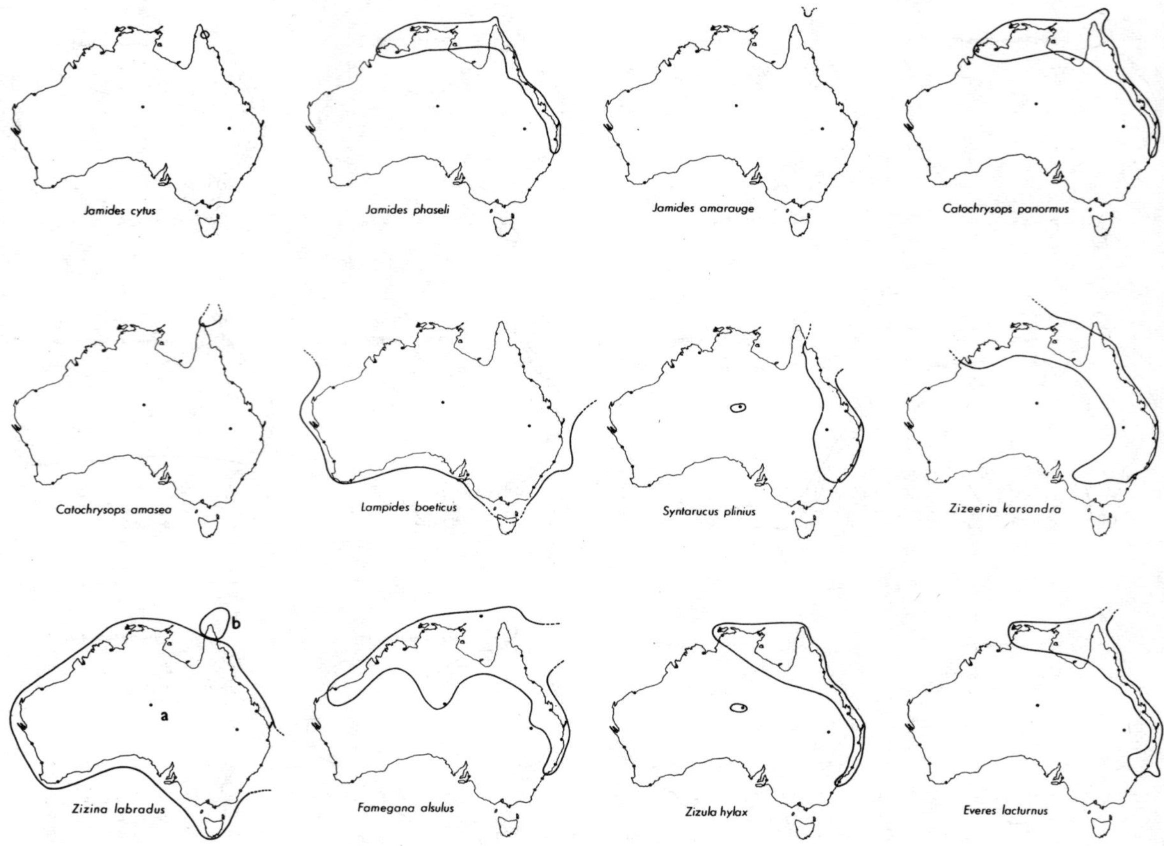
Jamides cytus
Jamides phaseli
Jamides amarauge
Catochrysops panormus
Catochrysops amasea
Lampides boeticus
Syntarucus plinius
Zizeeria karsandra
Zizina labradus
Famegana alsulus
Zizula hylax
Everes lacturnus

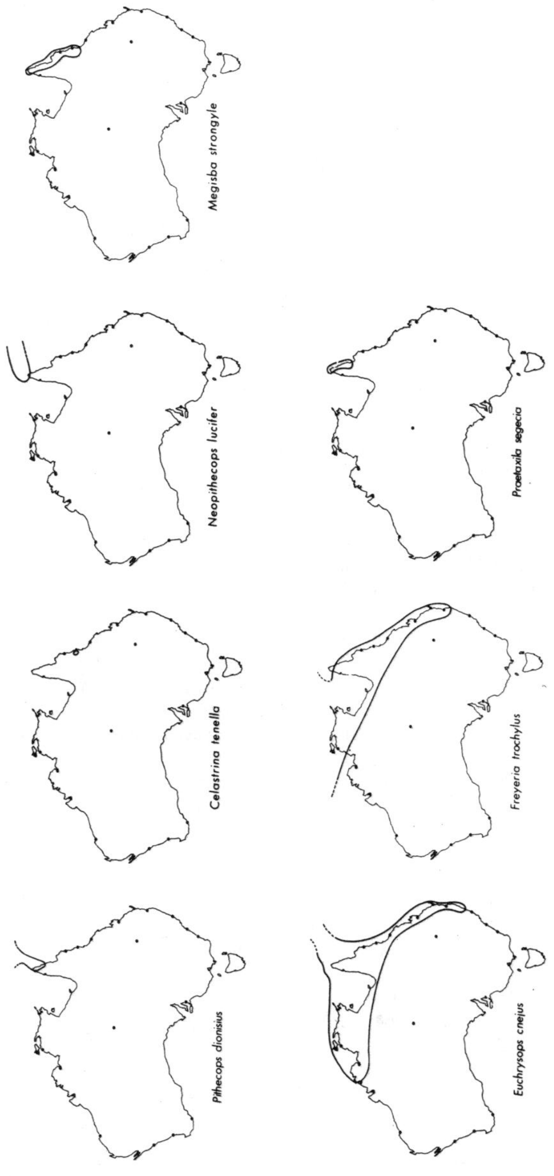

Megisba strongyle
Neopithecops lucifer
Praetaxila segecia
Celastrina tenella
Freyeria trochylus
Pithecops dionisius
Euchrysops cnejus

Index of Common Names

Blue moonbeam, see *Philiris nitens* 259
Blues 227
Blue tiger, see *Danaus hamatus* 156
Blue triangle, see *Graphium sarpedon* 115
Bright cerulean, see *Jamides aleuas* 335
Bright copper, see *Paralucia aurifera* 239
Bright-eyed brown, see *Heteronympha cordace* 183
Bright oakblue, see *Arhopala madytus* 262
Bright shield skipper, see *Signeta flammeata* 46
Broad-banded awl, see *Hasora hurama* 19
Broad-margined grass yellow, see *Eurema candida* 135
Brown
    Banks', see *Heteronympha banksii* 181
    bright-eyed, *Heteronympha cordace* 183
    common, see *Heteronympha merope* 176
    correa, see *Oreixenica correae* 193
    Cyril's, see *Argynnina cyrila* 170
    Elia, see *Nesoxenica leprea elia* 186
    evening, see *Melanitis leda* 164
    helena, see *Tisiphone helena* 196
    Hobart, see *Argynnina hobartia* 171
    Kershaw's, see *Oreixenica kershawi* 194
    leprea, see *Nesoxenica leprea* 185
    orichora, see *Oreixenica orichora* 187
    shouldered, see *Heteronympha penelope* 178
    Solander's, see *Heteronympha solandri* 182
    spotted, see *Heteronympha paradelpha* 177
    sword-grass, see *Tisiphone abeona* 196
    Tasmanian, see *Argynnina hobartia tasmanica* 171
    western, see *Heteronympha merope duboulayi* 177
    wonder, see *Heteronympha mirifica* 177
Brown awl, see *Badamia exclamationis* 20
Brown-eye, purple, see *Chaetocneme porphyropis* 24
Browns 162
Brown soldier, see *Junonia hedonia* 218
Brown tiger, see *Danaus philene* 156
Bushbrown
    cedar, see *Mycalesis sirius* 166
    dingy, see *Mycalesis perseus* 166
    orange, see *Mycalesis terminus* 166

Cabbage white, see *Pieris rapae* 151
Cairns hamadryad, see *Tellervo zoilus* 162
Canopus butterfly, see *Papilio canopus* 122
Capaneus butterfly, see *Papilio fuscus capaneus* 121
Caper white, see *Anaphaeis java* 147
Cape York aeroplane, see *Pantoporia venilia* 209
Cape York birdwing, see *Ornithoptera priamus* 127
Cape York hamadryad, see *Tellervo zoilus gelo* 162
Cedar bushbrown, see *Mycalesis sirius* 166
Cephenes blue, see *Pseudodipsas cephenes* 240
Cerulean
    bright, see *Jamides aleuas* 335
    dark, see *Jamides phaseli* 336
    pale, see *Jamides cytus* 335
Chaostola skipper, see *Hesperilla chaostola* 68
Chalk white, see *Elodina parthia* 139
Chequered blue, see *Theclinesthes serpentata* 331
Chequered swallowtail, see *Papilio demoleus* 123
Chrysotricha skipper, see *Hesperilla chrysotricha* 64
Common aeroplane, see *Phaedyma shepherdi* 206
Common albatross, see *Appias paulina* 149
Common Australian crow, see *Euploea core* 158
Common banded awl, see *Hasora chromus* 19
Common brown, see *Heteronympha merope* 176
Common brown ringlet, see *Hypocysta metirius* 168
Common dart, see *Ocybadistes flavovittatus* 89

412

White-fringed swift, see *Sabera fuliginosa*
105
White grassdart, see *Taractrocera papyria*
83
White lineblue, see *Nacaduba kurava* 315
White nymph, see *Mynes geoffroyi* 210
Whites 130
White-veined skipper, see *Anisynta albo-
venata* 39
Wonder brown, see *Heteronympha mirifica*
177
Wood white, see *Delias aganippe* 144

Xanthomera skipper, see *Neohesperilla
xanthomera* 52
Xenica
    alpine silver, see *Oreixenica latialis* 190
    common silver, see *Oreixenica lathoniella*
    191
    eastern ringed, see *Geitoneura acantha*
    172
    Klug's, see *Geitoneura klugii* 174
    western, see *Geitoneura minyas* 173
Xiphiphora skipper, see *Neohesperilla
xiphiphora* 51

Yellow
    broad-margined grass, see *Eurema
    candida* 135
    common grass, see *Eurema hecabe* 135
    line grass, see *Eurema laeta* 138
    Macleay's grass, see *Eurema herla* 138
    no-brand grass, see *Eurema brigitta* 134
    small grass, see *Eurema smilax* 137
    spotless grass, see *Eurema sana* 138
Yellow-banded dart, see *Ocybadistes
walkeri sothis* 90
Yellow migrant, see *Catopsilia gorgophone*
133
Yellow palmdart, see *Cephrenes trichopepla*
104
Yellowish skipper, see *Hesperilla flavescens*
63
Yellows 130
Yellow-spot blue, see *Candalides xan-
thospilos* 309
Yellow spot jewel, see *Hypochrysops byzos*
252
Yellow swift, see *Borbo impar* 108

Zebra blue, see *Syntarucus plinius* 338

Index of Scientific Names

417

caesina (Sabera) 105
calybe (Junonia) 219
Campomyrma 329
Camponotus 264, 265, 266, 267, 268, 269, 270, 271, 329, 344
Candalides 301
  absimilis 303
  acastus 307
  consimilis 304
    consimilis 304
    goodingi 305
    toza 304
  cyprotus 305
    cyprotus 305
    pallescens 306
  erinus 306
  geminus 307
  gilberti 301
  heathi 310
    aeratus 311
    alpinus 311
    doddi 311
    heathi 310
  helenita 303
  hyacinthinus 308
    eugenia 308
    hyacinthinus 308
    simplex 309
  margarita 302
  xanthospilos 309
candida (Eurema) 135
canopus (Papilio) 122
capaneus (Papilio) 121
capricornia (Theclinesthes) 325
caracalla (Ionolyce) 318
cardui (Vanessa) 216
caria (Appias) 151
cassandra (Cressida) 125
Catochrysops 336
  amasea 337
  panormus 337
    platissa 337
Catopsilia 131
  gorgophone 133
  pomona 132
  pyranthe 131

crokera 131
  scylla 133
    etesia 133
Catopyrops 319
  ancyra 319
    mysia 319
    florinda 319
      estrella 319
      halys 319
cela (Nacaduba) 314
Celastrina 342
  tenella 342
celestina (Appias) 150
centaurus (Arhopala) 261
cephenes (Pseudodipsas) 240
Cephrenes 104
  augiades 105
    sperthias 105
    trichopepla 104
Cepora 148
  perimale 148
    latilimbata 149
    scyllara 148
ceres (Ocybadistes) 90
Cethosia 220
  cydippe 220
    chrysippe 220
  penthesilea 220
    paksha 220
Chaetocneme 22
  beata 23
  critomedia 24
    sphinterifera 24
  denitza 23
  porphyropis 24
chaostola (Hesperilla) 67, 68
Charaxes 203
  latona 203
Charaxinae 202
chares (Hesperilla) 68
chlorinda (Pseudalmenus) 289, 291
chloris (Pseudalmenus) 293
choredon (Graphium) 115
chromus (Hasora) 19
chrysippe (Cethosia) 220
chrysippus (Danaus) 155